Birkhäuser

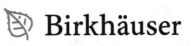

Lorenz Halbeisen • Regula Krapf

Gödel's Theorems and Zermelo's Axioms

A Firm Foundation of Mathematics

Birkhäuser

Lorenz Halbeisen
Departement Mathematik
ETH Zürich
Zürich, Switzerland

Regula Krapf
Institut für Mathematik
Universität Koblenz-Landau
Koblenz, Germany

ISBN 978-3-030-52281-0 ISBN 978-3-030-52279-7 (eBook)
https://doi.org/10.1007/978-3-030-52279-7

This book is published under the imprint Birkhäuser, www.birkhauser-science.com by the registered company Springer Nature Switzerland AG
The registered company address is: Gewerbestrasse 11, 6330 Cham, Switzerland

Preface

This book provides a self-contained introduction to the foundations of mathematics, where self-contained means that we assume as little prerequisites as possible. One such assumption is the notion of *finiteness*, which *cannot* be defined in mathematics.

The firm foundation of mathematics we provide is based on logic and models. In particular, it is based on Hilbert's axiomatisation of formal logic (including the notion of formal proofs), and on the notion of models of mathematical theories. On this basis, we first prove GÖDEL'S COMPLETENESS THEOREM and GÖDEL'S INCOMPLETENESS THEOREMS, and then we introduce Zermelo's Axioms of Set Theory. On the one hand, Gödel's Theorems set the framework within which mathematics takes place. On the other hand, using the example of Analysis, we shall see how mathematics can be developed within a model of Set Theory. So, Gödel's Theorems and Zermelo's Axioms are indeed a firm foundation of mathematics.

The book consists of four parts. The first part is an introduction to First-Order Logic from scratch. Starting with a set of symbols, the basic concepts of formal proofs and models are developed, where special care is given to the notion of *finiteness*.

The second part is concerned with GÖDEL'S COMPLETENESS THEOREM. Our proof follows Henkin's construction [22]. However, we modified Henkin's construction by working just with *potentially infinite* sets and avoid the use of *actually infinite* sets. Even though Henkin's construction works also for uncountable signatures, we prove the general COMPLETENESS THEOREM with an ultraproduct construction, using ŁOŠ'S THEOREM.

After a preliminary chapter on countable models of Peano Arithmetic, the third part is mainly concerned with GÖDEL'S INCOMPLETENESS THEOREMS, which will be proved from scratch (i.e., purely from the axioms of Peano Arithmetic) without any use of recursion theory. In Chapter 10 and 12 some weaker theories of arithmetic are investigated. In particular, in Chapter 10 it is shown that GÖDEL'S FIRST INCOMPLETENESS THEOREM also applies for Robinson Arithmetic and in Chapter 12 it is shown that Presburger Arithmetic is complete.

In the last part we present first Zermelo's axioms of Set Theory, including the Axiom of Choice. Then we discuss the consistency of this axiomatic system and provide standard as well as non-standard models of Set Theory (including Gödel's model **L**). After introducing the construction of models with ultraproducts, we prove the COMPLETENESS THEOREM for uncountable signatures as well the the LÖWENHEIM-SKOLEM THEOREMS. In the last two chapters, we construct several standard and non-standard models of Peano Arithmetic and of the real numbers, and give a brief introduction to Non-Standard Analysis.

Acknowledgement.

First of all, we would like to thank all our colleagues and students for many fruitful and inspiring discussions. In particular, we would like to thank Johann Birnick, Marius Furter, Adony Ghebressilasie, Norbert Hungerbühler, Marc Lischka, Daniel Paunovic, Philipp Provenzano, Michele Reho, Matthias Roshardt, Joel Schmitz, and especially Michael Yan for their careful reading and all their remarks and hints. Finally, we would like to thank Tobias Schwaibold from Birkhäuser Verlag for his numerous helpful remarks which have greatly improved the presentation of this book.

Zürich and Koblenz, 28 April 2020 *L. Halbeisen and R. Krapf*

Contents

Part II Gödel's Completeness Theorem

Part III Gödel's Incompleteness Theorems

Introduction: The Natural Numbers

In the late 19th and early 20th century, several unsuccessful attempts were made to develop the natural numbers from logic. The most promising approaches were the ones due to Frege and Russell, but also their approaches failed at the end. Even though it seems impossible to develop the natural numbers just from logic, it is still necessary to formalise them.

In fact, the problem with the natural numbers is, that we need the notion of finiteness in order to define them. This presupposes the existence of a kind of infinite list of objects, and it is not clear whether these objects are — in some sense — not already the natural numbers which we would like to define. However, in our opinion there is a subtle distinction between the infinite set of natural numbers and an arbitrarily long list of objects, since the set of natural numbers is an *actually infinite* set, whereas an arbitrarily long list is just *potentially infinite*. The difference between these two types of infinity is, that the actual infinity is something which is completed and definite and consists of infinitely many elements. On the other hand, the potential infinity — introduced by Aristotle — is something that is always finite, even though more and more elements can be added to make it arbitrarily large. For example, the set of prime numbers can be considered as an actually infinite set (as Cantor did), or just as a potentially infinite list of numbers without last element which is never completed (as Euclid did).

As mentioned above, it seems that there is no way to define the natural numbers just from logic. Hence, if we would like to define them, we have to make some assumptions which cannot be formalised within logic or mathematics in general. In other words, in order to define the natural numbers we have to presuppose some *metamathematical* notions like, e.g., the notion of F I N I T E N E S S. To emphasise this fact, we shall use a wider letter spacing for the metamathematical notions we suppose.

So, let us assume that we all have a notion of F I N I T E N E S S. Let us further assume that we have two characters, say 0 and s. With these characters, we build now the following finite strings:

$$0 \quad \mathsf{s}0 \quad \mathsf{ss}0 \quad \mathsf{sss}0 \quad \mathsf{ssss}0 \quad \mathsf{sssss}0 \quad \dots$$

The three dots on the right of the above expression mean that we always build the next string by appending on the left the character s to the string we just built. Proceeding this way, we get in fact a potentially infinite "list" $\mathbb{N}$ of different strings which is never completed. By introducing the primitive notion of "list", $\mathbb{N}$ is of the form

$$\mathbb{N} = [0, \mathsf{s}0, \mathsf{ss}0, \mathsf{sss}0, \mathsf{ssss}0, \mathsf{sssss}0, \ldots],$$

where each string in the list $\mathbb{N}$ is a so-called **natural number**. For each natural number n in the list $\mathbb{N}$ we have:

$$\text{either} \quad n \equiv 0 \quad \text{or} \quad n \equiv \sigma 0,$$

where the symbol $\equiv$ means "identical to" and σ is a non-empty finite string of the form $\mathsf{s} \cdots \mathsf{s}$ and hence $\sigma 0$ has the form

$$\underbrace{\mathsf{s} \cdots \mathsf{s}}_{\substack{\text{non-empty}\\ \text{finite string}}} 0 .$$

If σ and π are both (possibly empty) finite strings of the form $\mathsf{s} \cdots \mathsf{s}$, then we write $\sigma\pi$ for the concatenation of σ and π, i.e., for the string obtained by writing first the sequence σ followed by the sequence π.

Remark. For any (possibly empty) strings σ, π, ϱ of the form $\mathsf{s} \cdots \mathsf{s}$ we get

$$\sigma\pi 0 \;\equiv\; \pi\sigma 0, \qquad \mathsf{s}\sigma\pi 0 \;\equiv\; \mathsf{s}\pi\sigma 0, \qquad \mathsf{s}\sigma\pi 0 \;\equiv\; \sigma\mathsf{s}\pi 0,$$

and further we get:

$$\sigma 0 \equiv \pi 0 \iff \mathsf{s}\sigma 0 \equiv \mathsf{s}\pi 0$$
$$\sigma 0 \equiv \pi 0 \iff \sigma\varrho 0 \equiv \pi\varrho 0$$

If we order finite strings of the form

$$\underbrace{\mathsf{s} \cdots \mathsf{s}}_{\substack{\text{possibly empty}\\ \text{finite string}}} \; 0$$

by their length, we obtain that two strings are identical if and only if they have the same length. From this geometrical point of view, the facts above can be deduced from Euclid's *Elements* where he writes (see [9, p. 155]):

1. *Things which are equal to the same thing are also equal to one another.*
2. *If equals be added to equals, the wholes are equal.*
3. *If equals be subtracted from equals, the remainders are equal.*
4. *Things which coincide with one another are equal to one another.*

It is convenient to use Hindu-Arabic numerals to denote explicitly given natural numbers (e.g., we write the symbol 1 for $\mathsf{s}0$) and Latin letters like

$n, m, \ldots$ for non-specified natural numbers. If n and m denote different natural numbers, where n appears earlier than m in the list $\mathbb{N}$, then we write $n < m$ and the expression $n, \ldots, m$ means the natural numbers which belong to the sublist $[n, \ldots, m]$ of $\mathbb{N}$; if n appears later than m in $\mathbb{N}$, then we write $n > m$ and the expression $n, \ldots, m$ denotes the empty set.

We shall use natural numbers frequently as subscripts for finite lists of objects like $t_1, \ldots, t_n$. In this context we mean that for each natural number k in the list $[1, \ldots, n]$, there is an object t_k, where in the case when $n = 0$, the set of objects is empty.

If n is a natural number, then $n + 1$ denotes the natural number $\mathsf{s}n$ (i.e., the number which appears immediately after n in the list $\mathbb{N}$); and if $n \neq 0$, then $n - 1$ denotes the natural number which appears immediately before n in the list $\mathbb{N}$. Furthermore, for $\sigma 0, \pi 0$ in the list $\mathbb{N}$, we define

$$\sigma 0 + 0 :\equiv \sigma 0 \quad \text{and} \quad 0 + \pi 0 :\equiv \pi 0 \, ,$$

and in general, we define:

$$\sigma 0 + \pi 0 :\equiv \sigma \pi 0$$

Finally, by our construction of natural numbers we get the following fact:

> *If* a statement A holds for 0 and *if* whenever A holds for a natural number n in $\mathbb{N}$ then it also holds for $n + 1$, *then* the statement A holds for *all* natural numbers n in $\mathbb{N}$.

This fact is known as *Induction Priciple*, which is an important tool in proving statements about natural numbers.

A second principle which also uses that each natural number is either 0 or $n + 1$ for some natural number n in $\mathbb{N}$, is the *Recursion Principle*:

> *If* we define X_0 and *if* whenever X_n is defined then we can define X_{n+1}, *then* X_n *can* be defined for *all* natural numbers n in $\mathbb{N}$.

Part I
Introduction to First-Order Logic

First-Order Logic is the system of Symbolic Logic concerned not only to represent the logical relations between sentences or propositions as wholes (like *Propositional Logic*), but also to consider their internal structure in terms of subject and predicate. First-Order Logic can be considered as a kind of language which is distinguished from higher-orderlogic!higher-order languages in that it does not allow quantification over subsets of the domain of discourse or other objects of higher type (like statements of infinite length or statements about formulas). Nevertheless, First-Order Logic is strong enough to formalise all of Set Theory and thereby virtually all of Mathematics.

The goal of this brief introduction to First-Order Logic is to introduce the basic concepts of formal proofs and models, which shall be investigated further in Parts II & III.

Chapter 1
Syntax: The Grammar of Symbols

The goal of this chapter is to develop the formal language of First-Order Logic from scratch. At the same time, we introduce some terminology of the so-called metalanguage, which is the language we use when we speak *about* the formal language (e.g., when we want to express that two strings of symbols are equal). In the metalanguage, we shall use some notions of NAIVE SET THEORY like *sets* (which will always be FINITE), the *membership relation* $\in$, the *empty set* $\emptyset$, or the *subset relation* $\subseteq$. We would like to emphasise that these notions are not part of the language of formal logic and that they are just used in an informal way.

Alphabet

Like any other written language, First-Order Logic is based on an *alphabet*, which consists of the following *symbols*:

(a) **Variables** such as $x, y, v_0, v_1, \ldots$, which are place holders for objects of the *domain* under consideration (which can, e.g., be the elements of a group, natural numbers, or sets). We mainly use lower case Latin letters (with or without subscripts) for variables.

(b) **Logical operators** which are $\neg$ (*not*), $\wedge$ (*and*), $\vee$ (*or*), and $\rightarrow$ (*implies*).

(c) **Logical quantifiers** which are the *existential quantifier* $\exists$ (*there is* or *there exists*) and the *universal quantifier* $\forall$ (*for all* or *for each*), where quantification is restricted to objects only and not to formulae or sets of objects (but the objects themselves may be sets).

(d) **Equality symbol** $=$ which is a special binary *relation symbol* (see below).

(e) **Constant symbols** like the number $\mathsf{0}$ in Peano Arithmetic, or the neutral element e in Group Theory. Constant symbols stand for fixed individual objects in the domain.

(f) **Function symbols** such as $\circ$ (the operation in Group Theory), or $+, \cdot, \mathsf{s}$ (the operations in Peano Arithmetic). Function symbols stand for fixed

L. Halbeisen, R. Krapf, *Gödel's Theorems and Zermelo's Axioms*,
https://doi.org/10.1007/978-3-030-52279-7_1

functions taking objects as arguments and returning objects as values. With each function symbol we associate a positive natural number, its co-called *arity* (e.g., $\circ$ is a 2-ary or binary function, and the successor operation s is a 1-ary or unary function). More formally, to each function symbol F we adjoin a fixed F I N I T E string of place holders $\mathsf{x} \cdots \mathsf{x}$ and write $F\mathsf{x} \cdots \mathsf{x}$.

(g) **Relation symbols** or **predicate constants** (such as $\in$ in Set Theory) stand for fixed relations between (or properties of) objects in the domain. Again, we associate an arity with each relation symbol (e.g., $\in$ is a binary relation). More formally, to each relation symbol R we adjoin a fixed F I - N I T E string of place holders $\mathsf{x} \cdots \mathsf{x}$ and write $R\mathsf{x} \cdots \mathsf{x}$.

The symbols in (a)–(d) form the core of the alphabet and are called **logical symbols**. The symbols in (e)–(g) depend on the specific topic we are investigating and are called **non-logical symbols**. The set of non-logical symbols which are used in order to formalise a certain mathematical theory is called the **signature** (or **language**) of this theory, and *formulae* which are formulated in a language $\mathscr{L}$ are usually called $\mathscr{L}$-formulae. For example, if we investigate groups, then the only non-logical symbols we use are e and $\circ$, thus $\mathscr{L} = \{\mathsf{e}, \circ\}$ is the language of Group Theory.

Terms & Formulae

With the symbols of our alphabet, we can now start to compose names. In the language of First-Order Logic, these names are called called *terms*. Suppose that $\mathscr{L}$ is a signature.

Terms. A string of symbols is an **$\mathscr{L}$-term**, if it results from applying F I N I T E L Y many times the following rules:

(T0) Each variable is an $\mathscr{L}$-term.
(T1) Each constant symbol in $\mathscr{L}$ is an $\mathscr{L}$-term.
(T2) If $\tau_1, \ldots, \tau_n$ are any $\mathscr{L}$-terms which we have already built and $F\mathsf{x} \cdots \mathsf{x}$ is an n-ary function symbol in $\mathscr{L}$, then $F\tau_1 \cdots \tau_n$ is an $\mathscr{L}$-term (each place holder x is replaced by an $\mathscr{L}$-term).

When we write general statements which are independent of the signature $\mathscr{L}$, we omit the prefix $\mathscr{L}$ and simply write **term** rather than $\mathscr{L}$-term. Terms of the form (T0) or (T1) are the most basic terms we have, and since every term is built up from such terms, they are called **atomic terms**. In order to define the rule (T2) we had to use variables for terms, but since the variables of our alphabet stand just for objects of the domain and not for terms or other objects of the formal language, we had to introduce new symbols. For these new symbols, which do not belong to the alphabet of the formal language, we have chosen Greek letters. In fact, we shall mainly use Greek letters for variables which stand for objects of the formal language, also to emphasise the distinction between the formal language and the metalanguage. However, we shall use the Latin letters F and R as variables for function and relation symbols, respectively.

Note that this recursive definition of terms allows us to use the following principle: If we want to prove that all terms satisfy some property Φ, then one has to prove that

- all variables satisfy Φ;
- each constant symbol satisfies Φ;
- if some terms $\tau_1, \ldots, \tau_n$ satisfy Φ, then so does $F\tau_1, \cdots, \tau_n$ for every n-ary function symbol F.

We call this principle **induction on term construction**.

In order to make terms, relations, and other expressions in the formal language easier to read, it is convenient to introduce some more symbols, like brackets and commas, to our alphabet. For example, we usually write $F(\tau_1, \ldots, \tau_n)$ rather than $F\tau_1 \cdots \tau_n$.

To some extent, terms correspond to names, since they denote objects of the domain under consideration. Like real names, they are not statements and cannot express or describe possible relations between objects. So, the next step is to build sentences, or more precisely *formulae*, with these terms.

Formulae. A string of symbols is called an $\mathscr{L}$-**formula**, if it results from applying F I N I T E L Y many times the following rules:

(F0) If τ_1 and τ_2 are $\mathscr{L}$-terms, then $= \tau_1\tau_2$ is an $\mathscr{L}$-formula.
(F1) If $\tau_1, \ldots, \tau_n$ are any $\mathscr{L}$-terms and $R \times \cdots \times$ is any non-logical n-ary relation symbol in $\mathscr{L}$, then $R\tau_1 \cdots \tau_n$ is an $\mathscr{L}$-formula.
(F2) If φ is any $\mathscr{L}$-formula which we have already built, then $\neg\varphi$ is an $\mathscr{L}$-formula.
(F3) If φ and ψ are $\mathscr{L}$-formulae which we have already built, then $\wedge\varphi\psi$, $\vee\varphi\psi$, and $\rightarrow \varphi\psi$ are $\mathscr{L}$-formulae.
(F4) If φ is an $\mathscr{L}$-formula which we have already built, and ν is an arbitrary variable, then $\exists\nu\varphi$ and $\forall\nu\varphi$ are $\mathscr{L}$-formulae.

As in the case of terms, we usually write simply **formula** rather than $\mathscr{L}$-formula unless the statement in question refers to a specific language. Formulae of the form (F0) or (F1) are the most basic expressions we have, and since every formula is a logical connection or a quantification of these formulae, they are called **atomic formulae**.

In order to make formulae easier to read, we usually use *infix notation* instead of *Polish notation*, and use brackets if necessary. For example, we usually write $\varphi \wedge \psi$ instead of $\wedge\varphi\psi$, $\varphi \rightarrow (\psi \rightarrow \varphi)$ instead of $\rightarrow \varphi \rightarrow \psi\varphi$, and $(\varphi \rightarrow \psi) \rightarrow \varphi$ instead of $\rightarrow\rightarrow \varphi\psi\varphi$.

In the same way as for terms, a property Φ is satisfied by all formulae if we check the following:

- All atomic formulae satisfy Φ.
- If φ and ψ satisfy Φ and ν is a variable, then so do $\neg\varphi, \varphi \wedge \psi$, $\varphi \vee \psi$, $\varphi \rightarrow \psi$, $\exists\nu\varphi$ and $\forall\nu\varphi$.

In accordance with the corresponding principle for terms, we denote this as **induction on formula construction**.

For binary relation symbols $R\times\times$ and binary function symbols $F\times\times$, it is convenient to write xRy and xFy instead of $R(x, y)$ and $F(x, y)$, respectively. For example, we usually write $x = y$ instead of $= xy$.

If a formula φ is of the form $\exists\nu\psi$ or $\forall\nu\psi$ (for some variable ν and some formula ψ) and the variable ν occurs in ψ, then we say that ν is in the *range* of a logical quantifier. Every occurrence of a variable ν in a formula φ is said to be **bound** by the innermost quantifier in whose range it occurs. If an occurrence of the variable ν at a particular place is not in the range of a quantifier, it is said to be **free** at that particular place. Notice that it is possible that a variable occurs in a given formula at a certain place at bound and at another place at free. For example, in the formula $\exists z(x = z) \wedge \forall x(x = y)$, the variable x occurs bound and free, whereas z occurs just bound and y occurs just free. However, one can always rename the bound variables occurring in a given formula φ such that each variable in φ is either bound or free—the rules for this procedure are given later. For a formula φ, the set of variables occurring free in φ is denoted by free(φ). A formula φ is a **sentence** (or a **closed formula**) if it contains no free variables (i.e., free(φ) $= \emptyset$). For example, $\forall x(x = x)$ is a sentence but $x = x$ is just a formula.

In analogy to this definition, we say that a term is a **closed term** if it contains no variables. Obviously, the only terms which are closed are the constant symbols and the function symbols followed by closed terms.

Sometimes it is useful to indicate explicitly which variables occur free in a given formula φ, and for this we usually write $\varphi(x_1, \ldots, x_n)$ to indicate that $\{x_1, \ldots, x_n\} \subseteq$ free(φ).

If φ is a formula, ν a variable, and τ a term, then $\varphi(\nu/\tau)$ is the formula we get after replacing all *free* instances of the variable ν by τ. The process by which we obtain the formula $\varphi(\nu/\tau)$ is called **substitution**. Now, a substitution is **admissible** if and only if no free occurrence of ν in φ is in the range of a quantifier that binds any variable which appears in τ (i.e., for each variable $\tilde{\nu}$ appearing in τ, no place where ν occurs free in φ is in the range of $\exists\tilde{\nu}$ or $\forall\tilde{\nu}$). For example, if $x \notin$ free(φ), then $\varphi(x/\tau)$ is admissible for any term τ. In this case, the formulae φ and $\varphi(x/\tau)$ are identical, which we express by $\varphi \equiv \varphi(x/\tau)$. In general, we use the symbol $\equiv$ in the metalanguage to denote an equality of strings of symbols of the formal language. Furthermore, if φ is a formula and the substitution $\varphi(x/\tau)$ is admissible, then we write just $\varphi(\tau)$ instead of $\varphi(x/\tau)$. In order to express this, we write $\varphi(\tau) :\equiv \varphi(x/\tau)$, where we use the symbol $:\equiv$ in the metalanguage to define symbols (or strings of symbols) of the formal language.

So far, we have letters, and we can build names and sentences. However, these sentences are just strings of symbols without any inherent meaning. At a later stage, we shall interpret formulae in the intuitively natural way by giving the symbols their intended meaning (e.g., $\wedge$ meaning "and", $\forall x$ meaning "for all x", et cetera). But before we shall do so, let us stay a little bit longer on the syntactical side — nevertheless, one should consider the formulae from a semantical point of view as well.

Axioms

In what follows, we shall label certain formulae or types of formulae as **axioms**, which are used in connection with *inference rules* in order to derive further formulae. From a semantical point of view we can think of axioms as "true" statements from which we deduce or prove further results. We distinguish two types of axiom, namely *logical axioms* and *non-logical axioms* (which will be discussed later). A **logical axiom** is a sentence or formula φ which is universally valid (i.e., φ is true in any possible universe, no matter how the variables, constants, et cetera, occurring in φ are interpreted). Usually, one takes as logical axioms some minimal set of formulae that is sufficient for deriving all universally valid formulae — such a set is given below.

If a symbol, involved in an axiom, stands for an arbitrary relation, function, or even for a first-order formula, then we usually consider the statement as an **axiom schema** rather than a single axiom, since each instance of the symbol represents a single axiom. The following list of axiom schemata is a system of logical axioms.

Let φ, φ_1, φ_2, φ_3, and ψ, be arbitrary first-order formulae:

L_0: $\varphi \vee \neg\varphi$
L_1: $\varphi \rightarrow (\psi \rightarrow \varphi)$
L_2: $(\psi \rightarrow (\varphi_1 \rightarrow \varphi_2)) \rightarrow ((\psi \rightarrow \varphi_1) \rightarrow (\psi \rightarrow \varphi_2))$
L_3: $(\varphi \wedge \psi) \rightarrow \varphi$
L_4: $(\varphi \wedge \psi) \rightarrow \psi$
L_5: $\varphi \rightarrow (\psi \rightarrow (\psi \wedge \varphi))$
L_6: $\varphi \rightarrow (\varphi \vee \psi)$
L_7: $\psi \rightarrow (\varphi \vee \psi)$
L_8: $(\varphi_1 \rightarrow \varphi_3) \rightarrow ((\varphi_2 \rightarrow \varphi_3) \rightarrow ((\varphi_1 \vee \varphi_2) \rightarrow \varphi_3))$
L_9: $\neg\varphi \rightarrow (\varphi \rightarrow \psi)$

Let τ be a term, ν a variable, and assume that the substitution which leads to $\varphi(\nu/\tau)$ is admissible:

L_{10}: $\forall\nu\varphi(\nu) \rightarrow \varphi(\tau)$
L_{11}: $\varphi(\tau) \rightarrow \exists\nu\varphi(\nu)$

Let ψ be a formula and let ν a variable such that $\nu \notin \text{free}(\psi)$:

L_{12}: $\forall\nu(\psi \rightarrow \varphi(\nu)) \rightarrow (\psi \rightarrow \forall\nu\varphi(\nu))$
L_{13}: $\forall\nu(\varphi(\nu) \rightarrow \psi) \rightarrow (\exists\nu\varphi(\nu) \rightarrow \psi)$

What is not yet covered is the symbol $=$, so let us now have a closer look at the binary equality relation. The defining properties of equality can already be found in Book VII, Chapter 1 of Aristotle's *Topics* [2], where one of the rules to decide whether two things are the same is as follows: ... *you should look at every possible predicate of each of the two terms and at the things of which they are predicated and see whether there is any discrepancy anywhere. For anything which is predicated of the one ought also to be predicated of the other, and of anything of which the one is a predicate the other also ought to be a predicate.*

In our formal system, the binary equality relation is defined by the following three axioms. Let $\tau, \tau_1, \ldots, \tau_n, \tau_1', \ldots, \tau_n'$ be arbitrary terms, let R be an n-ary relation symbol (e.g., the binary relation symbol $=$), and let F be an n-ary function symbol:

L_{14}: $\tau = \tau$
L_{15}: $(\tau_1 = \tau_1' \wedge \cdots \wedge \tau_n = \tau_n') \rightarrow (R(\tau_1, \ldots, \tau_n) \rightarrow R(\tau_1', \ldots, \tau_n'))$
L_{16}: $(\tau_1 = \tau_1' \wedge \cdots \wedge \tau_n = \tau_n') \rightarrow (F(\tau_1, \ldots, \tau_n) = F(\tau_1', \ldots, \tau_n'))$

Finally, we define the logical operator $\leftrightarrow$, the quantifier $\exists!$ and the binary relation symbol $\neq$ by stipulating:

$$\begin{aligned} \varphi \leftrightarrow \psi \quad &:\Longleftrightarrow \quad (\varphi \rightarrow \psi) \wedge (\psi \rightarrow \varphi), \\ \exists! \nu \varphi \quad &:\Longleftrightarrow \quad \exists \nu (\varphi(\nu) \wedge \forall \mu (\varphi(\mu) \rightarrow \mu = \nu)) \\ \tau \neq \tau' \quad &:\Longleftrightarrow \quad \neg(\tau = \tau'), \end{aligned}$$

where we use the symbol $:\Longleftrightarrow$ in the metalanguage to define relations between symbols (or strings of symbols) of the formal language (i.e., $\leftrightarrow$, $\exists! \nu \varphi$ and $\neq$ are just abbreviations).

This completes the list of our logical axioms. In addition to these axioms, we are allowed to state arbitrarily many *sentences*. In logic, such a (possibly empty) set of sentences is also called a **theory**, or, when the signature $\mathscr{L}$ is specified, an **$\mathscr{L}$-theory**. The elements of a theory are called **non-logical axioms**. Notice that non-logical axioms are sentences (i.e., formulae without free variables). Examples of theories (i.e., of sets of non-logical axioms) which will be discussed in this book are the axioms of Set Theory (see Part IV), the axioms of Peano Arithmetic PA (also known as *Number Theory*), and the axioms of Group Theory GT, which we discuss first.

GT: The language of **Group Theory** is $\mathscr{L}_{\mathsf{GT}} = \{\mathsf{e}, \circ\}$, where e is a constant symbol and $\circ$ is a binary function symbol.

GT_0: $\forall x \forall y \forall z (x \circ (y \circ z) = (x \circ y) \circ z)$ (i.e., $\circ$ is *associative*)
GT_1: $\forall x (\mathsf{e} \circ x = x)$ (i.e., e is a *left-neutral* element)
GT_2: $\forall x \exists y (y \circ x = \mathsf{e})$ (i.e., every element has a *left-inverse*)

PA: The language of **Peano Arithmetic** is $\mathscr{L}_{\mathsf{PA}} = \{0, \mathsf{s}, +, \cdot\}$, where 0 is a constant symbol, s is a unary function symbol, and $+$, $\cdot$ are binary function symbols.

PA_0: $\neg \exists x (\mathsf{s}x = 0)$
PA_1: $\forall x \forall y (\mathsf{s}x = \mathsf{s}y \rightarrow x = y)$
PA_2: $\forall x (x + 0 = x)$
PA_3: $\forall x \forall y (x + \mathsf{s}y = \mathsf{s}(x + y))$

PA_4: $\forall x(x \cdot 0 = 0)$
PA_5: $\forall x \forall y(x \cdot \mathtt{s}y = (x \cdot y) + x)$

Let φ be an arbitrary $\mathscr{L}_{\mathsf{PA}}$-formula with $x \in \text{free}(\varphi)$:

PA_6: $\big(\varphi(0) \wedge \forall x(\varphi(x) \to \varphi(\mathtt{s}x))\big) \to \forall x \varphi(x)$

Notice that PA_6 is an axiom schema, known as the Induction Schema, and not just a single axiom like PA_0–PA_5.

It is often convenient to add certain *defined symbols* to a given language so that the expressions get shorter or are at least easier to read. For example, in Peano Arithmetic — which is an axiomatic system for the natural numbers — we usually replace the expression $\mathtt{s}0$ with 1 and $\mathtt{ss}0$ with 2. More formally, we define:

$$1 :\equiv \mathtt{s}0 \qquad \text{and} \qquad 2 :\equiv \mathtt{ss}0$$

where we use the symbol $:\equiv$ in the metalanguage to define new constant symbols or certain formulae. Obviously, all that can be expressed in the language $\mathscr{L}_{\mathsf{PA}} \cup \{1, 2\}$ can also be expressed in $\mathscr{L}_{\mathsf{PA}}$.

Formal Proofs

So far we have a set of logical and non-logical axioms in a certain language and can define, if we wish, as many new constants, functions, and relations as we like. However, we are still not able to deduce anything from the given axioms, since until now, we do not have *inference rules* which allow us, e.g., to infer a certain sentence from a given set of axioms.

Surprisingly, just two **inference rules** are sufficient, namely

$$\text{Modus Ponens (MP):} \qquad \frac{\varphi \to \psi,\ \varphi}{\psi}$$

and for variables ν which do not occur free in any non-logical axiom:

$$\text{Generalisation } (\forall): \qquad \frac{\varphi}{\forall \nu \varphi}$$

In the former case we say that the formula ψ is obtained from $\varphi \to \psi$ and φ by Modus Ponens, abbreviated (MP), and in the latter case we say that $\forall \nu \varphi$ (where ν can be any variable) is obtained from φ by Generalisation, abbreviated $(\forall)$. It is worth mentioning that the restriction on $(\forall)$ is not essential, but will simplify certain proofs (e.g., the proof of the DEDUCTION THEOREM 2.1).

Using these two inference rules, we are now able to define the notion of a formal proof: Let $\mathscr{L}$ be a signature (i.e., a possibly empty set of non-logical symbols) and let $\mathbf{\Phi}$ be a possibly empty set of $\mathscr{L}$-formulae (e.g., a set of axioms). An $\mathscr{L}$-formula ψ is **provable** from $\mathbf{\Phi}$ (or provable in $\mathbf{\Phi}$), denoted $\mathbf{\Phi} \vdash \psi$, if there is a F I N I T E sequence $\varphi_0, \ldots, \varphi_n$ of $\mathscr{L}$-formulae such that

$\varphi_n \equiv \psi$ (i.e., the formulae φ_n and ψ are identical), and for all i with $i \leq n$ we are in at least one of the following cases:

- φ_i is a logical axiom
- $\varphi_i \in \mathbf{\Phi}$
- there are $j, k < i$ such that $\varphi_j \equiv \varphi_k \to \varphi_i$
- there is a $j < i$ such that $\varphi_i \equiv \forall \nu\, \varphi_j$ for some variable ν

The sequence $\varphi_0, \ldots, \varphi_n$ is then called a **formal proof** of ψ.

In the case when $\mathbf{\Phi}$ is the empty set, we simply write $\vdash \psi$. If a formula ψ is not provable from $\mathbf{\Phi}$, i.e., if there is no formal proof for ψ which uses just formulae from $\mathbf{\Phi}$, then we write $\mathbf{\Phi} \nvdash \psi$.

Formal proofs, even of very simple statements, can get quite long and tricky. Nevertheless, we shall give two examples:

Example 1.1. For every formula φ we have:

$$\vdash \varphi \to \varphi$$

A formal proof of $\varphi \to \varphi$ is given by

φ_0:	$(\varphi \to ((\varphi \to \varphi) \to \varphi)) \to ((\varphi \to (\varphi \to \varphi)) \to (\varphi \to \varphi))$	instance of L_2
φ_1:	$\varphi \to ((\varphi \to \varphi) \to \varphi)$	instance of L_1
φ_2:	$(\varphi \to (\varphi \to \varphi)) \to (\varphi \to \varphi)$	from φ_0 and φ_1 by (MP)
φ_3:	$\varphi \to (\varphi \to \varphi)$	instance of L_1
φ_4:	$\varphi \to \varphi$	from φ_2 and φ_3 by (MP)

Example 1.2. We give a formal proof of $\mathsf{PA} \vdash 1 + 1 = 2$. Recall that we have defined $1 :\equiv \mathtt{s}0$ and $2 :\equiv \mathtt{ss}0$, so we need to prove $\mathsf{PA} \vdash \mathtt{s}0 + \mathtt{s}0 = \mathtt{ss}0$.

φ_0:	$\forall x \forall y (x + \mathtt{s}y = \mathtt{s}(x + y))$	instance of PA_3
φ_1:	$\forall x \forall y (x + \mathtt{s}y = \mathtt{s}(x + y)) \to \forall y (\mathtt{s}0 + \mathtt{s}y = \mathtt{s}(\mathtt{s}0 + y))$	instance of L_{10}
φ_2:	$\forall y (\mathtt{s}0 + \mathtt{s}y = \mathtt{s}(\mathtt{s}0 + y))$	from φ_1 and φ_0 by (MP)
φ_3:	$\forall y (\mathtt{s}0 + \mathtt{s}y = \mathtt{s}(\mathtt{s}0 + y)) \to \mathtt{s}0 + \mathtt{s}0 = \mathtt{s}(\mathtt{s}0 + 0)$	instance of L_{10}
φ_4:	$\mathtt{s}0 + \mathtt{s}0 = \mathtt{s}(\mathtt{s}0 + 0)$	from φ_3 and φ_2 by (MP)
φ_5:	$\forall x (x + 0 = x)$	instance of PA_2
φ_6:	$\forall x (x + 0 = x) \to \mathtt{s}0 + 0 = \mathtt{s}0$	instance of L_{10}
φ_7:	$\mathtt{s}0 + 0 = \mathtt{s}0$	from φ_6 and φ_5 by (MP)
φ_8:	$\mathtt{s}0 + 0 = \mathtt{s}0 \to \mathtt{s}(\mathtt{s}0 + 0) = \mathtt{ss}0$	instance of L_{16}
φ_9:	$\mathtt{s}(\mathtt{s}0 + 0) = \mathtt{ss}0$	from φ_8 and φ_7 by (MP)
φ_{10}:	$\mathtt{s}0 + \mathtt{s}0 = \mathtt{s}0 + \mathtt{s}0$	instance of L_{14}
φ_{11}:	$\varphi_{10} \to (\varphi_9 \to (\varphi_{10} \land \varphi_9))$	instance of L_5

φ_{12}:	$\varphi_9 \to (\varphi_{10} \wedge \varphi_9)$	from φ_{11} and φ_{10} by (MP)
φ_{13}:	$\varphi_{10} \wedge \varphi_9$	fromφ_{12} and φ_9 by (MP)
φ_{14}:	$(\varphi_{10} \wedge \varphi_9) \to (\mathtt{s0} + \mathtt{s0} = \mathtt{s}(\mathtt{s0} + 0) \to \mathtt{s0} + \mathtt{s0} = \mathtt{ss0})$	instance of L_{15}
φ_{15}:	$\mathtt{s0} + \mathtt{s0} = \mathtt{s}(\mathtt{s0} + 0) \to \mathtt{s0} + \mathtt{s0} = \mathtt{ss0}$	from φ_{14} and φ_{13} by (MP)
φ_{16}:	$\mathtt{s0} + \mathtt{s0} = \mathtt{ss0}$	from φ_{15} and φ_4 by (MP)

In Chapter 2, we will introduce some techniques which allow us to simplify formal proofs such as the one presented above.

Tautologies & Logical Equivalence

We say that two formulae φ and ψ are **logically equivalent** (or just **equivalent**), denoted $\varphi \Leftrightarrow \psi$, if $\vdash \varphi \leftrightarrow \psi$. More formally:

$$\varphi \Leftrightarrow \psi \quad :\Longleftrightarrow \quad \vdash \varphi \leftrightarrow \psi$$

In other words, if $\varphi \Leftrightarrow \psi$, then—from a logical point of view—φ and ψ state exactly the same, and therefore we could call $\varphi \leftrightarrow \psi$ a tautology, which means *saying the same thing twice*. Indeed, in logic, a formula φ is a **tautology** if $\vdash \varphi$. Thus, the formulae φ and ψ are equivalent if and only if $\varphi \leftrightarrow \psi$ is a tautology. More generally, if $\mathbf{\Phi}$ is a set of formulae, we write $\varphi \Leftrightarrow_{\mathbf{\Phi}} \psi$ to denote $\mathbf{\Phi} \vdash \varphi \leftrightarrow \psi$.

Example 1.3. For every formula φ we have:

$$\varphi \Leftrightarrow \varphi$$

This follows directly from Example 1.1, since $\varphi \leftrightarrow \varphi$ is simply an abbreviation for $(\varphi \to \varphi) \wedge (\varphi \to \varphi)$:

φ_0:	$\varphi \to \varphi$	Example 1.1
φ_1:	$(\varphi \to \varphi) \to ((\varphi \to \varphi) \to (\varphi \leftrightarrow \varphi))$	instance of L_5
φ_2:	$(\varphi \to \varphi) \to (\varphi \leftrightarrow \varphi)$	from φ_0 and φ_1 by (MP)
φ_3:	$\varphi \leftrightarrow \varphi$	from φ_0 and φ_2 by (MP)

Example 1.4. For every formula φ we have:

$$\varphi \Leftrightarrow \neg\neg\varphi$$

By applying L_5 as in Example 1.3, one can easily check that it suffices to prove separately that the formulae $\varphi \to \neg\neg\varphi$ and $\neg\neg\varphi \to \varphi$ are tautologies. We only prove the former statement, the latter one is proved in Example 2.2.

φ_0:	$(\neg\varphi \to (\varphi \to \neg\neg\varphi)) \to ((\neg\neg\varphi \to (\varphi \to \neg\neg\varphi)) \to ((\neg\varphi \vee \neg\neg\varphi) \to (\varphi \to \neg\neg\varphi)))$	instance of L_8
φ_1:	$\neg\varphi \to (\varphi \to \neg\neg\varphi)$	instance of L_9
φ_2:	$(\neg\neg\varphi \to (\varphi \to \neg\neg\varphi)) \to ((\neg\varphi \vee \neg\neg\varphi) \to (\varphi \to \neg\neg\varphi))$	from φ_0 and φ_1 by (MP)
φ_3:	$\neg\neg\varphi \to (\varphi \to \neg\neg\varphi)$	instance of L_1
φ_4:	$(\neg\varphi \vee \neg\neg\varphi) \to (\varphi \to \neg\neg\varphi)$	from φ_2 and φ_2 by (MP)
φ_5:	$\neg\varphi \vee \neg\neg\varphi$	instance of L_0
φ_6:	$\varphi \to \neg\neg\varphi$	from φ_4 and φ_5 by (MP)

Example 1.5. Commutativity and associativity of $\wedge$ and $\vee$ are tautological, i.e., for all formulae φ, ψ and χ we have $\varphi \wedge \psi \Leftrightarrow \psi \wedge \varphi$ and $\varphi \wedge (\psi \wedge \chi) \Leftrightarrow (\varphi \wedge \psi) \wedge \chi$; and similarly for $\vee$. Again, we omit the proof since it will be trivial once we have proved the DEDUCTION THEOREM (THEOREM 2.1). This result legitimizes the notations $\varphi_0 \wedge \ldots \wedge \varphi_n$ and $\varphi_0 \vee \ldots \vee \varphi_n$ respectively for $\varphi_0 \wedge (\varphi_1 \wedge (\ldots \wedge \varphi_n)\ldots)$ and $\varphi_0 \vee (\varphi_1 \vee (\ldots \vee \varphi_n)\ldots)$ respectively.

In Appendix 17, there is a list of tautologies which will be frequently used in formal proofs. Note that it follows from EXERCISE 1.4 that $\Leftrightarrow$ defines an equivalence relation on all $\mathscr{L}$-formulae for some given signature $\mathscr{L}$. Moreover, it even defines a congruence relation (i.e., equivalence is closed under all logical operations). More precisely, if $\varphi \Leftrightarrow \varphi'$ and $\psi \Leftrightarrow \psi'$, then:

$$\neg\varphi \Leftrightarrow \neg\varphi'$$
$$\varphi \circ \psi \Leftrightarrow \varphi' \circ \psi'$$
$$\forall\!\!\!\exists \nu\varphi \Leftrightarrow \forall\!\!\!\exists \nu\varphi'$$

where $\circ$ stands for either $\wedge, \vee$, or $\to$, and $\forall\!\!\!\exists$ stands for either $\exists$ or $\forall$. A proof of these statements will be easier once we have proved THEOREM 2.1.

The above observation enables us to replace subformulae of a given formula φ by equivalent formulae so that the resulting formula is equivalent to φ.

THEOREM 1.6 (SUBSTITUTION THEOREM). *Let φ be a formula and let α be a subformula of φ. Let ψ be the formula obtained from φ by replacing one or multiple occurrences of α by some formula β. Then we have:*

$$\alpha \Leftrightarrow \beta \quad \Longrightarrow \quad \varphi \Leftrightarrow \psi$$

Proof. We prove the theorem by induction on the recursive construction of the formula φ. If φ is an atomic formula or if α is φ, then the statement is trivial. If φ is a composite formula, then we use the observation that $\Leftrightarrow$ is a congruence relation: For example, if the formula φ is of the form $\neg\varphi'$, and ψ' is the formula obtained from φ' by replacing one or multiple occurrences of α by β, then by induction we may assume that $\varphi' \Leftrightarrow \psi'$. Consequently, we have $\neg\varphi' \Leftrightarrow \neg\psi'$ as desired. The other cases can be checked in a similar way. $\dashv$

THEOREM 1.7 (THREE-SYMBOLS). *For every formula φ there is an equivalent formula ψ which contains only the symbols $\neg$ and $\wedge$ as logical operators and $\exists$ as quantifier.*

Proof. By THEOREM 1.6, it suffices to prove the following equivalences:

$$\varphi \vee \psi \Leftrightarrow \neg(\neg\varphi \wedge \neg\psi)$$
$$\varphi \rightarrow \psi \Leftrightarrow \neg\varphi \vee \psi$$
$$\forall\nu\varphi \Leftrightarrow \neg\exists\nu\neg\varphi$$

The proof of these equivalences is left as an exercise (see EXERCISE 1.2). Note that THEOREM 2.1 as well as the methods of proof introduced in Chapter 2 will simplify the proofs to a great extent. ⊣

As a consequence of THEOREM 1.7, one could simplify both the alphabet and the logical axioms. Nevertheless, we do not wish to do so, since this would also decrease the readability of formulae.

NOTES

The logical axioms are essentially those given by Hilbert (see, e.g., [26]). However, Hilbert also introduced the axiom schemata $\big((\varphi \rightarrow \psi) \wedge (\neg\varphi \rightarrow \psi)\big) \rightarrow \psi$ and $\neg\neg\varphi \rightarrow \varphi$. On the other hand, he did not introduce the LAW OF EXCLUDED MIDDLE L_0, because it is not needed any more in this setting (see also EXERCISES 1.5 & 2.5.(c)).

EXERCISES

1.0 Show that $\vdash \exists x(x = x)$.

1.1 Show that the equality relation is transitive, i.e.,

$$\vdash \forall x\forall y\forall z\big((x = y \wedge y = z) \rightarrow x = z\big).$$

1.2 Prove the equivalences in the proof of THEOREM 1.7.
Hint: Prove the tautologies (K), (L.0), and (R) first.

1.3 (a) Show that quantifier-free formulae can be written with the only logical operator $\tilde{\wedge}$, where

$$\varphi \mathbin{\tilde{\wedge}} \psi :\Longleftrightarrow \neg(\varphi \wedge \psi).$$

(b) Show that quantifier-free formulae can be written with the only logical operator $\tilde{\vee}$, where

$$\varphi \mathbin{\tilde{\vee}} \psi :\Longleftrightarrow \neg(\varphi \vee \psi).$$

1.4 Show that logical equivalence $\Leftrightarrow$ defines an equivalence relation on the set of formulae.

1.5 Let $\mathsf{L}_{9^3/_4}$ be the axiom schema $(\varphi \rightarrow \psi) \rightarrow \big((\varphi \rightarrow \neg\psi) \rightarrow \neg\varphi\big)$.

(a) Show that $\{\mathsf{L}_0, \mathsf{L}_1, \mathsf{L}_2, \mathsf{L}_8, \mathsf{L}_9\} \vdash \mathsf{L}_{9^3/_4}$.

(b) Show that $\{\mathsf{L}_1, \mathsf{L}_2, \mathsf{L}_{9^3/_4}\} \vdash \varphi \rightarrow \neg\neg\varphi$.

(c) Show that $\{\mathsf{L}_1\text{–}\mathsf{L}_9\} \nvdash \mathsf{L}_{9\,3/4}$.

Hint: Define a mapping $|\cdot|$, which assigns to each formula φ a value $|\varphi| \in \{-1, 0, 1\}$ such that the following conditions are satisfied: $|\varphi \vee \psi| = \max\{|\varphi|, |\psi|\}$, $|\varphi \wedge \psi| = \min\{|\varphi|, |\psi|\}$, $|\neg\varphi| = -|\varphi|$, and the value of $|\varphi \to \psi|$ is given by the following table:

$\lvert\varphi\rvert \setminus \lvert\psi\rvert$	-1	0	1
-1	1	1	1
0	1	1	1
1	-1	0	1

$|\varphi \to \psi|$

Show that for every formula θ with $\{\mathsf{L}_1\text{–}\mathsf{L}_9\} \vdash \theta$ we have $|\theta| = 1$. On the other hand, for certain values of $|\varphi|$ and $|\psi|$ we have $|\mathsf{L}_{9\,3/4}| \neq 1$.

Chapter 2
The Art of Proof

In Example 1.2 we gave a proof of $1 + 1 = 2$ in 17 (!) proof steps. At that point you may have asked yourself: If it takes that much effort to prove such a simple statement, how can one ever prove any non-trivial mathematical result using formal proofs? This objection is of course justified; however, we will show in this chapter how one can simplify formal proofs using some methods of proof such as proofs by cases or by contradiction. It is crucial to note that the following results are not theorems of a formal theory, but theorems about formal proofs. In particular, they show how — under certain conditions — a formal proof can be transformed into another.

The Deduction Theorem

In common mathematics, one usually proves implications of the form

$$\text{IF } \Phi \text{ THEN } \Psi$$

by simply assuming the truth of Φ and deriving from this the truth of Ψ. When writing formal proofs, the so-called DEDUCTION THEOREM enables us to use a similar trick: Rather than proving $\mathbf{\Phi} \vdash \varphi \rightarrow \psi$, we simply add φ to our set of formulae $\mathbf{\Phi}$ and prove $\mathbf{\Phi} \cup \{\varphi\} \vdash \psi$.

If $\mathbf{\Phi}$ is a set of formulae and $\mathbf{\Phi}'$ is another set of formulae in the same language as $\mathbf{\Phi}$, then we write $\mathbf{\Phi} + \mathbf{\Phi}'$ for $\mathbf{\Phi} \cup \mathbf{\Phi}'$. In the case when $\mathbf{\Phi}' \equiv \{\varphi\}$ consists of a single formula, we write $\mathbf{\Phi} + \varphi$ for $\mathbf{\Phi} \cup \mathbf{\Phi}'$.

THEOREM 2.1 (DEDUCTION THEOREM). *If $\mathbf{\Phi}$ is a set of formulae and $\mathbf{\Phi} + \psi \vdash \varphi$, then $\mathbf{\Phi} \vdash \psi \rightarrow \varphi$; and vice versa, if $\mathbf{\Phi} \vdash \psi \rightarrow \varphi$, then $\mathbf{\Phi} + \psi \vdash \varphi$, i.e., we have:*

$$\mathbf{\Phi} + \psi \vdash \varphi \iff \mathbf{\Phi} \vdash \psi \rightarrow \varphi \tag{DT}$$

Proof. It is clear that $\mathbf{\Phi} \vdash \psi \rightarrow \varphi$ implies $\mathbf{\Phi} + \psi \vdash \varphi$. Conversely, suppose that $\mathbf{\Phi} + \psi \vdash \varphi$ holds and let the sequence $\varphi_0, \ldots, \varphi_n$ with $\varphi_n \equiv \varphi$ be a formal proof for φ from $\mathbf{\Phi} + \psi$. For each $i \leq n$ we will replace the formula φ_i

L. Halbeisen, R. Krapf, *Gödel's Theorems and Zermelo's Axioms*,
https://doi.org/10.1007/978-3-030-52279-7_2

by a sequence of formulae which ends with $\psi \to \varphi_i$. Let $i \leq n$ and assume $\mathbf{\Phi} \vdash \psi \to \varphi_j$ for every $j < i$.

- If φ_i is a logical axiom or $\varphi_i \in \mathbf{\Phi}$, we have:

$\varphi_{i,0}$:	φ_i	$\varphi_i \in \mathbf{\Phi}$ or φ_i is a logical axiom
$\varphi_{i,1}$:	$\varphi_i \to (\psi \to \varphi_i)$	instance of L_1
$\varphi_{i,2}$:	$\psi \to \varphi_i$	from $\varphi_{i,1}$ and $\varphi_{i,0}$ by (MP)

- The case $\varphi_i \equiv \psi$ follows directly from Example 1.1.

- If φ_i is obtained from φ_j and $\varphi_k \equiv (\varphi_j \to \varphi_i)$ by Modus Ponens, where $j, k < i$, we have:

$\varphi_{i,0}$:	$\psi \to \varphi_j$	since $j < i$
$\varphi_{i,1}$:	$\psi \to (\varphi_j \to \varphi_i)$	since $k < i$
$\varphi_{i,2}$:	$\varphi_{i,1} \to ((\psi \to \varphi_j) \to (\psi \to \varphi_i))$	instance of L_2
$\varphi_{i,3}$:	$(\psi \to \varphi_j) \to (\psi \to \varphi_i)$	from $\varphi_{i,2}$ and $\varphi_{i,1}$ by (MP)
$\varphi_{i,4}$:	$\psi \to \varphi_i$	from $\varphi_{i,3}$ and $\varphi_{i,0}$ by (MP)

- If φ_i is obtained from φ_j by Generalisation, where $j < i$, i.e., $\varphi_i \equiv \forall \nu \varphi_j$ for some variable ν, then, by the rules of Generalisation, the variable ν does not occur free in $\mathbf{\Phi} + \psi$. In particular, $\nu \notin \text{free}(\psi)$, and the claim follows from

$\varphi_{i,0}$:	$\psi \to \varphi_j$	since $j < i$
$\varphi_{i,1}$:	$\forall \nu(\psi \to \varphi_j)$	from $\varphi_{i,0}$ by ($\forall$)
$\varphi_{i,2}$:	$\forall \nu(\psi \to \varphi_j) \to (\psi \to \varphi_i)$	instance of L_{12}
$\varphi_{i,3}$:	$\psi \to \varphi_i$	from $\varphi_{i,2}$ and $\varphi_{i,1}$ by (MP)

Hence we have $\mathbf{\Phi} \vdash \psi \to \varphi$. ⊣

As a first application, note that $\vdash \varphi \to \varphi$ is a trivial consequence of the Deduction Theorem, whereas its formal proof in Example 1.1 has five steps.

As a further application of the Deduction Theorem, we show that the equality relation is symmetric. We first show that $\{x = y\} \vdash y = x$:

φ_0:	$(x = y \wedge x = x) \to (x = x \to y = x)$	instance of L_{15}
φ_1:	$x = x$	instance of L_{14}
φ_2:	$x = y$	$x = y$ belongs to $\{x = y\}$
φ_3:	$x = x \to (x = y \to (x = y \wedge x = x))$	instance of L_5
φ_4:	$x = y \to (x = y \wedge x = x)$	from φ_3 and φ_1 by (MP)
φ_5:	$x = y \wedge x = x$	from φ_4 and φ_2 by (MP)
φ_6:	$x = x \to y = x$	from φ_0 and φ_5 by (MP)
φ_7:	$y = x$	from φ_6 and φ_1 by (MP)

Thus, we have $\{x = y\} \vdash y = x$, and by the DEDUCTION THEOREM 2.1 we see that $\vdash x = y \to y = x$. Finally, by Generalisation we get

$$\vdash \forall x \forall y(x = y \to y = x).$$

We leave it as an exercise to the reader to show that the equality relation is also transitive (see EXERCISE 1.1).

Example 2.2. We prove the tautology $\neg\neg\varphi \to \varphi$. By the DEDUCTION THEOREM it suffices to prove $\{\neg\neg\varphi\} \vdash \varphi$.

φ_0:	$\neg\neg\varphi \to (\neg\varphi \to \varphi)$	instance of L_9
φ_1:	$\neg\neg\varphi$	$\neg\neg\varphi \in \{\neg\neg\varphi\}$
φ_2:	$\neg\varphi \to \varphi$	from φ_0 and φ_1 by (MP)
φ_3:	$(\varphi \to \varphi) \to ((\neg\varphi \to \varphi) \to ((\varphi \vee \neg\varphi) \to \varphi))$	instance of L_8
φ_4:	$\varphi \to \varphi$	by Example 1.1
φ_5:	$(\neg\varphi \to \varphi) \to ((\varphi \vee \neg\varphi) \to \varphi)$	from φ_3 and φ_4 by (MP)
φ_6:	$(\varphi \vee \neg\varphi) \to \varphi$	from φ_5 and φ_2 by (MP)
φ_7:	$\varphi \vee \neg\varphi$	instance of L_0
φ_8:	φ	from φ_6 and φ_7 by (MP)

Natural Deduction

We have introduced predicate logic in such a way that there are many logical axioms and only two inference rules. However, it is also possible to introduce calculi with an opposite approach: few axioms and many inference rules. In the calculus of **natural deduction** there are, in fact, no axioms at all. Its inference rules essentially state how to transform a given formal proof to another one. We write $\mathbf{\Phi} \vdash \varphi$ to state that there is a formal proof of φ in the calculus of natural deduction with the non-logical axioms given by $\mathbf{\Phi}$.

Let $\mathbf{\Phi}$ be a set of formulae and let φ, ψ, χ be any formulae. The first rule states how formal proofs can be initialized.

$$\text{INITIAL RULE (IR):} \quad \frac{}{\mathbf{\Phi} \vdash \varphi} \quad \text{for } \varphi \in \mathbf{\Phi}.$$

In the calculus of natural deduction there are so-called **introduction rules** and **elimination rules** for each logical symbol.

$$(\mathsf{I}\wedge): \quad \frac{\mathbf{\Phi} \vdash \varphi \text{ and } \mathbf{\Phi} \vdash \psi}{\mathbf{\Phi} \vdash \varphi \wedge \psi} \qquad (\mathsf{E}\wedge): \quad \frac{\mathbf{\Phi} \vdash \varphi \wedge \psi}{\mathbf{\Phi} \vdash \varphi \text{ and } \mathbf{\Phi} \vdash \psi}$$

$$(\mathsf{I}\vee): \quad \frac{\mathbf{\Phi} \vdash \varphi \text{ or } \mathbf{\Phi} \vdash \psi}{\mathbf{\Phi} \vdash \varphi \vee \psi} \qquad (\mathsf{E}\vee): \quad \frac{\mathbf{\Phi} \vdash \varphi \vee \psi, \mathbf{\Phi} + \varphi \vdash \chi \text{ and } \mathbf{\Phi} + \psi \vdash \chi}{\mathbf{\Phi} \vdash \chi}$$

(I→): $\dfrac{\mathbf{\Phi} + \{\varphi\} \vdash \psi}{\mathbf{\Phi} \vdash \varphi \to \psi}$ (E→): $\dfrac{\mathbf{\Phi} \vdash \varphi \to \psi \text{ and } \mathbf{\Phi} \vdash \varphi}{\mathbf{\Phi} \vdash \psi}$

(I¬): $\dfrac{\mathbf{\Phi} + \varphi \vdash \psi \wedge \neg\psi}{\mathbf{\Phi} \vdash \neg\varphi}$ (E¬): $\dfrac{\mathbf{\Phi} \vdash \neg\neg\varphi}{\mathbf{\Phi} \vdash \varphi}$

Let τ be a term and ν be a variable such that the substitution $\varphi(\nu/\tau)$ is admissible and $\nu \notin \text{free}(\chi)$ for any formula $\chi \in \mathbf{\Phi}$ and — in the case of (E∃) — $\nu \notin \text{free}(\psi)$. Now we can state the corresponding introduction and elimination rules for quantifiers:

(I∃): $\dfrac{\mathbf{\Phi} \vdash \varphi(\tau)}{\mathbf{\Phi} \vdash \exists\nu\varphi(\nu)}$ (E∃): $\dfrac{\mathbf{\Phi} \vdash \exists\nu\varphi(\nu) \text{ and } \mathbf{\Phi} + \varphi(\nu) \vdash \psi}{\mathbf{\Phi} \vdash \psi}$

(I∀): $\dfrac{\mathbf{\Phi} \vdash \varphi(\nu)}{\mathbf{\Phi} \vdash \forall\nu\varphi(\nu)}$ (E∀): $\dfrac{\mathbf{\Phi} \vdash \forall\nu\varphi(\nu)}{\mathbf{\Phi} \vdash \varphi(\tau)}$

Finally, we need to deal with equality and atomic formulae. Let τ, τ_1 and τ_2 be terms and φ an atomic formula. The following introduction and elimination rules for equality are closely related to the logical axioms L_{14}–L_{16}:

(I=): $\dfrac{}{\tau = \tau}$ (E=): $\dfrac{\mathbf{\Phi} \vdash \tau_1 = \tau_2 \text{ and } \mathbf{\Phi} \vdash \varphi}{\mathbf{\Phi} \vdash \varphi(\tau_1/\tau_2)}$

Formal proofs in the calculus of natural deduction are defined in a similar way as in our usual calculus: There is a formal proof of a formula φ from a set of formulae $\mathbf{\Phi}$, denoted by $\mathbf{\Phi} \vdash \varphi$, if there is a F I N I T E sequence of of pairs $(\mathbf{\Phi}_0, \varphi_0), \dots, (\mathbf{\Phi}_n, \varphi_n)$ such that $\mathbf{\Phi}_n \equiv \mathbf{\Phi}$, $\varphi_n \equiv \varphi$ and for each $i \leq n$, $\mathbf{\Phi}_i \vdash \varphi_i$ is obtained by the application of an inference rule

$$\frac{\mathbf{\Phi}_{j_0} \vdash \varphi_{j_0}, \dots, \mathbf{\Phi}_{j_k} \vdash \varphi_{j_k}}{\mathbf{\Phi}_i \vdash \varphi_i}$$

with $k \leq 3$ and $j_0, \dots, j_k < i$. Note that the the case $k = 0$ is permitted, which corresponds to an application of the INITIAL RULE. In the case when $\mathbf{\Phi}$ is the empty set, we simply write $\vdash \varphi$.

We have now described two ways of introducing formal proofs. It is therefore natural to ask whether the two systems prove the same theorems. Fortunately, this question turns out to have a positive answer.

THEOREM 2.3. *Let $\mathbf{\Phi}$ be a set of formulae and let φ be a formula. Then we have*

$$\mathbf{\Phi} \vdash \varphi \iff \mathbf{\Phi} \vdash \varphi.$$

Proof. We need to verify that every formal proof in the usual sense can be turned into a formal proof in the calculus of natural deduction and vice versa. In order to prove that $\mathbf{\Phi} \vdash \varphi$ implies $\mathbf{\Phi} \vdash \varphi$ for every formula φ, we need to derive all introduction and elimination rules from our logical axioms and

(MP) and ($\forall$). We focus on some of the rules only and leave the others as an exercise.

Formal proofs of the form $\mathbf{\Phi} \vdash \varphi$ with $\varphi \in \mathbf{\Phi}$ using only (IR) obviously correspond to trivial formal proofs of the form $\mathbf{\Phi} \vdash \varphi$. We consider the more interesting elimination rule (E$\vee$). Suppose that $\mathbf{\Phi} \vdash \varphi \vee \psi, \mathbf{\Phi} + \varphi \vdash \chi$ and $\mathbf{\Phi} + \psi \vdash \chi$. We verify that $\mathbf{\Phi} \vdash \chi$.

φ_0:	$\varphi \to \chi$	from $\mathbf{\Phi} + \varphi \vdash \chi$ by (DT)
φ_1:	$\psi \to \chi$	from $\mathbf{\Phi} + \psi \vdash \chi$ by (DT)
φ_2:	$(\varphi \to \chi) \to ((\psi \to \chi) \to ((\varphi \vee \psi) \to \chi))$	instance of L_8
φ_3:	$(\psi \to \chi) \to ((\varphi \vee \psi) \to \chi)$	from φ_2 and φ_0 by (MP)
φ_4:	$(\varphi \vee \psi) \to \chi$	from φ_3 and φ_1 by (MP)
φ_5:	$\varphi \vee \psi$	by assumption
φ_6:	χ	from φ_4 and φ_5 by (MP)

The corresponding introduction rule (I$\vee$) follows directly from L_6 and L_7 using (DT). Note that (I$\to$) follows directly from (DT) and (E$\to$) from (MP).

We further prove the rules for negation. For (I$\neg$) suppose that $\mathbf{\Phi} + \varphi \vdash \psi \wedge \neg\psi$. It follows from (E$\wedge$) that $\mathbf{\Phi} + \varphi \vdash \psi$ and $\mathbf{\Phi} + \varphi \vdash \neg\psi$. We prove that $\mathbf{\Phi} + \varphi \vdash \neg\varphi$, since then $\mathbf{\Phi} \vdash \neg\varphi$ by (E$\vee$) and L_0. We have:

φ_0:	$\neg\psi \to (\psi \to \neg\varphi)$	instance of L_9
φ_1:	$\neg\psi$	by assumption
φ_2:	$\psi \to \neg\varphi$	from φ_0 and φ_1 by (MP)
φ_3:	ψ	by assumption
φ_4:	$\neg\varphi$	from φ_2 and φ_3 by (MP)

The corresponding elimination rule (E$\neg$) follows from Example 2.2. Finally, we prove (I$\exists$) and (E$\exists$). Note that (I$\exists$) follows directly from L_{11} using (DT) and (MP). For (E$\exists$), let ν be a variable such that $\nu \notin \text{free}(\chi)$ for any $\chi \in \mathbf{\Phi}$ and suppose that $\mathbf{\Phi} \vdash \exists\nu\varphi(\nu)$ and $\mathbf{\Phi} + \varphi(\nu) \vdash \psi$. An application of (DT) then yields $\mathbf{\Phi} \vdash \varphi(\nu) \to \psi$. Then we obtain $\mathbf{\Phi} \vdash \psi$ by the following formal proof:

φ_0:	$\forall\nu(\varphi(\nu) \to \psi) \to (\exists\nu\varphi(\nu) \to \psi)$	instance of L_{13}
φ_1:	$\varphi(\nu) \to \psi$	by assumption
φ_2:	$\forall\nu(\varphi(\nu) \to \psi)$	from φ_1 by ($\forall$)
φ_3:	$\exists\nu\varphi(\nu) \to \psi$	from φ_0 and φ_2 by (MP)
φ_4:	$\exists\nu\varphi(\nu)$	by assumption
φ_5:	ψ	from φ_3 and φ_4 by (MP)

This completes the proof of (E$\exists$). The verification of the other rules of the calculus of natural deduction are left to the reader (see Exercise 2.0).

Conversely, we need to check that the calculus of natural deduction proves the logical axioms L_0–L_{16} as well as the inference rules (MP) and ($\forall$). Observe that (MP) corresponds to (E$\to$) and ($\forall$) corresponds to (I$\forall$). As before, we only present the proof for some axioms and leave the others to the reader. We consider first L_9. We need to check that $\vdash \neg\varphi \to (\varphi \to \psi)$.

$$
\begin{array}{rll}
\{\neg\varphi,\varphi,\neg\psi\} & \vdash \varphi & \text{by (IR)} \\
\{\neg\varphi,\varphi,\neg\psi\} & \vdash \neg\varphi & \text{by (IR)} \\
\{\neg\varphi,\varphi,\neg\psi\} & \vdash \varphi\wedge\neg\varphi & \text{by (I}\wedge\text{)} \\
\{\neg\varphi,\varphi\} & \vdash \neg\neg\psi & \text{by (I}\neg\text{)} \\
\{\neg\varphi,\varphi\} & \vdash \psi & \text{by (E}\neg\text{)} \\
\{\neg\varphi\} & \vdash \varphi\to\psi & \text{by (I}\to\text{)} \\
 & \vdash \neg\varphi\to(\varphi\to\psi) & \text{by (I}\to\text{)}
\end{array}
$$

Secondly, we derive Axiom L_{13} using the calculus of natural deduction, i.e., we show $\vdash \forall\nu(\varphi(\nu)\to\psi)\to(\exists\nu\varphi(\nu)\to\psi)$.

$$
\begin{array}{rll}
\{\forall\nu(\varphi(\nu)\to\psi),\exists\nu\varphi(\nu),\varphi(\nu)\} & \vdash \varphi(\nu) & \text{by (IR)} \\
\{\forall\nu(\varphi(\nu)\to\psi),\exists\nu\varphi(\nu),\varphi(\nu)\} & \vdash \exists\nu\varphi(\nu) & \text{by (IR)} \\
\{\forall\nu(\varphi(\nu)\to\psi),\exists\nu\varphi(\nu),\varphi(\nu)\} & \vdash \forall\nu(\varphi(\nu)\to\psi) & \text{by (IR)} \\
\{\forall\nu(\varphi(\nu)\to\psi),\exists\nu\varphi(\nu),\varphi(\nu)\} & \vdash \varphi(\nu)\to\psi & \text{by (E}\forall\text{)} \\
\{\forall\nu(\varphi(\nu)\to\psi),\exists\nu\varphi(\nu),\varphi(\nu)\} & \vdash \psi & \text{by (E}\to\text{)} \\
\{\forall\nu(\varphi(\nu)\to\psi),\exists\nu\varphi(\nu)\} & \vdash \psi & \text{by (E}\exists\text{)} \\
\{\forall\nu(\varphi(\nu)\to\psi) & \vdash \exists\nu\varphi(\nu)\to\psi & \text{by (I}\to\text{)} \\
 & \vdash \forall\nu(\varphi(\nu)\to\psi)\to(\exists\nu\varphi(\nu)\to\psi) & \text{by (I}\to\text{).}
\end{array}
$$

The other axioms can be verified in a similar way. ⊣

Methods of Proof

The inference rules of the calculus of natural deduction are very useful because they resemble methods of proof which are commonly used in mathematics. For example, the elimination rule (E∨) mimicks proofs by case distinction: Under the assumption that $\mathbf{\Phi}\vdash\varphi\vee\psi$, one can prove a formula χ by separately proving $\mathbf{\Phi}\cup\{\varphi\}\vdash\chi$ and $\mathbf{\Phi}\cup\{\psi\}\vdash\chi$.

In the following, we list several methods of proof such as proofs by contradiction, contraposition and case distinction.

PROPOSITION 2.4 (PROOF BY CASES). *Let $\mathbf{\Phi}$ be a set of formulae and let φ, ψ, χ be some formulae. Then the following two statements hold:*

$$\mathbf{\Phi}\vdash\varphi\vee\psi \text{ and } \mathbf{\Phi}+\varphi\vdash\chi \text{ and } \mathbf{\Phi}+\psi\vdash\chi \implies \mathbf{\Phi}\vdash\chi \quad (\vee 0)$$

$$\mathbf{\Phi}+\varphi\vdash\chi \text{ and } \mathbf{\Phi}+\neg\varphi\vdash\chi \implies \mathbf{\Phi}\vdash\chi \quad (\vee 1)$$

Proof. Note that (∨0) is exactly the statement of (E∨) and (∨1) is a special case of (∨0), since $\mathbf{\Phi}\vdash\varphi\vee\neg\varphi$ by L_0. ⊣

COROLLARY 2.5 (GENERALISED PROOF BY CASES). *Let $\mathbf{\Phi}$ be a set of formulae and let $\psi_0,\dots,\psi_n,\varphi$ be some formulae. Then we have:*

$$\mathbf{\Phi}\vdash\psi_0\vee\dots\vee\psi_n \text{ and } \mathbf{\Phi}+\psi_i\vdash\varphi \text{ for all } i\le n \implies \mathbf{\Phi}\vdash\varphi$$

Since COROLLARY 2.5 is just a generalization of ($\vee 0$), we will denote all instances of this form by ($\vee 0$) as well.

Proof of COROLLARY 2.5. We proceed by induction on $n \geq 1$. For $n = 1$ the statement is exactly ($\vee 0$). Now assume that $\mathbf{\Phi} \vdash \psi_0 \vee \ldots \vee \psi_n \vee \psi_{n+1}$ and $\mathbf{\Phi} + \psi_i \vdash \varphi$ for all $i \leq n + 1$. Let $\mathbf{\Phi}' :\equiv \mathbf{\Phi} + \psi_0 \vee \ldots \vee \psi_n$ and observe that $\mathbf{\Phi}' \vdash \psi_0 \vee \ldots \vee \psi_n$ and $\mathbf{\Phi}' + \psi_i \vdash \varphi$, so by induction hypothesis $\mathbf{\Phi}' \vdash \varphi$. By the DEDUCTION THEOREM this implies $\mathbf{\Phi} \vdash \psi_0 \vee \ldots \vee \psi_n \to \varphi$. Moreover, by another application of (DT) we also have $\mathbf{\Phi} \vdash \psi_{n+1} \to \varphi$. Using L_8 and twice (DT), we obtain $\mathbf{\Phi} \vdash \psi_0 \vee \ldots \vee \psi_n \vee \psi_{n+1} \to \varphi$, hence (DT) yields the claim. ⊣

PROPOSITION 2.6 (EX FALSO QUODLIBET). *Let $\mathbf{\Phi}$ be a set of formulae and let φ an arbitrary formula. Then for every $\mathscr{L}$-formula ψ we have:*

$$\mathbf{\Phi} \vdash \varphi \wedge \neg\varphi \quad \Longrightarrow \quad \mathbf{\Phi} \vdash \psi \qquad (\boxtimes)$$

Proof. Let ψ be any formula and assume that $\mathbf{\Phi} \vdash \varphi \wedge \neg\varphi$ for some formula φ. By (E$\wedge$) we have $\mathbf{\Phi} \vdash \varphi$ and $\mathbf{\Phi} \vdash \neg\varphi$. Now the instance $\neg\varphi \to (\varphi \to \psi)$ of the logical axiom L_9 and two applications of Modus Ponens imply $\mathbf{\Phi} \vdash \psi$. ⊣

Notice that PROPOSITION 2.6 implies that if we can derive a contradiction from $\mathbf{\Phi}$, we can derive *every* formula we like, even the impossible, denoted by the symbol

$$\boxtimes.$$

This is closely related to proofs by contradiction:

COROLLARY 2.7 (PROOF BY CONTRADICTION). *Let $\mathbf{\Phi}$ be a set of formulae, and let φ be an arbitrary formula. Then the following statements hold:*

$$\mathbf{\Phi} + \neg\varphi \vdash \boxtimes \quad \Longrightarrow \quad \mathbf{\Phi} \vdash \varphi$$

$$\mathbf{\Phi} + \varphi \vdash \boxtimes \quad \Longrightarrow \quad \mathbf{\Phi} \vdash \neg\varphi$$

Proof. Note that the second statement is exactly the introduction rule (I$\neg$). For the first statment, note that by ($\vee 1$) it is enough to check $\mathbf{\Phi} + \varphi \vdash \varphi$ and $\mathbf{\Phi} + \neg\varphi \vdash \varphi$. The first condition is clearly satisfied and the second one follows directly from (I$\wedge$) and ($\boxtimes$). ⊣

PROPOSITION 2.8 (PROOF BY CONTRAPOSITIVE). *Let $\mathbf{\Phi}$ be a set of formulae and φ and ψ two arbitrary formulae. Then we have:*

$$\mathbf{\Phi} + \varphi \vdash \psi \quad \Longleftrightarrow \quad \mathbf{\Phi} + \neg\psi \vdash \neg\varphi \qquad (\mathsf{CP})$$

Proof. Suppose first that $\mathbf{\Phi} + \varphi \vdash \psi$. Then by (I$\wedge$), $\mathbf{\Phi} \cup \{\neg\psi, \varphi\} \vdash \psi \wedge \neg\psi$ and hence by (I$\neg$) we obtain $\mathbf{\Phi} + \neg\psi \vdash \neg\varphi$. Conversely, assume that $\mathbf{\Phi} + \neg\psi \vdash \neg\varphi$. A similar argument as above yields $\mathbf{\Phi} + \varphi \vdash \neg\neg\psi$. An application of (E$\neg$) completes the proof. ⊣

Note that PROPOSITION 2.8 proves the logical equivalence

$$\varphi \to \psi \Leftrightarrow \neg\psi \to \neg\varphi.$$

THEOREM 2.9 (GENERALISED DEDUCTION THEOREM). *If $\mathbf{\Phi}$ is an arbitrary set of formulae and $\mathbf{\Phi} \cup \{\psi_1, \ldots, \psi_n\} \vdash \varphi$, then $\mathbf{\Phi} \vdash (\psi_1 \wedge \cdots \wedge \psi_n) \to \varphi$; and vice versa:*

$$\mathbf{\Phi} \cup \{\psi_1, \ldots, \psi_n\} \vdash \varphi \quad \Longleftrightarrow \quad \mathbf{\Phi} \vdash (\psi_1 \wedge \cdots \wedge \psi_n) \to \varphi \qquad \text{(GDT)}$$

Proof. This follows immediately from the DEDUCTION THEOREM and Part (c) of DEMORGAN'S LAWS (see EXERCISE 2.2). ⊣

The Normal Forms NNF & DNF

In many proofs it is convenient to convert a formula into an equivalent formula in some normal form. The simplest normal form is the following: A formula is said to be in **Negation Normal Form**, denoted by NNF, if it does not contain the implication symbol $\to$ and if the negation symbol $\neg$ only occurs directly in front of atomic subformulae.

THEOREM 2.10 (NEGATION NORMAL FORM THEOREM). *Every formula is equivalent to some formula in* NNF.

Proof. We successively apply the following transformations to every non-atomic negated subformula ψ of φ, starting with the outermost negation symbols.

- Using (K), we may replace subformulae of the form $\psi_1 \to \psi_2$ by $\neg\psi_1 \vee \psi_2$.
- If $\psi \equiv \neg\neg\psi'$ for some formula ψ', we replace ψ with ψ' using (F).
- By the DEMORGAN'S LAWS (see EXERCISE 2.2), we replace subformulae of the form $\neg(\psi_1 \wedge \psi_2)$ and $\neg(\psi_1 \vee \psi_2)$, respectively, with $\neg\psi_1 \vee \neg\psi_2$ and $\neg\psi_1 \wedge \neg\psi_2$, respectively.
- If $\psi \equiv \neg\exists\nu\psi'$ then it follows from (Q.0) that $\psi \Leftrightarrow \forall\nu\neg\psi'$, and hence we replace ψ with $\forall\nu\neg\psi'$. Similarly, using (Q.1), we replace subformulae of the form $\neg\forall\nu\psi'$ with the equivalent formula $\exists\nu\neg\psi'$.

⊣

A quantifier-free formula φ is said to be in **Disjunctive Normal Form**, denoted by DNF, if it is a disjunction of conjunctions of atomic formulae or negated atomic formulae, i.e., if it is of the form

$$(\varphi_{1,1} \wedge \ldots \wedge \varphi_{1,k_1}) \vee \cdots \vee (\varphi_{m,1} \wedge \cdots \wedge \varphi_{m,k_m})$$

for some quantifier-free formulae $\varphi_{i,j}$ which are either atomic or the negation of an atomic formula. In particular, each formula in DNF is also in NNF.

THEOREM 2.11 (DISJUNCTIVE NORMAL FORM THEOREM). *Every quantifier-free formula φ is equivalent to some formula in DNF.*

Proof. By the NEGATION NORMAL FORM THEOREM we may assume that φ is in NNF. Starting with the outermost conjunction symbol, we successively apply the distributive laws

$$\psi \wedge (\varphi_1 \vee \varphi_2) \Leftrightarrow (\psi \wedge \varphi_1) \vee (\psi \wedge \varphi_2) \quad \text{and}$$
$$(\varphi_1 \vee \varphi_2) \wedge \psi \Leftrightarrow (\varphi_1 \wedge \psi) \vee (\varphi_2 \wedge \psi)$$

until all conjunction symbols occur between atomic or negated atomic formulae. This process ends after F I N I T E L Y many steps, since there are only F I N I T E L Y many conjunction symbols. ⊣

A similar result holds for the so-called *Conjunctive Normal Form*, denoted by CNF, see EXERCISE 2.4.

Substitution of Variables and the Prenex Normal Form

In Part II & III, we shall encode formulae by strings of certain symbols and by natural numbers, respectively. In order to do so, we have to make sure that the variables are among a well-defined set of symbols, namely among $v_0, v_1, \ldots$, where the index n of v_n is a natural number (i.e., a member of $\mathbb{N}$).

THEOREM 2.12 (VARIABLE SUBSTITUTION THEOREM). *For every sentence σ there is an equivalent sentence $\tilde{\sigma}$ which contains just variables among $v_0, v_1, \ldots$, where for any $m, n \in \mathbb{N}$ with $m < n$, if v_n appears in $\tilde{\sigma}$, then also v_m appears in $\tilde{\sigma}$.*

Proof. Let σ be an arbitrary sentence and let m be such that no variable v_k with $k \geq m$ appears in σ. Assume that $\exists \nu \varphi(\nu)$ is a sub-sentence of σ. Then $\exists \nu \varphi(\nu) \Leftrightarrow \exists v_k \varphi(\nu / v_k)$ for any $k \geq m$. To see this, first notice that since v_k does not appear in σ, the substitution $\varphi(\nu / v_k)$ is admissible. Furthermore, we have:

φ_0:	$\varphi(v_k) \to \exists \nu \varphi(\nu)$	instance of L_{11}
φ_1:	$\forall v_k \big(\varphi(v_k) \to \exists \nu \varphi(\nu)\big)$	from φ_0 by $(\forall)$
φ_2:	$\forall v_k \big(\varphi(v_k) \to \exists \nu \varphi(\nu)\big) \to \big(\exists v_k \varphi(v_k) \to \exists \nu \varphi(\nu)\big)$	instance of L_{13}
φ_3:	$\exists v_k \varphi(v_k) \to \exists \nu \varphi(\nu)$	from φ_2 and φ_1 by (MP)

Similarly, we obtain $\exists \nu \varphi(\nu) \to \exists v_k \varphi(v_k)$, which shows that $\exists \nu \varphi(\nu) \Leftrightarrow \exists v_k \varphi(v_k)$.

Assume now that $\forall \nu \varphi(\nu)$ is a sub-sentence of σ. Then $\forall \nu \varphi(\nu) \Leftrightarrow \forall v_k \varphi(\nu / v_k)$. Since the substitution $\varphi(\nu / v_k)$ is admissible, we have:

φ_0:	$\forall \nu \varphi(\nu) \rightarrow \varphi(v_k)$	instance of L_{10}
φ_1:	$\forall v_k \big(\forall \nu \varphi(\nu) \rightarrow \varphi(v_k)\big)$	from φ_0 by $(\forall)$
φ_2:	$\forall v_k \big(\forall \nu \varphi(\nu) \rightarrow \varphi(v_k)\big) \rightarrow \big(\forall \nu \varphi(\nu) \rightarrow \forall v_k \varphi(v_k)\big)$	instance of L_{13}
φ_3:	$\forall \nu \varphi(\nu) \rightarrow \forall v_k \varphi(v_k)$	from φ_2 and φ_1 by (MP)

Similarly we obtain $\forall v_k \varphi(v_k) \rightarrow \forall \nu \varphi(\nu)$, which shows that $\forall \nu \varphi(\nu) \Leftrightarrow \forall v_k \varphi(v_k)$.

Therefore, we can replace all the variables $\nu_0, \nu_1, \ldots$ appearing in σ step by step with variables $v_m, v_{m+1}, \ldots$ and obtain a sentence σ' which is equivalent to σ. In a last step, we replace the variables $v_m, v_{m+1}, \ldots$ with $v_0, v_1, \ldots$ and finally obtain $\tilde{\sigma}$. ⊣

In Chapter 5, we will use the fact that every sentence can be transformed into a semantically equivalent sentence in the so-called *special Prenex Normal Form*:

A sentence σ is said to be in **Prenex Normal Form**, denoted by PNF, if it is of the form

$$\exists\!\!\!\forall_0 \nu_0 \ldots \exists\!\!\!\forall_n \nu_n \tilde{\sigma},$$

where the variables $\nu_0, \ldots, \nu_n$ are pairwise distinct, each $\exists\!\!\!\forall_m$ (for $0 \leq m \leq n$) stands for either $\exists$ or for $\forall$, and $\tilde{\sigma}$ is a quantifier-free formula. Furthermore, a sentence σ is in **special Prenex Normal Form**, denoted by sPNF, if σ is in PNF and

$$\sigma \equiv \exists\!\!\!\forall_0 v_0 \exists\!\!\!\forall_1 v_1 \ldots \exists\!\!\!\forall_n v_n \tilde{\sigma}\,,$$

where each $\exists\!\!\!\forall_m$ (for $0 \leq m \leq n$) stands for either $\exists$ or $\forall$, $\tilde{\sigma}$ is quantifier-free, and, in addition, each variable $v_0, \ldots, v_n$ appears free in $\tilde{\sigma}$.

THEOREM 2.13 (PRENEX NORMAL FORM THEOREM). *For every sentence σ there is an equivalent sentence $\tilde{\sigma}$ in* sPNF.

Sketch of the Proof. Using the NEGATION NORMAL FORM THEOREM we may suppose that φ is in NNF. Moreover, by TAUTOLOGY (O.1) and (O.2) (see in the proof of the VARIABLE SUBSTITUTION THEOREM) we may additionally suppose that no variable is quantified more than once. The crucial part is now to show that for all formulae φ and ψ, where $\nu \notin \text{free}(\psi)$, the following formulae are tautologies:

$$\exists \nu \varphi \circ \psi \;\Leftrightarrow\; \exists \nu \big(\varphi \circ \psi\big),$$

$$\forall x \varphi \circ \psi \;\Leftrightarrow\; \forall \nu \big(\varphi \circ \psi\big),$$

where $\circ$ stands for either $\vee$ or $\wedge$.

We just prove that the formula

$$\exists\nu\varphi \vee \psi \;\rightarrow\; \exists\nu(\varphi \vee \psi)\,, \quad \text{where } \nu \notin \text{free}(\psi),$$

is a tautology; all other cases are proved similarly.

By L_8, we obtain

$$\vdash\; \big(\exists\nu\varphi \rightarrow \exists\nu(\varphi \vee \psi)\big) \rightarrow \Big(\big(\psi \rightarrow \exists\nu(\varphi \vee \psi)\big) \rightarrow \big((\exists\nu\varphi \vee \psi) \rightarrow \exists\nu(\varphi \vee \psi)\big)\Big).$$

Therefore, it is enough to show:

$$\vdash \exists\nu\varphi \rightarrow \exists\nu(\varphi \vee \psi) \quad \text{and} \quad \vdash \psi \rightarrow \exists\nu(\varphi \vee \psi)$$

We first prove $\vdash \exists\nu\varphi \rightarrow \exists\nu(\varphi \vee \psi)$:

φ_0:	$\varphi \rightarrow \varphi \vee \psi$	instance of L_6
φ_1:	$\varphi \vee \psi \rightarrow \exists\nu(\varphi \vee \psi)$	instance of L_{11}
φ_2:	$\varphi \rightarrow \exists\nu(\varphi \vee \psi)$	by TAUTOLOGY (D.0)
φ_3:	$\forall\nu\big(\varphi \rightarrow \exists\nu(\varphi \vee \psi)\big)$	from φ_2 by $(\forall)$
φ_4:	$\forall\nu\big(\varphi \rightarrow \exists\nu(\varphi \vee \psi)\big) \rightarrow \big(\exists\nu\varphi \rightarrow \exists\nu(\varphi \vee \psi)\big)$	instance of L_{13}
φ_5:	$\exists\nu\varphi \rightarrow \exists\nu(\varphi \vee \psi)$	from φ_4 and φ_3 by (MP)

Now we show $\vdash \psi \rightarrow \exists\nu(\varphi \vee \psi)$:

φ_0:	$\psi \rightarrow \varphi \vee \psi$	instance of L_7
φ_1:	$\varphi \vee \psi \rightarrow \exists\nu(\varphi \vee \psi)$	instance of L_{11}
φ_2:	$\psi \rightarrow \exists\nu(\varphi \vee \psi)$	by TAUTOLOGY (D.0)

After F I N I T E L Y many applications of the above tautologies, we obtain a sentence in PNF. Moreover, the VARIABLE SUBSTITUTION THEOREM yields a formula in sPNF. ⊣

Consistency & Compactness

We say that a set of formulae $\mathbf{\Phi}$ is **consistent**, denoted by $\text{Con}(\mathbf{\Phi})$, if $\mathbf{\Phi} \nvdash$ ⅅ, i.e., if there is *no* formula φ such that $\mathbf{\Phi} \vdash (\varphi \wedge \neg\varphi)$. Otherwise $\mathbf{\Phi}$ is called **inconsistent**, denoted by $\neg\,\text{Con}(\mathbf{\Phi})$.

FACT 2.14. *Let $\mathbf{\Phi}$ be a set of formulae.*

(a) *If $\neg\,\text{Con}(\mathbf{\Phi})$, then for all formulae ψ we have $\mathbf{\Phi} \vdash \psi$.*

(b) *If $\text{Con}(\mathbf{\Phi})$ and $\mathbf{\Phi} \vdash \varphi$ for some formula φ, then $\mathbf{\Phi} \nvdash \neg\varphi$.*

(c) *If* $\neg\,\mathrm{Con}(\mathbf{\Phi} + \varphi)$*, for some formula* φ*, then* $\mathbf{\Phi} \vdash \neg\varphi$*.*

(d) *If* $\mathbf{\Phi} \vdash \neg\varphi$*, for some formula* φ*, then* $\neg\,\mathrm{Con}(\mathbf{\Phi} + \varphi)$*.*

Proof. Condition (a) is just PROPOSITION 2.6. For (b), notice that if $\mathbf{\Phi} \vdash \varphi$ and $\mathbf{\Phi} \vdash \neg\varphi$, then by (I∧) we get $\mathbf{\Phi} \vdash$ ⧉ and thus also $\neg\,\mathrm{Con}(\mathbf{\Phi})$. Moreover, (c) coincides with the second statement of COROLLARY 2.7. Finally, for (d) note that if $\mathbf{\Phi} \vdash \neg\varphi$, then $\mathbf{\Phi} + \varphi \vdash \varphi \wedge \neg\varphi$ and hence $\mathbf{\Phi} + \varphi$ is inconsistent. ⊣

If we choose a set of formulae $\mathbf{\Phi}$ as the basis of a theory (e.g., a set of axioms), we have to make sure that $\mathbf{\Phi}$ is consistent. However, as we shall see later, in many cases this task is impossible.

We conclude this chapter with the COMPACTNESS THEOREM, which is a powerful tool for constructing non-standard models of Peano Arithmetic or Set Theory. On the one hand, it is just a consequence of the fact that formal proofs are F I N I T E sequences of formulae. On the other hand, the COMPACTNESS THEOREM is the main tool for proving that a given set of sentences is consistent with some given set of formulae $\mathbf{\Phi}$.

THEOREM 2.15 (COMPACTNESS THEOREM). *Let* $\mathbf{\Phi}$ *be an arbitrary set of formulae. Then* $\mathbf{\Phi}$ *is consistent if and only if every finite subset* $\mathbf{\Phi}'$ *of* $\mathbf{\Phi}$ *is consistent.*

Proof. Obviously, if $\mathbf{\Phi}$ is consistent, then every finite subset $\mathbf{\Phi}'$ of $\mathbf{\Phi}$ must be consistent. On the other hand, if $\mathbf{\Phi}$ is inconsistent, then there is a formula φ such that $\mathbf{\Phi} \vdash \varphi \wedge \neg\varphi$. In other words, there is a proof of $\varphi \wedge \neg\varphi$ from $\mathbf{\Phi}$. Now, since every proof is finite, there are only finitely many formulae of $\mathbf{\Phi}$ involved in this proof, and if $\mathbf{\Phi}'$ is this finite set of formulae, then $\mathbf{\Phi}' \vdash \varphi \wedge \neg\varphi$, which shows that $\mathbf{\Phi}'$, a finite subset of $\mathbf{\Phi}$, is inconsistent. ⊣

Semi-formal Proofs

Previously, we have shown that formal proofs can be simplified by applying methods of proof such as case distinctions, proofs by contradiction or proofs by contraposition. However, in order to make proofs even more natural, it is beneficial to use natural language for describing a proof step as in an "informal" mathematical proof.

Example 2.16. We want to prove the tautology $\vdash \varphi \to \neg\neg\varphi$. Instead of writing out the whole formal proof, which is quite tedious, we can apply the methods of proof which we introduced above. The first modification we make is to use (DT) to obtain the new goal

$$\{\varphi\} \vdash \neg\neg\varphi.$$

The easiest way to proceed is to make a proof by contradiction; hence it remains to show

$$\{\varphi, \neg\varphi\} \vdash \text{⧉}$$

which by (I$\wedge$) is again a consequence of the trivial goals

$$\{\varphi, \neg\varphi\} \vdash \varphi \quad \text{and} \quad \{\varphi, \neg\varphi\} \vdash \neg\varphi.$$

To sum up, this procedure can actually be transformed back into a formal proof, so it suffices as a proof of $\vdash \varphi \rightarrow \neg\neg\varphi$. Now this is still not completely satisfactory, since we would like to write the proof in natural language. A possible translation could thus be the following:

Proof. We want to prove that φ implies $\neg\neg\varphi$. Assume φ. Suppose for a contradiction that $\neg\varphi$. But then we have φ and $\neg\varphi$. Contradiction. ⊣

We will now show in a systematic way how formal proofs can — in principle — be replaced by semi-formal proofs, which make use of a **controlled natural language**, i.e., a limited vocabulary consisting of natural language phrases such as "assume that" which are often used in mathematical proof texts. This language is controlled in the sense that its allowed vocabulary is only a subset of the entire English vocabulary and that every word and every phrase, respectively, has a unique precisely defined interpretation. However, for the sake of a nice proof style, we will not always stick to this limited vocabulary. Moreover, this section should be considered as a hint of how formal proofs can be formulated using a controlled natural language as well as a justification for working with natural language proofs rather than formal ones.

Every statement which we would like to prove formally is of the form $\mathbf{\Phi} \vdash \varphi$, where $\mathbf{\Phi}$ is a set of formulae and φ is a formula. Note that as in Example 2.16, in order to prove $\mathbf{\Phi} \vdash \varphi$ — which is actually a meta-proof — we perform operations both on the set of formulae $\mathbf{\Phi}$ and on the formula to be formally proved. We call a statement of the form $\mathbf{\Phi} \vdash \varphi$ a **goal**, the set $\mathbf{\Phi}$ is called **premises**, and the formula φ to be verified as **target**. Now instead of listing a formal proof, we can step by step reduce our current goal to a simpler one using the methods of proof from the previous section, until the target is tautological as in the case of Example 2.16.

In that sense, methods of proof are simply operations on the premises and the targets. For example, the proof by contraposition adds the negation of the target to the premises and replaces the original target by the negation of the premise from which it shall be derived: If we want to show $\mathbf{\Phi} + \psi \vdash \varphi$ we can prove $\mathbf{\Phi} + \neg\varphi \vdash \neg\psi$ instead. A slightly different example is the proof of a conjunction $\mathbf{\Phi} \vdash \varphi \wedge \psi$, which is usually split into the goals given by $\mathbf{\Phi} \vdash \varphi$ and $\mathbf{\Phi} \vdash \psi$. Thus we have to revise our first attempt and interpret methods of proof as operations on F I N I T E lists of goals consisting of premises and targets.

We distinguish between two types of operations on goals: **Backward reasoning** means performing operations on targets, whereas **forward reasoning** denotes operations on the premises. We give some examples of both backward and forward reasoning and indicate how such proofs can be phrased in a semi-formal way.

Backward reasoning

- Targets are often of the universal conditional form $\forall\nu(\varphi(\nu) \to \psi(\nu))$. In particular, this pattern includes the purely universal formulae $\forall\nu\psi(\nu)$ by taking φ to be a tautology as well as simple conditionals of the form $\varphi \to \psi$. Now the usual procedure is to reduce $\mathbf{\Phi} \vdash \forall\nu(\varphi(\nu) \to \psi(\nu))$ to $\mathbf{\Phi} + \varphi(\nu) \vdash \psi(\nu)$ using $(\forall)$ and (DT). This can be rephrased as

  Assume $\varphi(\nu)$. *Then ... This shows* $\psi(\nu)$.

- As already mentioned above, if the target is a conjunction $\varphi \wedge \psi$, then one can show the conjuncts separately using (I$\wedge$). This step is usually executed without mentioning it explicitly.
- If the target is a negation $\neg\varphi$, one often uses a proof by contradiction or by contraposition: In the first case, we transform $\mathbf{\Phi} \vdash \neg\varphi$ to $\mathbf{\Phi} + \varphi \vdash \lightning$ and use the natural language notation

  Suppose for a contradiction that φ. *Then ... Contradiction.*

  In the latter case, we want to go from $\mathbf{\Phi} + \neg\psi \vdash \neg\varphi$ to $\mathbf{\Phi} + \varphi \vdash \psi$ or, in its positive version, from $\mathbf{\Phi} + \psi \vdash \neg\varphi$ to $\mathbf{\Phi} + \varphi \vdash \neg\psi$, respectively. In both cases, we can mark this with the keyword *contraposition*, e.g., as

  We proceed by contraposition ... This shows $\neg\varphi$.

Forwards reasoning

- By (E$\wedge$), conjunctive premises $\varphi \wedge \psi$ can be split into two premises φ, ψ; i.e., $\mathbf{\Phi} + \varphi \wedge \psi \vdash \chi$ can be reduced to $\mathbf{\Phi} \cup \{\varphi, \psi\} \vdash \chi$. This is usually performed automatically.
- Disjunctive premises are used for proofs by case distinction: If a goal of the form $\mathbf{\Phi} + \varphi \vee \psi \vdash \chi$ is given, we can reduce it to the new goals $\mathbf{\Phi} + \varphi \vdash \chi$ and $\mathbf{\Phi} + \psi \vdash \chi$. We can write this in a semi-formal way as

  Case 1: Assume φ*.... This proves* χ.
  Case 2: Assume ψ*.... This proves* χ.

- Intermediate proof steps: Often we first want to prove some intermediate statement which shall then be applied in order to resolve the target. Formally, this means that we want to show $\mathbf{\Phi} \vdash \varphi$ by first showing $\mathbf{\Phi} \vdash \psi$ and then adding ψ to the list of premises and checking $\mathbf{\Phi} + \psi \vdash \varphi$. Clearly, if we have $\mathbf{\Phi} \vdash \psi$ and $\mathbf{\Phi} + \psi \vdash \varphi$, using (DT) and (MP) we obtain that $\mathbf{\Phi} \vdash \varphi$. In a semi-formal proof, this can be described by

We first show ψ... This proves ψ.

Note that it is important to mark where the proof of the intermediate statement ψ ends, since from this point on, ψ can be used as a new premise.

Observe that in any case, once a goal $\mathbf{\Phi} \vdash \varphi$ is reduced to a tautology, it can be removed from the list of goals. This should be marked by a phrase like

This shows/proves φ.

so that it is clear that we move on to the next goal. The proof is complete as soon as no unresolved goals remain.

What is the use of such a formalised natural proof language? First of all, it increases readibility. Secondly, by giving a precise formal definition to some of the common natural language phrases which appear in proof texts, we show how — in principle — one could write formal proofs with a controlled natural language input. This input could then be parsed into a formal proof and subsequently be verified by a proof checking system.

We would like to emphasize that this section should only be considered as a motivation rather than a precise description of how formal proofs can be translated into semi-formal ones and vice versa. Nevertheless, it suffices to understand how this can theoretically be achieved. Therefore, in subsequent chapters, especially in Chapters 8 and 9, we will often present semi-formal proofs rather than formal ones.

Notes

Natural deduction in its modern form was developed by the German mathematician Gentzen in 1934 (see [11, 12]).

Exercises

2.0 Complete the proof of Theorem 2.3.

2.1 Formalise the method of proof by counterexample and prove that it works.

2.2 Let $\varphi_0, \dots, \varphi_n$ be formulae. Prove the DeMorgan's Laws:

(a) $\neg(\varphi_0 \wedge \dots \wedge \varphi_n) \Leftrightarrow (\neg\varphi_0 \vee \dots \vee \neg\varphi_n)$

(b) $\neg(\varphi_0 \vee \dots \vee \varphi_n) \Leftrightarrow (\neg\varphi_0 \wedge \dots \wedge \neg\varphi_n)$

(c) $\varphi_0 \to \big(\varphi_1 \to (\dots \to \varphi_n)\cdots\big) \Leftrightarrow \neg(\varphi_0 \wedge \dots \wedge \varphi_n)$

2.3 Prove the following generalisation of L_{15} to an arbitrary formula φ:

$$\vdash (\tau_1 = \tau_1' \wedge \dots \wedge \tau_n = \tau_n') \to \big(\varphi(\tau_1, \dots, \tau_n) \to \varphi(\tau_1', \dots, \tau_n')\big)$$

where $\tau, \tau_1, \dots, \tau_n, \tau_1', \dots, \tau_n'$ are terms and φ is a formula with n free variables.

2.4 A quantifier-free formula φ is said to be in **Conjunctive Normal Form**, denoted by CNF, if it is a conjunction of disjunctions of atomic formulae or negated atomic formulae, i.e., it is of the form

$$(\varphi_{1,1} \vee \ldots \vee \varphi_{1,k_1}) \wedge \cdots \wedge (\varphi_{m,1} \vee \cdots \vee \varphi_{m,k_m})$$

for some quantifier-free formulae $\varphi_{i,j}$ which are either atomic or the negation of an atomic formula.

Show that every quantifier-free formula φ is equivalent to some formula in CNF.

2.5 Let $\mathsf{L}_{9^3/_4}$ be the axiom schema $(\varphi \to \psi) \to \big((\varphi \to \neg\psi) \to \neg\varphi\big)$, and let $\mathsf{L}_{9^1/_4}$ be the axiom schema $\neg\neg\varphi \to \varphi$.

(a) Show that $\{\mathsf{L}_0, \mathsf{L}_1, \mathsf{L}_2, \mathsf{L}_8, \mathsf{L}_9\} \vdash \mathsf{L}_{9^1/_4}$.

Hint: Notice that every proof of φ from $\boldsymbol{\Phi}$ which involves (DT), but not L_{12}, can be transformed with the help of L_1 and L_2 into a formal proof $\boldsymbol{\Phi} \vdash \varphi$.

(b) Show that $\{\mathsf{L}_1, \mathsf{L}_2, \mathsf{L}_6, \mathsf{L}_7, \mathsf{L}_{9^3/_4}\} \vdash \neg\neg(\varphi \vee \neg\varphi)$.

(c) Show that $\{\mathsf{L}_1, \mathsf{L}_2, \mathsf{L}_6, \mathsf{L}_7, \mathsf{L}_{9^3/_4}, \mathsf{L}_{9^1/_4}\} \vdash \mathsf{L}_0$.

(d) Show that $\{\mathsf{L}_1\text{–}\mathsf{L}_9, \mathsf{L}_{9^3/_4}\} \nvdash \mathsf{L}_{9^1/_4}$ (compare with EXERCISE 1.5.(b)).

(e) Show that $\{\mathsf{L}_1\text{–}\mathsf{L}_9, \mathsf{L}_{9^3/_4}\} \nvdash \mathsf{L}_0$ (compare with EXERCISE 2.5.(c)).

Hint for parts (d) & (e): As in EXERCISE 1.5.(c), define a mapping $|\cdot|$, which assigns to each formula φ a value $|\varphi| \in \{-1, 0, 1\}$ such that the following conditions are satisfied: $|\varphi \vee \psi| = \max\{|\varphi|, |\psi|\}$, $|\varphi \wedge \psi| = \min\{|\varphi|, |\psi|\}$, and the values of $|\neg\varphi|$ and $|\varphi \to \psi|$ are given by the following tables:

$\lvert\varphi\rvert$	-1	0	1
$\lvert\neg\varphi\rvert$	1	-1	-1

$\lvert\varphi\rvert$ \ $\lvert\psi\rvert$	-1	0	1
-1	1	1	1
0	-1	1	1
1	-1	0	1

$|\varphi \to \psi|$

Show that for every formula θ with $\{\mathsf{L}_1\text{–}\mathsf{L}_9, \mathsf{L}_{9^3/_4}\} \vdash \theta$ we have $|\theta| = 1$. On the other hand, for certain values of $|\varphi|$ we have $|\mathsf{L}_0| \neq 1$ and $|\mathsf{L}_{9^1/_4}| \neq 1$, respectively.

2.6 Prove GLIVENKO'S THEOREM, which states that

$$\{\mathsf{L}_0\text{–}\mathsf{L}_9\} \vdash \varphi \quad \textit{if and only if} \quad \{\mathsf{L}_1\text{–}\mathsf{L}_9, \mathsf{L}_{9^3/_4}\} \vdash \neg\neg\varphi\,.$$

Chapter 3
Semantics: Making Sense of the Symbols

There are two different views on a given set of formulae $\mathbf{\Phi}$, namely the *syntactical* view and the *semantical* view.

From the syntactical point of view (presented in the previous chapters), we consider the set $\mathbf{\Phi}$ just as a set of well-formed formulae — regardless of their intended sense or meaning — from which we can prove some formulae. So, from a formal point of view there is no need to assign real objects (whatever this means) to our strings of symbols.

In contrast to this very formal syntactical view, there is also the semantical point of view according to which we consider the intended meaning of the formulae in $\mathbf{\Phi}$ and then seek for a *model* in which all formulae of $\mathbf{\Phi}$ become true. For this, we have to explain some basic notions of Model Theory like *structure* and *interpretation*, which we will do in a natural, informal language. In this language, we will use words like "or", "and", or phrases like "if...then". These words and phrases have the usual meaning. Furthermore, we assume that in our normal world, which we describe with our informal language, the basic rules of *common logic* apply. For example, a statement φ is true or false, and if φ is true, then $\neg\varphi$ is false; and vice versa. Hence, the statement "φ or $\neg\varphi$" is always true, which means that we tacitly assume the LAW OF EXCLUDED MIDDLE, also known as TERTIUM NON DATUR, which corresponds to the logical axiom L_0. Furthermore, we assume DEMORGAN'S LAWS and apply MODUS PONENS as an inference rule.

Structures & Interpretations

In order to define structures and interpretations, we have to assume some notions of NAIVE SET THEORY like *subset*, *cartesian product*, or *relation*, which shall be properly defined in Part IV. On this occasion, we also make use of the set theoretical symbol $\in$, which stands for the binary *membership relation*.

L. Halbeisen, R. Krapf, *Gödel's Theorems and Zermelo's Axioms*,
https://doi.org/10.1007/978-3-030-52279-7_3

Let $\mathscr{L}$ be an arbitrary but fixed language. An **$\mathscr{L}$-structure M** consists of a non-empty set A, called the **domain** of **M**, together with a mapping which assigns to each constant symbol $c \in \mathscr{L}$ an element $c^{\mathbf{M}} \in A$, to each n-ary relation symbol $R \in \mathscr{L}$ a set of n-tuples $R^{\mathbf{M}}$ of elements of A, and to each n-ary function symbol $F \in \mathscr{L}$ a function $F^{\mathbf{M}}$ from n-tuples of A to A. In other word, the constant symbols denote elements of A, n-ary relation symbols denote subsets of A^n (i.e., subsets of the n-fold cartesian product of A), and n-ary functions symbols denote n-ary functions from A^n to A.

The interpretation of variables is given by a so-called assignment: An **assignment** in an $\mathscr{L}$-structure **M** is a mapping j which assigns to each variable an element of the domain A.

Finally, an **$\mathscr{L}$-interpretation I** is a pair $(\mathbf{M}, j)$ consisting of an $\mathscr{L}$-structure **M** and an assignment j in **M**. For a variable ν, an element $a \in A$, and an assignment j in **M**, we define the assignment $j\frac{a}{\nu}$ by stipulating

$$j\tfrac{a}{\nu}(\nu') = \begin{cases} a & \text{if } \nu' \equiv \nu, \\ j(\nu') & \text{otherwise.} \end{cases}$$

Furthermore, for elements $a, a' \in A$ and variables ν, ν', we shall write $j\frac{a}{\nu}\frac{a'}{\nu'}$ instead of $\left(j\frac{a}{\nu}\right)\frac{a'}{\nu'}$.

For an interpretation $\mathbf{I} = (\mathbf{M}, j)$ and an element $a \in A$, we define:

$$\mathbf{I}\tfrac{a}{\nu} := (\mathbf{M}, j\tfrac{a}{\nu})$$

We associate with every interpretation $\mathbf{I} = (\mathbf{M}, j)$ and every $\mathscr{L}$-term τ an element $\mathbf{I}(\tau) \in A$ as follows:

- For a variable ν, let $\mathbf{I}(\nu) := j(\nu)$.
- For a constant symbol $c \in \mathscr{L}$, let $\mathbf{I}(c) := c^{\mathbf{M}}$.
- For an n-ary function symbol $F \in \mathscr{L}$ and terms $\tau_1, \ldots, \tau_n$, let

$$\mathbf{I}\big(F(\tau_1, \ldots, \tau_n)\big) := F^{\mathbf{M}}\big(\mathbf{I}(\tau_1), \ldots, \mathbf{I}(\tau_n)\big) .$$

Now, we are able to define precisely when a formula φ becomes *true* under an interpretation $\mathbf{I} = (\mathbf{M}, j)$; in which case we write $\mathbf{I} \vDash \varphi$ and say that φ is **true** in **I** (or that φ **holds** in **I**). The definition is by induction on the complexity of the formula φ, where the truth value of expressions involving "NOT", "AND", "IF ... THEN", et cetera, is explained later. By the rules (F0)–(F4), φ must be of the form $\tau_1 = \tau_2$, $R(\tau_1, \ldots, \tau_n)$, $\neg\psi$, $\psi_1 \wedge \psi_2$, $\psi_1 \vee \psi_2$, $\psi_1 \rightarrow \psi_2$, $\exists\nu\psi$, or $\forall\nu\psi$:

$$\mathbf{I} \vDash \tau_1 = \tau_2 \quad :\Longleftrightarrow \quad \mathbf{I}(\tau_1) \text{ IS THE SAME OBJECT AS } \mathbf{I}(\tau_2)$$

$$\mathbf{I} \vDash R(\tau_1, \ldots, \tau_n) \quad :\Longleftrightarrow \quad \big\langle \mathbf{I}(\tau_1), \ldots, \mathbf{I}(\tau_n)\big\rangle \text{ BELONGS TO } R^{\mathbf{M}}$$

$$\begin{aligned}
\mathbf{I} \vDash \neg\psi \quad &:\Longleftrightarrow \quad \text{NOT } \mathbf{I} \vDash \psi \\
\mathbf{I} \vDash \psi_1 \wedge \psi_2 \quad &:\Longleftrightarrow \quad \mathbf{I} \vDash \psi_1 \text{ AND } \mathbf{I} \vDash \psi_2 \\
\mathbf{I} \vDash \psi_1 \vee \psi_2 \quad &:\Longleftrightarrow \quad \mathbf{I} \vDash \psi_1 \text{ OR } \mathbf{I} \vDash \psi_2 \\
\mathbf{I} \vDash \psi_1 \rightarrow \psi_2 \quad &:\Longleftrightarrow \quad \text{IF } \mathbf{I} \vDash \psi_1 \text{ THEN } \mathbf{I} \vDash \psi_2 \\
\mathbf{I} \vDash \exists\nu\psi \quad &:\Longleftrightarrow \quad \text{THERE EXISTS } a \text{ IN } A: \ \mathbf{I}\tfrac{a}{\nu} \vDash \psi \\
\mathbf{I} \vDash \forall\nu\psi \quad &:\Longleftrightarrow \quad \text{FOR ALL } a \text{ IN } A: \ \mathbf{I}\tfrac{a}{\nu} \vDash \psi
\end{aligned}$$

Notice that by the logical rules in our informal language, for *every* $\mathscr{L}$-formula φ we have either $\mathbf{I} \vDash \varphi$ or $\mathbf{I} \vDash \neg\varphi$. So, every $\mathscr{L}$-formula is either true or false in $\mathbf{I}$. On the syntactical level, however, we do not necessarily have $\mathbf{\Phi} \vdash \varphi$ or $\mathbf{\Phi} \vdash \neg\varphi$ for each set $\mathbf{\Phi}$ of $\mathscr{L}$-formulae and each $\mathscr{L}$-formula φ.

The following fact summarises a few immediate consequences of the above definitions:

FACT 3.1. (a) *If φ is a formula and $\nu \notin \text{free}(\varphi)$, then:*

$$\mathbf{I}\tfrac{a}{\nu} \vDash \varphi \quad \textit{if and only if} \quad \mathbf{I} \vDash \varphi$$

(b) *If $\varphi(\nu)$ is a formula and the substitution $\varphi(\nu/\tau)$ is admissible, then:*

$$\mathbf{I}\tfrac{\mathbf{I}(\tau)}{\nu} \vDash \varphi(\nu) \quad \textit{if and only if} \quad \mathbf{I} \vDash \varphi(\tau)$$

Basic Notions of Model Theory

Let $\mathbf{\Phi}$ be an arbitrary set of $\mathscr{L}$-formulae. Then an $\mathscr{L}$-structure $\mathbf{M}$ is a **model of $\mathbf{\Phi}$** if for every assignment j and for each $\mathscr{L}$-formula $\varphi \in \mathbf{\Phi}$ we have $(\mathbf{M}, j) \vDash \varphi$, i.e., φ is true in the $\mathscr{L}$-interpretation $\mathbf{I} = (\mathbf{M}, j)$. Instead of saying "$\mathbf{M}$ is a model of $\mathbf{\Phi}$" we just write $\mathbf{M} \vDash \mathbf{\Phi}$. If φ fails in $\mathbf{M}$, then we write $\mathbf{M} \nvDash \varphi$. Notice that in the case when φ is a sentence, $\mathbf{M} \nvDash \varphi$ is equivalent to $\mathbf{M} \vDash \neg\varphi$, since for any $\mathscr{L}$-sentence φ we have *either* $\mathbf{M} \vDash \varphi$ *or* $\mathbf{M} \vDash \neg\varphi$.

Example 3.2. Let $\mathscr{L} = \{\mathsf{c}, f\}$, where c is a constant symbol and f is a unitary function symbol. Furthermore, let $\mathbf{\Phi}$ consist of the following two $\mathscr{L}$-sentences:

$$\underbrace{\forall x\big(x = \mathsf{c} \vee x = f(\mathsf{c})\big)}_{\varphi_1} \quad \text{and} \quad \underbrace{\exists x(x \neq \mathsf{c})}_{\varphi_2}$$

We construct two models $\mathbf{M}_1$ and $\mathbf{M}_2$ with the same domain A, such that $\mathbf{M}_1 \vDash \boldsymbol{\Phi}$ and $\mathbf{M}_2 \nvDash \boldsymbol{\Phi}$: For this, let $A := \{0, 1\}$, and let

$$\mathsf{c}^{\mathbf{M}_1} := 0, \quad f^{\mathbf{M}_1}(0) := 1, \quad f^{\mathbf{M}_1}(1) := 0,$$

$$\mathsf{c}^{\mathbf{M}_2} := 0, \quad f^{\mathbf{M}_2}(0) := 0, \quad f^{\mathbf{M}_2}(1) := 1.$$

We leave it as an exercise to the reader to show that φ_2 holds in both models, whereas φ_1 holds just in the model $\mathbf{M}_1$. In fact, we have $\mathbf{M}_1 \vDash \varphi_1 \wedge \varphi_2$ and $\mathbf{M}_2 \vDash \varphi_1 \wedge \neg\varphi_2$.

As an immediate consequence of the definition of models, we get:

FACT 3.3. *If φ is an $\mathscr{L}$-formula, ν a variable, and $\mathbf{M}$ a model, then $\mathbf{M} \vDash \varphi$ if and only if $\mathbf{M} \vDash \forall\nu\varphi$.*

This leads to the following definition: Let $\langle \nu_1, \dots, \nu_n \rangle$ be the sequence of variables which appear free in the $\mathscr{L}$-formula φ, where the variables appear in the sequence in the same order as they appear for the first time in φ if one reads φ from left to right. Then the **universal closure** of φ, denoted $\overline{\varphi}$, is defined by stipulating

$$\overline{\varphi} :\equiv \forall\nu_1 \cdots \forall\nu_n\, \varphi\,.$$

As a generalisation of FACT 3.3, we get:

FACT 3.4. *If φ is an $\mathscr{L}$-formula and $\mathbf{M}$ a model, then*

$$\mathbf{M} \vDash \varphi \quad \Longleftrightarrow \quad \mathbf{M} \vDash \overline{\varphi}$$

The following notation will be used later on to simplify the arguments when we investigate the truth-value of sentences in some model $\mathbf{M}$: Suppose that $\mathbf{M}$ is a model with domain A. Let $\varphi(\nu_1, \dots, \nu_n)$ be an $\mathscr{L}$-formula whose free variables are $\nu_1, \dots, \nu_n$ and let $a_1, \dots, a_n \in A$. Then we write

$$\mathbf{M} \vDash \varphi(a_1, \dots, a_n)$$

to denote that for every assignment j in $\mathbf{M}$ we have:

$$(\mathbf{M}, j\tfrac{a_1}{\nu_1} \cdots \tfrac{a_n}{\nu_n}) \vDash \varphi(\nu_1, \dots, \nu_n)$$

Let us now have a closer look at models: For this, we fix a signature $\mathscr{L}$ (i.e., we fix a possibly empty set of constant symbols c, n-ary function symbols F, and n-ary relation symbols R). Two $\mathscr{L}$-structures $\mathbf{M}$ and $\mathbf{N}$ with domains A and B are **isomorphic**, denoted by $\mathbf{M} \cong \mathbf{N}$, if there is a bijection $f : A \to B$ such that for all constant symbols $c \in \mathscr{L}$ we have

$$f\big(c^{\mathbf{M}}\big) = c^{\mathbf{N}}\,,$$

and for all natural numbers n, all n-ary function symbols $F \in \mathscr{L}$, all n-ary relation symbols $R \in \mathscr{L}$, and any $a_1, \dots, a_n \in A$ we have:

$$f\big(F^{\mathbf{M}}(a_1,\ldots,a_n)\big) \;=\; F^{\mathbf{N}}\big(f(a_1),\ldots,f(a_n)\big)$$

$$\langle a_1,\ldots,a_n\rangle \in R^{\mathbf{M}} \;\Leftrightarrow\; \big\langle f(a_1),\ldots,f(a_n)\big\rangle \in R^{\mathbf{N}}$$

Since models are just $\mathscr{L}$-structures, we can extend the notion of an isomorphism to models and obtain the following:

FACT 3.5. *If* $\mathbf{M}$ *and* $\mathbf{N}$ *are isomorphic models of some given set of* $\mathscr{L}$-*formulae and* φ *is an* $\mathscr{L}$-*formula, then:*

$$\mathbf{M} \vDash \varphi \quad \Longleftrightarrow \quad \mathbf{N} \vDash \varphi$$

It may happen that although two $\mathscr{L}$-structures $\mathbf{M}$ and $\mathbf{N}$ are not isomorphic there is no $\mathscr{L}$-sentence that can distinguish between them. In this case, we say that $\mathbf{M}$ and $\mathbf{N}$ are elementarily equivalent. More formally, we say that $\mathbf{M}$ is **elementarily equivalent** to $\mathbf{N}$, denoted by $\mathbf{M} \equiv_e \mathbf{N}$, if each $\mathscr{L}$-sentence σ which is true in $\mathbf{M}$ is also true in $\mathbf{N}$. The following lemma shows that $\equiv_e$ is symmetric:

LEMMA 3.6. *If* $\mathbf{M}$ *and* $\mathbf{N}$ *are* $\mathscr{L}$-*structures and* $\mathbf{M} \equiv_e \mathbf{N}$, *then for each* $\mathscr{L}$-*sentence* σ *we have:*

$$\mathbf{M} \vDash \sigma \quad \Longleftrightarrow \quad \mathbf{N} \vDash \sigma$$

Proof. One direction follows immediately from the definition. For the other direction, assume that σ is not true in $\mathbf{M}$, i.e., $\mathbf{M} \nvDash \sigma$. Then $\mathbf{M} \vDash \neg\sigma$, which implies $\mathbf{N} \vDash \neg\sigma$, and hence, σ is not true in $\mathbf{N}$. ⊣

As a consequence of FACT 3.4, we get:

FACT 3.7. *If* $\mathbf{M}$ *and* $\mathbf{N}$ *are elementarily equivalent models of some given set of* $\mathscr{L}$-*formulae and* φ *is an* $\mathscr{L}$-*formula, then we have:*

$$\mathbf{M} \vDash \varphi \quad \Longleftrightarrow \quad \mathbf{N} \vDash \varphi$$

In what follows, we investigate the relationship between syntax and semantic. In particular, we investigate the relationship between a formal proof of a formula φ from a set of formulae $\mathbf{\Phi}$ and the truth-value of φ in a model of $\mathbf{\Phi}$. In this context, two questions arise naturally:

- Is each formula φ which is provable from some set of formulae $\mathbf{\Phi}$ valid in every model $\mathbf{M}$ of $\mathbf{\Phi}$?
- Is every formula φ which is valid in each model $\mathbf{M}$ of $\mathbf{\Phi}$ provable from $\mathbf{\Phi}$?

Soundness Theorem

In this section, we give an answer to the former question in the previous paragraph; the answer to the latter question is postponed to Part II.

A logical calculus is called **sound** if all that we can prove is valid (i.e., true), which implies that we cannot derive a contradiction. The following theorem shows that First-Order Logic is sound.

THEOREM 3.8 (SOUNDNESS THEOREM). *Let* $\mathbf{\Phi}$ *be a set of* $\mathscr{L}$*-formulae and* $\mathbf{M}$ *a model of* $\mathbf{\Phi}$. *Then for every* $\mathscr{L}$*-formula* φ_0 *we have:*

$$\mathbf{\Phi} \vdash \varphi_0 \quad \Longrightarrow \quad \mathbf{M} \vDash \varphi_0$$

Somewhat shorter, we could say

$$\forall \varphi_0 : \ \mathbf{\Phi} \vdash \varphi_0 \quad \Longrightarrow \quad \forall \mathbf{M}\big(\mathbf{M} \vDash \mathbf{\Phi} \Longrightarrow \mathbf{M} \vDash \varphi_0\big)\,,$$

where the symbol $\forall$ *stands for* "FOR ALL".

Proof. First we show that all logical axioms are valid in $\mathbf{M}$. For this, we have to define truth-values of composite statements in the metalanguage. In the previous chapter, e.g., we defined:

$$\underbrace{\mathbf{M} \vDash \varphi \wedge \psi}_{\Theta} \quad \Longleftrightarrow \quad \underbrace{\mathbf{M} \vDash \varphi}_{\Phi} \quad \text{AND} \quad \underbrace{\mathbf{M} \vDash \psi}_{\Psi}$$

Thus, in the metalanguage the statement Θ is true if and only if the statement "Φ AND Ψ" is true. So, the truth-value of Θ depends on the truth-values of Φ and Ψ. In order to determine truth-values of composite statement like "Φ AND Ψ" or "IF Φ THEN Ψ", where the latter statement will get the same truth-value as "NOT Φ OR Ψ", we introduce so called *truth-tables*, in which **1** stands for **true** and **0** stands for **false**:

Φ	Ψ	NOT Φ	Φ AND Ψ	Φ OR Ψ	IF Φ THEN Ψ
0	0	1	0	0	1
0	1	1	0	1	1
1	0	0	0	1	0
1	1	0	1	1	1

With these truth-tables, one can show that all logical axioms are valid in $\mathbf{M}$. As an example, we show that every instance of L_1 is valid in $\mathbf{M}$. For this, let φ_1 be an instance of L_1, i.e., $\varphi_1 \equiv \varphi \to (\psi \to \varphi)$ for some $\mathscr{L}$-formulae φ and ψ. Then $\mathbf{M} \vDash \varphi_1$ if and only if $\mathbf{M} \vDash \varphi \to (\psi \to \varphi)$:

$$\begin{aligned}\underbrace{\mathbf{M} \vDash \varphi \to (\psi \to \varphi)}_{\Theta} \quad &\Longleftrightarrow \quad \text{IF } \underbrace{\mathbf{M} \vDash \varphi}_{\Phi} \text{ THEN } \mathbf{M} \vDash \psi \to \varphi \\ &\Longleftrightarrow \quad \text{IF } \Phi \text{ THEN } \big(\text{IF } \underbrace{\mathbf{M} \vDash \psi}_{\Psi} \text{ THEN } \underbrace{\mathbf{M} \vDash \varphi}_{\Phi}\big)\end{aligned}$$

This shows that

$$\Theta \iff \text{IF } \Phi \text{ THEN } (\text{IF } \Psi \text{ THEN } \Phi).$$

Writing the truth-table of Θ, we see that the statement Θ is always true (i.e., φ_1 is valid in $\mathbf{M}$):

Φ	Ψ	IF Ψ THEN Φ	IF Φ THEN (IF Ψ THEN Φ)
0	0	1	1
0	1	0	1
1	0	1	1
1	1	1	1

Therefore, $\mathbf{M} \vDash \varphi_1$, and since φ_1 was an arbitrary instance of L_1, every instance of L_1 is valid in $\mathbf{M}$.

In order to show that the logical axioms L_{10}–L_{16} are also valid in $\mathbf{M}$, we need somewhat more than just truth-tables. For this purpose, let A be the domain of $\mathbf{M}$, let j be an arbitrary assignment, and let $\mathbf{I} = (\mathbf{M}, j)$ be the corresponding $\mathscr{L}$-interpretation. Now we show that every instance of L_{10} is valid in $\mathbf{M}$. For this, let φ_{10} be an instance of L_{10}, i.e., $\varphi_{10} \equiv \forall\nu\varphi(\nu) \to \varphi(\tau)$ for some $\mathscr{L}$-formula φ, where ν is a variable, τ an $\mathscr{L}$-term, and the substitution $\varphi(\nu/\tau)$ is admissible. We work with $\mathbf{I}$ and show that $\mathbf{I} \vDash \varphi_{10}$. By definition, we have

$$\mathbf{I} \vDash \forall\nu\varphi(\nu) \to \varphi(\tau) \iff \text{IF } \mathbf{I} \vDash \forall\nu\varphi(\nu) \text{ THEN } \mathbf{I} \vDash \varphi(\tau),$$

and again by definition, we have

$$\mathbf{I} \vDash \forall\nu\varphi(\nu) \iff \text{FOR ALL } a \text{ IN } A: \ \mathbf{I}\tfrac{a}{\nu} \vDash \varphi(\nu).$$

In particular, we obtain

$$\mathbf{I} \vDash \forall\nu\varphi(\nu) \implies \mathbf{I}\tfrac{I(\tau)}{\nu} \vDash \varphi(\nu).$$

Furthermore, by FACT 3.1.(b) we get

$$\mathbf{I} \vDash \varphi(\tau) \iff \mathbf{I}\tfrac{I(\tau)}{\nu} \vDash \varphi(\nu).$$

Hence, we have

$$\text{IF } \mathbf{I} \vDash \forall\nu\varphi(\nu) \text{ THEN } \mathbf{I} \vDash \varphi(\tau)$$

which shows that

$$(\mathbf{M}, j) \vDash \forall\nu\varphi(\nu) \to \varphi(\tau).$$

Since the assignment j was arbitrary, we finally have:

$$\mathbf{M} \vDash \forall\nu\varphi(\nu) \rightarrow \varphi(\tau)$$

Therefore, $\mathbf{M} \vDash \varphi_{10}$, and since φ_{10} was an arbitrary instance of L_{10}, every instance of L_{10} is valid in $\mathbf{M}$.

With similar arguments, one can show that every instance of L_{11}, L_{12}, or L_{13} is also valid in $\mathbf{M}$ (see EXERCISE 3.6.(a)). Furthermore, one can show that L_{14}, L_{15}, and L_{16} are also valid in $\mathbf{M}$ (see EXERCISE 3.6.(b)).

Let $\mathbf{\Phi}$ be a set of formulae, let $\mathbf{M}$ be a model of $\mathbf{\Phi}$, and assume that $\mathbf{\Phi} \vdash \varphi_0$ for some $\mathscr{L}$-formula φ_0. We shall show that $\mathbf{M} \vDash \varphi_0$. For this, we first notice the following facts:

- As we have seen above, each instance of a logical axiom is valid in $\mathbf{M}$.
- Since $\mathbf{M} \vDash \mathbf{\Phi}$, each formula of $\mathbf{\Phi}$ is valid in $\mathbf{M}$.
- By the truth-tables, we get

$$\text{IF } (\mathbf{M} \vDash \varphi \rightarrow \psi \text{ AND } \mathbf{M} \vDash \varphi) \text{ THEN } \mathbf{M} \vDash \psi$$

  and therefore, every application of Modus Ponens in the proof of φ_0 from $\mathbf{\Phi}$ yields a valid formula (if the premises are valid).
- Since, by FACT 3.3,

$$\mathbf{M} \vDash \varphi \iff \mathbf{M} \vDash \forall\nu\varphi(\nu)$$

  every application of the Generalisation in the proof of φ_0 from $\mathbf{\Phi}$ yields a valid formula.

From these facts, it follows immediately that *each* formula in the proof of φ_0 from $\mathbf{\Phi}$ is valid in $\mathbf{M}$. In particular, we get

$$\mathbf{M} \vDash \varphi_0$$

which completes the proof. ⊣

The following fact summarises a few consequences of the SOUNDNESS THEOREM.

FACT 3.9.

(a) *Every tautology is valid in each model:*

$$\forall \varphi : \ \vdash \varphi \implies \forall \mathbf{M} : \mathbf{M} \vDash \varphi$$

(b) *If a set of formulae $\mathbf{\Phi}$ has a model, then $\mathbf{\Phi}$ is consistent:*

$$\exists \mathbf{M} : \mathbf{M} \vDash \mathbf{\Phi} \implies \mathrm{Con}(\mathbf{\Phi})$$

Here, the symbol $\exists$ *stands for* "IT EXISTS".

(c) *The logical axioms are consistent:*

$$\mathrm{Con}(\mathsf{L}_0\text{-}\mathsf{L}_{16})$$

(d) *If a sentence σ is not valid in $\mathbf{M}$, where $\mathbf{M}$ is a model of $\mathbf{\Phi}$, then σ is not provable from $\mathbf{\Phi}$:*

$$\text{IF } (\mathbf{M} \nvDash \sigma \text{ AND } \mathbf{M} \vDash \mathbf{\Phi}) \text{ THEN } \mathbf{\Phi} \nvdash \sigma$$

Completion of Theories

A set of $\mathscr{L}$-sentences is called an **$\mathscr{L}$-theory**, denoted by T. An $\mathscr{L}$-theory T is called **complete**, if for every $\mathscr{L}$-sentence σ we have *either* $\mathsf{T} \vdash \sigma$ *or* $\mathsf{T} \vdash \neg\sigma$. Notice that, by definition, an inconsistent theory is always **incomplete** (i.e., not complete). Furthermore, for an $\mathscr{L}$-theory T let $\mathbf{Th}(\mathsf{T})$ be the set of all $\mathscr{L}$-sentences σ, such that $\mathsf{T} \vdash \sigma$. By these definitions, we get that a consistent $\mathscr{L}$-theory T is complete if and only if for every $\mathscr{L}$-sentence σ we have *either* $\sigma \in \mathbf{Th}(\mathsf{T})$ *or* $\neg\sigma \in \mathbf{Th}(\mathsf{T})$.

PROPOSITION 3.10. *If T is an $\mathscr{L}$-theory which has a model, then there exists a complete $\mathscr{L}$-theory $\bar{\mathsf{T}}$ which contains T. In particular, every $\mathscr{L}$-theory which has a model can be completed (i.e., can be extended to a complete theory).*

Proof. Let $\mathbf{M}$ be a model of some $\mathscr{L}$-theory T and let $\bar{\mathsf{T}}$ be the set of $\mathscr{L}$-sentences σ such that $\mathbf{M} \vDash \sigma$. Since for each $\mathscr{L}$-sentence σ_0 we have either $\mathbf{M} \vDash \sigma_0$ or $\mathbf{M} \vDash \neg\sigma_0$, we get either $\sigma_0 \in \bar{\mathsf{T}}$ or $\neg\sigma_0 \in \bar{\mathsf{T}}$, which shows that $\bar{\mathsf{T}}$ is complete, and since $\mathbf{M} \vDash \mathsf{T}$, we get that $\bar{\mathsf{T}}$ contains T. ⊣

Let $\mathbf{M}$ be a model of some $\mathscr{L}$-theory T and let $\bar{\mathsf{T}}$ be the set of $\mathscr{L}$-sentences σ such that $\mathbf{M} \vDash \sigma$. Then $\bar{\mathsf{T}}$ is called the **theory of M**, denoted by $\mathbf{Th}(\mathbf{M})$. By definition, the theory $\mathbf{Th}(\mathbf{M})$ is always complete.

It is natural to ask whetherthe converse of PROPOSITION 3.10 also holds, i.e., whether every $\mathscr{L}$-theory which can be completed has a model. Notice that if an $\mathscr{L}$-theory T can be completed, then T must be consistent. So, one may ask whether every consistent theory has a model. An affirmative answer to this question together with FACT 3.9.(b) would imply that an $\mathscr{L}$-theory T is consistent if and only if T has a model — which is indeed the case, as we shall see below.

NOTES

The history of Model Theory can be traced back to the 19th century, when semantics began to play a role in Logic. However, one of the earliest results in modern Model Theory is Gödel's COMPLETENESS THEOREM (see Chapter 5). In the 1950's and 1960's, Model Theory was further developed, e.g., by Jerry Łoś (see Chapter 15) and Abraham Robinson (see Chapter 17).

Exercises

3.0 Show that the domain of a model is never empty.
Hint: Use Exercise 1.0.

3.1 Let R be a binary relation symbol and let the three sentences $\varphi_1, \varphi_2, \varphi_3$ be defined as follows:

$$\varphi_1 :\equiv \forall x(xRx), \quad \varphi_2 :\equiv \forall x \forall y(xRy \to yRx), \quad \varphi_3 :\equiv \forall x \forall y \forall z\big((xRy \land yRz) \to xRz\big)$$

Find three models $\mathbf{M}_1, \mathbf{M}_2, \mathbf{M}_3$ with domains as small as possible, such that

$$\mathbf{M}_1 \vDash \neg\varphi_1 \land \varphi_2 \land \varphi_3, \qquad \mathbf{M}_2 \vDash \varphi_1 \land \neg\varphi_2 \land \varphi_3, \qquad \mathbf{M}_3 \vDash \varphi_1 \land \varphi_2 \land \neg\varphi_3.$$

3.2 Let T be a set of $\mathscr{L}'$-sentences (for some signature $\mathscr{L}'$) and let $\mathbf{M}'$ be an $\mathscr{L}'$-structure such that $\mathbf{M}' \vDash \mathsf{T}$. Furthermore, let $\mathscr{L}$ be an extension of $\mathscr{L}'$ (i.e., $\mathscr{L}$ is a signature which contains $\mathscr{L}'$). Then there is an $\mathscr{L}$-structure $\mathbf{M}$ with the same domain as $\mathbf{M}'$, such that $\mathbf{M} \vDash \mathsf{T}$.
Hint: Let a_0 be an arbitrary but fixed element of the domain A of $\mathbf{M}'$. For each constant symbol $c \in \mathscr{L}$ which does not belong to $\mathscr{L}'$, let $c^{\mathbf{M}} := a_0$. Similarly, for each n-ary function symbol $F \in \mathscr{L}$ which does not belong to $\mathscr{L}'$, let $F^{\mathbf{M}} : A^n \to A$ be such that $F^{\mathbf{M}}$ maps each element of A^n to a_0. Finally, for each n-ary relation symbol $R \in \mathscr{L}$ which does not belong to $\mathscr{L}'$, let $R^{\mathbf{M}} := A^n$.

3.3 If an $\mathscr{L}$-theory T has, up to isomorphisms, a unique model, then T is complete.

3.4 If two structures $\mathbf{M}$ and $\mathbf{N}$ are isomorphic, then they are elementarily equivalent.

3.5 Let DLO be the theory of dense linearly ordered sets without endpoints. More precisely, the signature $\mathscr{L}_{\mathsf{DLO}}$ contains just the binary relation symbol $<$, and the non-logical axioms of DLO are the following sentences:

DLO_0 $\forall x \neg(x < x)$
DLO_1 $\forall x \forall y \forall z\big((x < y \land y < z) \to x < z\big)$
DLO_2 $\forall x \forall y\big(x < y \lor x = y \lor y < x\big)$
DLO_3 $\forall x \forall y \exists z\big(x < y \to (x < z \land z < y)\big)$
DLO_4 $\forall x \exists y \exists z\big(y < x \land x < z\big)$

Show that every countable model of DLO is isomorphic to $(\mathbb{Q}, <)$.
Hint: Enumerate both $\mathbb{Q}$ and some model $\mathbf{M}$ of DLO, and construct an isomorphism by recursion in such a way that in the n-th step the n-th element of M is mapped to an element of $\mathbb{Q}$ with the order being preserved.

3.6 Let $\mathscr{L}$ be an arbitrary signature and let $\mathbf{M}$ be an arbitrary $\mathscr{L}$-structure.

(a) Show that L_{11}-L_{13} are valid in $\mathbf{M}$.
(b) Show that L_{14}-L_{16} are valid in $\mathbf{M}$.

3.7 We say that two $\mathscr{L}$-formulae φ and ψ are **semantically equivalent** if for all $\mathscr{L}$-structures $\mathbf{M}$ and every assignment j we have

$$(\mathbf{M}, j) \vDash \varphi \iff (\mathbf{M}, j) \vDash \psi$$

(a) Show that for every sentence σ there is a semantically equivalent sentence $\tilde{\sigma}$ which contains just variables among $v_0, v_1, \ldots$, where for any $m, n \in \mathbb{N}$ with $m < n$, if v_n appears in $\tilde{\sigma}$, then also v_m appears in $\tilde{\sigma}$ (compare with Theorem 2.12).
(b) Show that for every $\mathscr{L}$-sentence σ there is a semantically equivalent $\mathscr{L}$-sentence in sPNF. (compare with Theorem 2.13).

Part II
Gödel's Completeness Theorem

In this part of the book, we shall prove GÖDEL'S COMPLETENESS THEOREM and show several consequences of it. Roughly speaking, GÖDEL'S COMPLETENESS THEOREM states that every consistent $\mathscr{L}$-theory T has a model $\mathbf{M} \vDash \mathsf{T}$. With respect to the model $\mathbf{M}$, every $\mathscr{L}$-sentence σ is either true or false (i.e., either σ or $\neg\sigma$ is true in $\mathbf{M}$). Hence, the set of $\mathscr{L}$-sentences which are true in $\mathbf{M}$ is with this respect *complete*. Therefore, as a consequence of GÖDEL'S COMPLETENESS THEOREM we obtain that every consistent theory is contained in a complete theory. However, this result should not be confused with GÖDEL'S FIRST INCOMPLETENESS THEOREM (presented in Part III), which states that the theory of Peano Arithmetik PA is incomplete and cannot be completed constructively. The reason for the latter is that the completion of PA is not feasible in a constructive way (notice the metamathematical assumption in the proof of LINDENBAUM'S LEMMA 4.5).

Gödel proved his famous theorem in his doctoral dissertation *Über die Vollständigkeit des Logikkalküls* [14] which he completed in 1929. In 1930, he published the same material as in the doctoral dissertation in a rewritten and shortened form in [15]. However, instead of presenting Gödel's original proof we decided to follow Henkin's construction, which can be found in [22] (see also [24]), since it fits better in the logical framework as developed in Part I. Even though Henkin's construction also works for uncountable signatures, we shall prove in Chapter 15 the general COMPLETENESS THEOREM with an ultraproduct construction, using ŁOŚ'S THEOREM.

We would like to mention that in our proof of GÖDEL'S COMPLETENESS THEOREM — in contrast to Henkin's proof — we only have to assume the existence of *potentially infinite* sets, but no instance of an *actually infinite* set is required (see also the introductory chapter).

Chapter 4
Maximally Consistent Extensions

Throughout this chapter, we require that all formulae are written in Polish notation and that the variables are among $v_0, v_1, v_2, \ldots$ Notice that the former requirement is just another notation which does not involve brackets, and that by the VARIABLE SUBSTITUTION THEOREM 2.12, the latter requirement gives us semantically equivalent formulae.

Maximally Consistent Theories

Let $\mathscr{L}$ be an arbitrary signature and let T be an $\mathscr{L}$-theory (i.e., a set of $\mathscr{L}$-sentences). We say that T is **maximally consistent** if T is consistent and for every $\mathscr{L}$-sentence σ we have *either* $\sigma \in \mathsf{T}$ *or* $\neg\operatorname{Con}(\mathsf{T} + \sigma)$. In other words, a consistent theory T is maximally consistent if no proper extension of T is consistent. The following fact is just a reformulation of this definition.

FACT 4.1. *Let $\mathscr{L}$ be a signature and let T be a consistent $\mathscr{L}$-theory. Then T is maximally consistent if and only if for every $\mathscr{L}$-sentence σ, either $\sigma \in \mathsf{T}$ or $\mathsf{T} \vdash \neg\sigma$.*

Proof. By FACT 2.14(c) & (d) we have:

$$\neg\operatorname{Con}(\mathsf{T} + \sigma) \iff \mathsf{T} \vdash \neg\sigma$$

Hence, an $\mathscr{L}$-theory is maximally consistent if and only if for every $\mathscr{L}$-sentence σ, either $\sigma \in \mathsf{T}$ or $\mathsf{T} \vdash \neg\sigma$. ⊣

As a consequence of FACT 4.1, we get

LEMMA 4.2. *Let $\mathscr{L}$ be a signature and let T be a consistent $\mathscr{L}$-theory. Then T is maximally consistent if and only if for every $\mathscr{L}$-sentence σ, either $\sigma \in \mathsf{T}$ or $\neg\sigma \in \mathsf{T}$.*

L. Halbeisen, R. Krapf, *Gödel's Theorems and Zermelo's Axioms*,
https://doi.org/10.1007/978-3-030-52279-7_4

Proof. We have to show that the following equivalence holds:

$$\forall\sigma(\sigma \in \mathsf{T} \text{ or } \mathsf{T} \vdash \neg\sigma) \iff \forall\sigma(\sigma \in \mathsf{T} \text{ or } \neg\sigma \in \mathsf{T})$$

($\Rightarrow$) Assume that for every $\mathscr{L}$-sentence σ we have $\sigma \in \mathsf{T}$ or $\mathsf{T} \vdash \neg\sigma$. If $\sigma \in \mathsf{T}$, then the implication obviously holds. If $\sigma \notin \mathsf{T}$, then $\mathsf{T} \vdash \neg\sigma$, and since T is consistent, this implies $\mathsf{T} \nvdash \sigma$. Now, by TAUTOLOGY (F), this implies $\mathsf{T} \nvdash \neg\neg\sigma$ and by our assumption we finally get $\neg\sigma \in \mathsf{T}$.

($\Leftarrow$) Assume that for every $\mathscr{L}$-sentence σ we have $\sigma \in \mathsf{T}$ or $\neg\sigma \in \mathsf{T}$. If $\sigma \in \mathsf{T}$, then the implication obviously holds. Now, if $\sigma \notin \mathsf{T}$, then by our assumption we have $\neg\sigma \in \mathsf{T}$, which obviously implies $\mathsf{T} \vdash \neg\sigma$. ⊣

Maximally consistent theories have similar features as complete theories: Recall that an $\mathscr{L}$-theory T is complete if for every $\mathscr{L}$-sentence σ we have *either* $\mathsf{T} \vdash \sigma$ *or* $\mathsf{T} \vdash \neg\sigma$.

As an immediate consequence of the definitions, we get

FACT 4.3. *Let $\mathscr{L}$ be a signature, let T be a consistent $\mathscr{L}$-theory, and let* $\mathbf{Th}(\mathsf{T})$ *be the set of all $\mathscr{L}$-sentences which are provable from T.*

(a) *If T is complete, then $\mathbf{Th}(\mathsf{T})$ is maximally consistent.*

(b) *If T is maximally consistent, then $\mathbf{Th}(\mathsf{T})$ is the same as T.*

The next result gives a condition under which a theory can be extended to a maximally consistent theory. In fact, it is just a reformulation of PROPOSITION 3.10.

FACT 4.4. *If an $\mathscr{L}$-theory T has a model, then T has a maximally consistent extension.*

Proof. Let $\mathbf{M}$ be a model of the $\mathscr{L}$-theory T and let $\mathbf{Th}(\mathbf{M})$ be the set of $\mathscr{L}$-sentences σ such that $\mathbf{M} \vDash \sigma$. Then $\mathbf{Th}(\mathbf{M})$ is obviously a maximally consistent theory which contains T. ⊣

Later we shall see that every consistent theory has a model. For this, we first show how a consistent theory can be extended to a maximally consistent theory.

Universal List of Sentences

Let $\mathscr{L}$ be an arbitrary but fixed countable signature, where by "countable" we mean that the symbols in $\mathscr{L}$ can be listed in a F I N I T E or P O T E N - T I A L L Y I N F I N I T E list $L_{\mathscr{L}}$.

First, we encode the symbols of $\mathscr{L}$ corresponding to the order in which they appear in the list $L_{\mathscr{L}}$: The first symbol is encoded with "2", the second with "22", the third with "222", and so on. For every symbol $\zeta \in L_{\mathscr{L}}$, let $\#\zeta$ denote the code of ζ. Therefore, the code of a symbol of $\mathscr{L}$ is just a sequence of 2's.

Furthermore, we encode the logical symbols as follows:

Symbol ζ	Code $\#\zeta$
$=$	11
$\neg$	1111
$\wedge$	111111
$\vee$	11111111
$\rightarrow$	1111111111
$\exists$	111111111111
$\forall$	11111111111111
v_0	1
v_1	111
$\vdots$	$\vdots$
v_n	$\underbrace{1111\ \ldots\ 11111}_{(2n+1)\ 1\text{'s}}$

In the next step, we encode strings of symbols: Let $\bar{\zeta} \equiv \zeta_0\zeta_1\zeta_2\ldots\zeta_n$ be a finite string of symbols, then

$$\#\bar{\zeta} := \#\zeta_0 0\#\zeta_1 0\#\zeta_2\ldots 0\#\zeta_n$$

For a string $\#\zeta$ (i.e., a string of 0's, 1's, and 2's), let $|\#\zeta|$ be the length of $\#\zeta$ (i.e., the number of 0's, 1's, and 2's which appear in $\#\zeta$).

Now, we order the codes of strings of symbols by their length and strings of the same length lexicographically, where $0 < 1 < 2$. If, with respect to this ordering, $\#\zeta$ is less than $\#\zeta'$, then we write $\zeta \prec \zeta'$.

Finally, let

$$\Lambda_{\mathscr{L}} := [\sigma_1, \sigma_2, \ldots]$$

be the potentially infinite list of all $\mathscr{L}$-sentences written in Polish notation (notice that we did not encode brackets), where we require

$$\#\sigma_i \prec \#\sigma_j \iff i < j\,.$$

We call $\Lambda_{\mathscr{L}}$ the **universal list of $\mathscr{L}$-sentences.**

Lindenbaum's Lemma

In this section, we show that every consistent set of $\mathscr{L}$-sentences T can be extended to a maximally consistent set of $\mathscr{L}$-sentences $\bar{\mathsf{T}}$. Since the universal list of $\mathscr{L}$-sentences contains all possible $\mathscr{L}$-sentences, every set T of $\mathscr{L}$-sentences can be listed in a finite or potentially infinite list.

THEOREM 4.5 (LINDENBAUM'S LEMMA). *Let $\mathscr{L}$ be a countable signature and let T be a consistent set of $\mathscr{L}$-sentences. Furthermore, let σ_0 be an $\mathscr{L}$-sentence which cannot be proved from T, i.e., $\mathsf{T} \nvdash \sigma_0$. Then there exists a maximally consistent set $\bar{\mathsf{T}}$ of $\mathscr{L}$-sentences which contains $\neg\sigma_0$ as well as all the sentences of T.*

Proof. Let $\Lambda_{\mathscr{L}} = [\sigma_1, \sigma_2, \ldots]$ be the universal list of all $\mathscr{L}$-sentences. First we extend $\Lambda_{\mathscr{L}}$ with the $\mathscr{L}$-sentence $\neg\sigma_0$; let $\Lambda^0_{\mathscr{L}} = [\neg\sigma_0, \sigma_1, \sigma_2, \ldots]$.

Now, we go through the list $\Lambda^0_{\mathscr{L}}$ and define step by step a list $\bar{\mathsf{T}}$ of $\mathscr{L}$-sentences. For this, we define T_0 as the list which contains just $\neg\sigma_0$, i.e., $\mathsf{T}_0 := [\neg\sigma_0]$. If T_n is already defined, then

$$\mathsf{T}_{n+1} := \begin{cases} \mathsf{T}_n + [\sigma_{n+1}] & \text{if } \operatorname{Con}(\mathsf{T} + \mathsf{T}_n + \sigma_{n+1}), \\ \mathsf{T}_n & \text{otherwise.} \end{cases}$$

Let $\bar{\mathsf{T}} = [\neg\sigma_0, \sigma_{i_1}, \ldots]$ be the resulting list, i.e., $\bar{\mathsf{T}}$ is the potentially infinite list which contains each finite list T_n as an initial list. Notice that this construction only works if we assume the metamathematical LAW OF EXCLUDED MIDDLE or a similar principle like the WEAK KÖNIG'S LEMMA (see EXERCISE 4.2): Even in the case when we cannot decide whether $\mathsf{T} + T_n + \sigma_n$ is consistent or not, we assume, from a metamathematical point of view, that *either* $\mathsf{T} + T_n + \sigma_n$ is consistent *or* $\mathsf{T} + T_n + \sigma_n$ is inconsistent (and *neither* both, *nor* none).

The following claim states that we cannot derive a contradiction from finitely many $\mathscr{L}$-sentences in $\bar{\mathsf{T}}$ and that we cannot add any new $\mathscr{L}$-sentence to $\bar{\mathsf{T}}$ without destroying this property. However, in order to simplify our terminology, we shall consider the potentially infinite list $\bar{\mathsf{T}}$ as an actual infinite set — notice that this is just a "façon-de-parler" since we do not have to assume the existence of actual infinite sets.

CLAIM. *$\bar{\mathsf{T}}$ is a maximally consistent set of $\mathscr{L}$-sentences which contains $\neg\sigma_0$ as well as all the sentences of T.*

Proof of Claim. First we show that $\bar{\mathsf{T}}$ contains $\mathsf{T} + \neg\sigma_0$, then we show that $\bar{\mathsf{T}}$ is consistent, and finally we show that for every $\mathscr{L}$-sentence σ we have either $\sigma \in \bar{\mathsf{T}}$ or $\neg \operatorname{Con}(\bar{\mathsf{T}} + \sigma)$.

$\bar{\mathsf{T}}$ contains all sentences of $\mathsf{T} + \neg\sigma_0$: By definition, $\mathsf{T}_0 = [\neg\sigma_0]$, and since T_0 is an initial segment of the list $\bar{\mathsf{T}}$, $\neg\sigma_0$ belongs to $\bar{\mathsf{T}}$. For every $\sigma \in \mathsf{T}$, there is $n \in \mathbb{N}$ with $n \geq 1$ such that $\sigma \equiv \sigma_n$. By induction, we show that if $\sigma_n \in \mathsf{T}$ then $\sigma_n \in \bar{\mathsf{T}}$. For this, suppose that the claim holds for all $m \leq n$ and that

$\sigma_{n+1} \in \mathsf{T}$. Let m be the largest number $m \leq n$ such that $\sigma_m \in \mathsf{T}$; if no such m exists, we set $m = 0$. Then we have $\mathsf{T}_n = \mathsf{T}_m$. If $\neg\,\mathrm{Con}(\mathsf{T} + \mathsf{T}_n + \sigma_{n+1})$, then since $\sigma_{n+1} \in \mathsf{T}$, we have $\neg\,\mathrm{Con}(\mathsf{T} + \mathsf{T}_m)$, contradicting $\sigma_m \in \mathsf{T}$ (respectively $\mathsf{T} \nvdash \sigma_0$ in the case of $m = 0$). Hence we have $\mathrm{Con}(\mathsf{T} + \mathsf{T}_n + \sigma_{n+1})$ and therefore $\sigma_{n+1} \in \mathsf{T}_{n+1}$.

$\bar{\mathsf{T}}$ *is consistent*: By the COMPACTNESS THEOREM 2.15 it is enough to show that every finite subset of $\bar{\mathsf{T}}$ is consistent. Since every finite subset of $\bar{\mathsf{T}}$ is contained in T_n for some n, it suffices to prove by induction that T_n is consistent for every $n \in \mathbb{N}$. Since $\mathsf{T} \nvdash \sigma_0$, $\mathsf{T}_0 = [\sigma_0]$ is consistent. Now suppose $\mathrm{Con}(\mathsf{T}_m)$ for all $m \leq n$. If $\neg\,\mathrm{Con}(\mathsf{T}+\mathsf{T}_n+\sigma_{n+1})$, then $\mathsf{T}_{n+1} = \mathsf{T}_n$ is consistent by our induction hypothesis. Otherwise, we have $\mathrm{Con}(\mathsf{T} + \mathsf{T}_n + \sigma_{n+1})$ and hence T_{n+1} is also consistent.

For every σ, either $\sigma \in \bar{\mathsf{T}}$ or $\neg\,\mathrm{Con}(\bar{\mathsf{T}} + \sigma)$: For every $\mathscr{L}$-sentence σ, there is a $n \in \mathbb{N}$ with $n \geq 1$ such that $\sigma \equiv \sigma_n$. By the LAW OF EXCLUDED MIDDLE, we have *either* $\mathrm{Con}(\mathsf{T}+\mathsf{T}_{n-1}+\sigma_n)$ *or* $\neg\,\mathrm{Con}(\mathsf{T}+\mathsf{T}_{n-1}+\sigma_n)$. In the former case we obtain $\sigma_n \in \mathsf{T}_n$, which implies $\sigma \in \bar{\mathsf{T}}$. In the latter case we obtain $\neg\,\mathrm{Con}(\bar{\mathsf{T}} + \sigma_n)$, which is the same as $\neg\,\mathrm{Con}(\bar{\mathsf{T}} + \sigma)$. $\dashv_{\text{Claim}}$

Thus, the list $\bar{\mathsf{T}}$ has all the required properties, which completes the proof. $\dashv$

The following fact summarises the main properties of $\bar{\mathsf{T}}$.

FACT 4.6. *Let T, $\bar{\mathsf{T}}$, and σ_0 be as above, and let σ and σ' be any $\mathscr{L}$-sentences.*

(a) $\neg\sigma_0 \in \bar{\mathsf{T}}$.

(b) *Either* $\sigma \in \bar{\mathsf{T}}$ *or* $\neg\sigma \in \bar{\mathsf{T}}$.

(c) *If* $\mathsf{T} \vdash \sigma$, *then* $\sigma \in \bar{\mathsf{T}}$.

(d) $\bar{\mathsf{T}} \vdash \sigma$ *if and only if* $\sigma \in \bar{\mathsf{T}}$.

(e) *If* $\sigma \Leftrightarrow \sigma'$, *then* $\sigma \in \bar{\mathsf{T}}$ *if and only if* $\sigma' \in \bar{\mathsf{T}}$.

Proof. (a) follows by construction of $\bar{\mathsf{T}}$.

Since $\bar{\mathsf{T}}$ is maximally consistent, (b) follows by LEMMA 4.2.

For (c), notice that $\mathsf{T} \vdash \sigma$ implies $\neg\,\mathrm{Con}(\mathsf{T} + \neg\sigma)$, hence $\neg\sigma \notin \bar{\mathsf{T}}$ and by (b) we get $\sigma \in \bar{\mathsf{T}}$.

For (d), let us first assume $\bar{\mathsf{T}} \vdash \sigma$, where $\sigma \equiv \sigma_n$. This implies $\mathrm{Con}(\bar{\mathsf{T}} + \sigma)$, hence $\mathrm{Con}(\mathsf{T} + T_n + \sigma_n)$, and by construction of $\bar{\mathsf{T}}$ we get $\sigma_n \in \bar{\mathsf{T}}$. On the other hand, if $\sigma \in \bar{\mathsf{T}}$, then we obviously have $\bar{\mathsf{T}} \vdash \sigma$.

For (e), recall that $\sigma \Leftrightarrow \sigma'$ is just an abbreviation for $\vdash \sigma \leftrightarrow \sigma'$. Thus, (e) follows immediately from (d). $\dashv$

FACT 4.6 shows that the $\mathscr{L}$-sentences in $\bar{\mathsf{T}}$ "behave" like valid sentences in a model, which is indeed the case—as the following proposition shows.

PROPOSITION 4.7. *Let* $\bar{\mathsf{T}}$ *be as above, and let* $\sigma, \sigma_1, \sigma_2$ *be any* $\mathscr{L}$*-sentences in Polish notation.*

(a) $\neg\sigma \in \bar{\mathsf{T}} \iff$ NOT $\sigma \in \bar{\mathsf{T}}$

(b) $\wedge\sigma_1\sigma_2 \in \bar{\mathsf{T}} \iff \sigma_1 \in \bar{\mathsf{T}}$ AND $\sigma_2 \in \bar{\mathsf{T}}$

(c) $\vee\sigma_1\sigma_2 \in \bar{\mathsf{T}} \iff \sigma_1 \in \bar{\mathsf{T}}$ OR $\sigma_2 \in \bar{\mathsf{T}}$

(d) $\rightarrow\sigma_1\sigma_2 \in \bar{\mathsf{T}} \iff$ IF $\sigma_1 \in \bar{\mathsf{T}}$ THEN $\sigma_2 \in \bar{\mathsf{T}}$

Proof. (a) Follows immediately from FACT 4.6.(b).

(b) First notice that by FACT 4.6.(d), $\wedge\sigma_1\sigma_2 \in \bar{\mathsf{T}}$ if and only if $\bar{\mathsf{T}} \vdash \wedge\sigma_1\sigma_2$. Thus, by L_3 and L_4 and (MP) we get $\bar{\mathsf{T}} \vdash \sigma_1$ and $\bar{\mathsf{T}} \vdash \sigma_2$. Therefore, by FACT 4.6.(d), we get $\sigma_1 \in \bar{\mathsf{T}}$ AND $\sigma_2 \in \bar{\mathsf{T}}$. On the other hand, if $\sigma_1 \in \bar{\mathsf{T}}$ AND $\sigma_2 \in \bar{\mathsf{T}}$, then, by FACT 4.6.(d), we get $\bar{\mathsf{T}} \vdash \sigma_1$ and $\bar{\mathsf{T}} \vdash \sigma_2$. Now, by TAUTOLOGY (B), this implies $\bar{\mathsf{T}} \vdash \wedge\sigma_1\sigma_2$, and by FACT 4.6.(d) we finally get $\wedge\sigma_1\sigma_2 \in \bar{\mathsf{T}}$.

(c) and (d) follow from FACT 4.6.(e) and from the 3-SYMBOLS THEOREM 1.7 which states that for each formula σ there is an equivalent formula σ' which contains neither $\vee$ nor $\rightarrow$. ⊣

EXERCISES

4.0 (a) Show that Group Theory GT is incomplete.

(b) Let $\psi \equiv \forall x \forall y\, (x \circ y = y \circ x)$. Show that $\mathsf{GT} + \psi$ is incomplete.

(c) Extend GT with a single axiom φ, such that $\mathsf{GT} + \varphi$ is complete.

4.1 Show that all the logical axioms of propositional logic (i.e., L_0–L_9) were used in the proofs of FACT 4.1, LEMMA 4.2, FACT 4.6, and PROPOSITION 4.7. Notice that in the proof of FACT 4.1, we used FACT 2.14.(c) & (d).

4.2 The WEAK KÖNIG'S LEMMA is a very weak choice principle: A tree T is a 0-1-*tree* if it is a sub-tree of a binary tree in which the two successors of a node are always labelled with 0 and 1, respectively. Now, the WEAK KÖNIG'S LEMMA states that

every infinite 0-1-*tree contains an infinite branch.*

Show that in the proof of LINDENBAUM'S LEMMA 4.5, the LAW OF EXCLUDED MIDDLE can be replaced by the metamathematical WEAK KÖNIG'S LEMMA.

Hint: First, consider the set Λ of finite lists $\lambda = [\sigma_{i_0}, \dots, \sigma_{i_n}]$ of $\mathscr{L}$-sentences, where $k < l$ implies $i_k < i_l$. Next, encode formal proofs, which are finite sequences of $\mathscr{L}$-formulae, with natural numbers. Finally, construct the tree T consisting of all lists $\lambda \in \Lambda$, such that there is no formal proof of an inconsistency from $\mathsf{T} + \lambda$ with a code-number less than the length of λ. Then T corresponds to an infinite 0-1-tree with the property that each infinite branch through T corresponds to a maximally consistent set of $\mathscr{L}$-sentences.

4.3 Show that in the case when the theory $\mathsf{T} + \neg\sigma_0$ already has a model $\mathbf{M}$, then we can just set $\bar{\mathsf{T}} = \mathbf{Th}(\mathbf{M})$. Therefore, we do not need the LAW OF EXCLUDED MIDDLE in this case.

Chapter 5
The Completeness Theorem

As in the previous chapter, we require that all formulae are written in Polish notation and that the variables are among $v_0, v_1, v_2, \ldots$ Furthermore, let $\mathscr{L}$ be a countable signature, let T be a consistent $\mathscr{L}$-theory, and let σ_0 be an $\mathscr{L}$-sentence which is not provable from T. Finally, let $\bar{\mathsf{T}}$ be the maximally consistent extension of $\mathsf{T} + \neg\sigma_0$ obtained with LINDENBAUM'S LEMMA 4.5.

We shall construct a model of $\bar{\mathsf{T}}$ as follows: In a first step, we extend the signature $\mathscr{L}$ to a signature $\mathscr{L}_c$ by adding countably many new constant symbols, so-called *special constants*. In a second the step, we extend the $\mathscr{L}$-theory $\bar{\mathsf{T}}$ to an $\mathscr{L}_c$-theory $\bar{\mathsf{T}}_c$ by adding so-called *witnesses* to existential sentences in $\bar{\mathsf{T}}$. In particular, for each sentence $\exists\nu\sigma(\nu) \in \bar{\mathsf{T}}$ we add an $\mathscr{L}_c$-sentence $\sigma(c)$, where c is some special constant. In a third step, we extend the $\mathscr{L}_c$-theory $\bar{\mathsf{T}}_c$ to a maximally consistent $\mathscr{L}_c$-theory $\tilde{\mathsf{T}}$, and in a last step, we build the domain of the model of $\tilde{\mathsf{T}}$ as a list of lists of closed $\mathscr{L}_c$-terms.

Extending the Language

A string of symbols is called a **term-constant**, if it results from applying F I N I T E L Y many times the following rules:

(C0) Each closed (i.e., variable-free) $\mathscr{L}$-term is a term-constant.

(C1) If $\tau_0, \ldots, \tau_{n-1}$ are any term-constants which we have already built and F is an n-ary function symbol, then $F\tau_0 \cdots \tau_{n-1}$ is a term-constant.

(C2) For any natural numbers i, n, if $\tau_0, \ldots, \tau_{n-1}$ are any term-constants which we have already built, then $(i, \tau_0, \ldots, \tau_{n-1}, n)$ is a term-constant.

The strings $(i, \tau_0, \ldots, \tau_{n-1}, n)$ which are built with rule (C2) are called **special constants**. Notice that for $n = 0$, $(i, \tau_0, \ldots, \tau_{n-1}, n)$ becomes $(i, 0)$.

Let $\mathscr{L}_c$ be the signature $\mathscr{L}$ extended by the countably many special constants. In order to write the special constants in a list, we first encode them and then define an ordering on the set of codes.

First we encode closed $\mathscr{L}$-terms with strings of 0's and 2's as in Chapter 4. Now, let $c \equiv (i, \tau_0, \ldots, \tau_{n-1}, n)$ be a special constant, where the codes

L. Halbeisen, R. Krapf, *Gödel's Theorems and Zermelo's Axioms*,
https://doi.org/10.1007/978-3-030-52279-7_5

$\#\tau_1, \ldots, \#\tau_{n-1}$ of $\tau_0, \ldots, \tau_{n-1}$ are already defined. Then we encode c as follows:

$$\begin{array}{rcccccccccc} c & \equiv & (& i & , & \tau_0 & , & \ldots & , & \tau_{n-1} & , & n &) \\ & & \downarrow & \downarrow & \downarrow & \downarrow & \downarrow & & \downarrow & \downarrow & \downarrow & \downarrow & \downarrow \\ \#c & \equiv & 6 & \underbrace{1\ldots1}_{i\text{-times } 1} & 8 & \#\tau_0 & 8 & \ldots & 8 & \#\tau_{n-1} & 8 & \underbrace{1\ldots1}_{n\text{-times } 1} & 9 \end{array}$$

The codes of special constants are ordered by their length and lexicographically, where $0 < 1 < 2 < 6 < 8 < 9$.

Finally, let $\Lambda_c = [c_0, c_1, \ldots]$ be the potentially infinite list of all special constants, ordered with respect to the ordering of their codes.

Extending the Theory

In this section, we shall add witnesses for certain existential $\mathscr{L}_c$-sentences σ_i in the list $\bar{\mathsf{T}} = [\neg\sigma_0, \sigma_1, \ldots, \sigma_i, \ldots]$, where an $\mathscr{L}_c$-sentence is existential if it is of the form $\exists\nu\varphi$. We choose the witnesses from the list Λ_c of special constants. In order to make sure that we have a witness for each existential $\mathscr{L}_c$-sentence (and not just for $\mathscr{L}$-sentences), and also to make sure that the choice of witnesses does not lead to a contradiction, we have to choose the witnesses carefully.

Let $\sigma_i \in \bar{\mathsf{T}}$ and let $c_j \equiv (i, t_0, \ldots, t_{n-1}, n)$ be a special constant. Then we say that c_j witnesses σ_i or that c_j is a **witness** for σ_i, if

- $i \geq 1$ and σ_i is in special Prenex Normal Form sPNF (see Chapter 3),
- the symbol $\exists v_n$ appears in σ_i,
- for all $m < n$: if $\exists v_m$ appears in σ_i, then $t_m \equiv (i, t_0, \ldots, t_{m-1}, m)$.

On the one hand, we have only witnesses c_j for $\mathscr{L}$-sentences σ_i with $i \geq 1$. On the other hand, notice that since $\neg\sigma_0$ is not necessarily in sPNF, by construction of $\bar{\mathsf{T}}$ there exists an $i \geq 1$ such that σ_i and $\neg\sigma_0$ are semantically equivalent, which will be sufficient for our purposes.

If an $\mathscr{L}$-sentence $\sigma_i \in \bar{\mathsf{T}}$ is in sPNF and either $\exists v_n$ or $\forall v_n$ appears in σ_i, then

$$\sigma_i \equiv \forall\!\!\!/_0 v_0 \forall\!\!\!/_1 v_1 \cdots \forall\!\!\!/_n v_n \sigma_{i,n}(v_0, \ldots, v_n),$$

where $\sigma_{i,n}(v_0, \ldots, v_n)$ is an $\mathscr{L}$-formula in which each variable among $v_0, \ldots, v_n$ appears free. In particular, if $c_j \equiv (i, t_0, \ldots, t_{n-1}, n)$ witnesses σ_i, then

$$\sigma_i \equiv \forall\!\!\!/_0 v_0 \forall\!\!\!/_1 v_1 \cdots \forall\!\!\!/_{n-1} v_{n-1} \exists v_n \sigma_{i,n}(v_0, \ldots, v_n),$$

i.e., $\exists v_n$ appears in σ_i. Furthermore, if $\sigma_i \in \bar{\mathsf{T}}$ is in sPNF, $c_j \equiv (i, t_0, \ldots, t_{n-1}, n)$ is a special constant, and c_j witnesses σ_i, then let

$$\sigma_{i,n}[c_j] :\equiv \sigma_{i,n}(v_0/t_0, \ldots, v_{n-1}/t_{n-1}, v_n/c_j).$$

Now, we go through the list $\Lambda_c = [c_0, c_1, \ldots]$ of special constants and extend step by step the list $\bar{\mathsf{T}} = [\sigma_0, \sigma_1, \ldots]$. For this, we first stipulate $\mathsf{T}_0 := \bar{\mathsf{T}}$. Now assume that T_j is already defined and that $c_j \equiv (i, t_0, \ldots, t_{n-1}, n)$ for some natural numbers i, n and term-constants $t_0, \ldots, t_{n-1}$. Then we have the following two cases:

Case 1. The special constant c_j does not witness the $\mathscr{L}$-sentence $\sigma_i \in \bar{\mathsf{T}}$. In this case, we set $\mathsf{T}_{j+1} := \mathsf{T}_j$.

Case 2. The special constant c_j witnesses $\sigma_i \in \bar{\mathsf{T}}$. In this case, we insert the $\mathscr{L}_c$-sentence $\sigma_{i,n}[c_j]$ into the list T_j at the place which corresponds to the code $\#\sigma_{i,n}[c_j]$. The extended list is then T_{j+1}.

Finally, let $\bar{\mathsf{T}}_c$ be the resulting list, i.e., $\bar{\mathsf{T}}_c$ is the union of all the T_j's.

LEMMA 5.1. *$\bar{\mathsf{T}}_c$ is consistent.*

Proof. By construction of $\bar{\mathsf{T}}$ we have $\mathrm{Con}(\bar{\mathsf{T}})$ with respect to the signature $\mathscr{L}$. We first show that $\bar{\mathsf{T}}$ is also consistent with respect to the signature $\mathscr{L}_c$: Assume towards a contradiction that $\bar{\mathsf{T}} \vdash \text{⎅}$ with respect to the signature $\mathscr{L}_c$. In that proof, we replace each special constant c by a variable ν_c which does not occur in any of the finitely many formulae of the proof, such that ν_c and $\nu_{c'}$ are distinct variables whenever c and c' are distinct special constants. Notice that every logical axiom becomes a logical axiom of the same type. Moreover, notice that all $\mathscr{L}$-sentences of $\bar{\mathsf{T}}$ remain unchanged since they do not contain special constants. Furthermore, each application of Modus Ponens or Generalisation becomes a new application of the same inference rule. To see this, notice that we do not apply Generalisation to any of the ν_c's, since otherwise, we would have applied Generalisation to a special constant c, but c is a term-constant and not a variable. Since the obtained proof does not contain any special constants, we get $\bar{\mathsf{T}} \vdash \text{⎅}$ (with respect to $\mathscr{L}$), which contradicts the fact that $\bar{\mathsf{T}}$ is consistent (with respect to $\mathscr{L}$). Therefore we have $\mathrm{Con}(\bar{\mathsf{T}})$ with respect to $\mathscr{L}_c$.

Now, assume towards a contradiction that $\bar{\mathsf{T}}_c$ is inconsistent, i.e., $\neg\,\mathrm{Con}(\bar{\mathsf{T}}_c)$. Then, by the COMPACTNESS THEOREM 2.15, we find finitely many pairwise distinct $\mathscr{L}_c$-sentences $\sigma_{i,n}[c_j]$ in $\bar{\mathsf{T}}_c$ such that

$$\neg\,\mathrm{Con}\left(\bar{\mathsf{T}} + \left\{\sigma_{i_1,n_1}[c_{j_1}], \ldots, \sigma_{i_k,n_k}[c_{j_k}]\right\}\right).$$

Notice that since the $\mathscr{L}_c$-sentences $\sigma_{i_1,n_1}[c_{j_1}], \ldots, \sigma_{i_k,n_k}[c_{j_k}]$ are pairwise distinct, so are the special constants $c_{j_1}, \ldots, c_{j_k}$. Without loss of generality we may assume that $\sigma_{i_1,n_1}[c_{j_1}], \ldots, \sigma_{i_k,n_k}[c_{j_k}]$ are such that the sum $n_1 + \ldots + n_k + k$ is minimal.

For term-constants τ we define the **height** $h(\tau)$ as follows: If τ is a closed $\mathscr{L}$-term, then $h(\tau) := 0$; if $\tau_0, \ldots, \tau_{n-1}$ are term-constants and $F \in \mathscr{L}$ is an n-ary function symbol, then

$$h(F\tau_0 \cdots \tau_{n-1}) := \max\left\{h(\tau_0), \ldots, h(\tau_{n-1})\right\};$$

and finally, if $\tau \equiv (i, \tau_0, \ldots, \tau_{n-1}, n)$ is a special constant, then

$$h(\tau) := 1 + \max\{h(\tau_0), \ldots, h(\tau_{n-1})\},$$

where $\max\emptyset := 0$. Without loss of generality we may assume that

$$h(c_{j_k}) = \max\{h(c_{j_1}), \ldots, h(c_{j_k})\},$$

i.e., for each special constant c_j occurring in $c_{j_1}, \ldots, c_{j_k}$ we have $h(c_j) < h(c_{j_k})$. In particular, it follows that c_{j_k} does not occur in each such special constant c_j.

Let us now consider the formula $\sigma_{i_k,n_k}[c_{j_k}]$. In order to simplify the notation, we write i, n, j instead of i_k, n_k, j_k, respectively; in particular, $\sigma_{i_k,n_k}[c_{j_k}]$ becomes $\sigma_{i,n}[c_j]$. Furthermore, let

$$\Sigma := \{\sigma_{i_1,n_1}[c_{j_1}], \ldots, \sigma_{i_{k-1},n_{k-1}}[c_{j_{k-1}}]\},$$

and let $c_j \equiv (i, t_0, \ldots, t_{n-1}, n)$, i.e.,

$$\sigma_{i,n}[c_j] \equiv \sigma_{i,n}(v_0/t_0, \ldots, v_{n-1}/t_{n-1}, v_n/c_j).$$

Since c_j witnesses σ_i, $\exists v_n$ appears in σ_i, i.e.,

$$\sigma_{i,n-1}(v_0, \ldots, v_{n-1}) \equiv \exists v_n \sigma_{i,n}(v_0, \ldots, v_{n-1}, v_n).$$

In order to simplify the notation again, we set

$$\tilde{\sigma}(v_n) :\equiv \sigma_{i,n}(v_0/t_0, \ldots, v_{n-1}/t_{n-1}, v_n).$$

Notice that v_n is the only variable which appears free in $\tilde{\sigma}$.

Claim. $\neg\operatorname{Con}(\bar{\mathsf{T}} + \Sigma + \sigma_{i,n}[c_j]) \quad \Longrightarrow \quad \neg\operatorname{Con}(\bar{\mathsf{T}} + \Sigma + \exists v_n \tilde{\sigma}(v_n))$

Proof of Claim. If $\bar{\mathsf{T}} + \Sigma + \sigma_{i,n}[c_j]$ is inconsistent, then

$$\bar{\mathsf{T}} + \Sigma + \sigma_{i,n}[c_j] \vdash \boxtimes, \qquad (\vdash_1)$$

and with the Deduction Theorem we get

$$\bar{\mathsf{T}} + \Sigma \vdash \sigma_{i,n}[c_j] \to \boxtimes. \qquad (\vdash_2)$$

In the latter proof $(\vdash_2)$ we replace the special constant c_j throughout the proof by a variable ν which does not occur in $\sigma_{i,n}$ and which does not occur in any of the finitely many formulae of the former proof $(\vdash_1)$. Notice that every logical axiom becomes a logical axiom of the same type. Moreover, notice that $\mathscr{L}$-sentences of $\bar{\mathsf{T}}$ are not affected (since they do not contain special constants). Furthermore, $\mathscr{L}_c$-sentences of Σ are not affected either, since they do not contain the special constant c_j (recall that the special constants $c_{j_1}, \ldots, c_{j_k}$ are pairwise distinct). Finally, each application of Modus Ponens or Generalisation becomes a new application of the same inference rule (notice that we do not apply Generalisation to ν, since otherwise we would have applied Generalisation to c_j, but c_j is a term-constant). Now, we construct a proof of $\exists v_n \tilde{\sigma}(v_n) \to \boxtimes$ from $\bar{\mathsf{T}} + \Sigma$ as follows:

$$
\begin{array}{lll}
\bar{\mathsf{T}} + \Sigma \vdash \tilde{\sigma}(\nu) \to \boxtimes & & \text{by assumption} \\
\bar{\mathsf{T}} + \Sigma \vdash \forall \nu \big(\tilde{\sigma}(\nu) \to \boxtimes\big) & & \text{by Generalisation} \\
\bar{\mathsf{T}} + \Sigma \vdash \forall \nu \big(\tilde{\sigma}(\nu) \to \boxtimes\big) \to \big(\exists \nu \tilde{\sigma}(\nu) \to \boxtimes\big) & & \mathsf{L}_{13} \\
\bar{\mathsf{T}} + \Sigma \vdash \exists \nu \tilde{\sigma}(\nu) \to \boxtimes & & \text{by Modus Ponens} \\
\bar{\mathsf{T}} + \Sigma \vdash \exists v_n \tilde{\sigma}(v_n) \to \boxtimes & & \text{TAUTOLOGY (Q.1)}
\end{array}
$$

Therefore, we finally have $\neg \operatorname{Con}\big(\bar{\mathsf{T}} + \Sigma + \exists v_n \tilde{\sigma}(v_n)\big)$. $\dashv_{\text{Claim}}$

Let us now consider σ_i. Let $m \leq n$ be the largest natural number such that for each l with $1 \leq l \leq m$ we have that $\forall v_{n-l}$ appears in σ_i. For example, if $m = 0$, then $\forall\!\!\!\exists_{n-1}$ is the quantifier $\exists$, and if $m = n$, then for no $n' < n$, $\forall\!\!\!\exists_{n'}$ is the quantifier $\exists$. In general, σ_i is of the form

$$\sigma_i \equiv \underbrace{\forall\!\!\!\exists_0 v_0 \quad \cdots \quad \exists v_{n-m-1}}_{\exists \text{ or } \forall} \underbrace{\forall v_{n-m} \quad \cdots \quad \forall v_{n-1}}_{\text{only } \forall} \exists v_n \, \sigma_{i,n}(v_0, \ldots, v_n)\,.$$

Consider now the formula

$$\tilde{\sigma}_m :\equiv \sigma_{i,n-m-1}(v_0/t_0, \ldots, v_{n-m-1}/t_{n-m-1})\,.$$

Then either $\tilde{\sigma}_m \in \bar{\mathsf{T}}$ (in case $m = n$), or $\exists v_{n-m-1}$ appears in σ_i (in case $m < n$), and therefore, we are in one of following two cases:

Case 1. $\tilde{\sigma}_m \in \bar{\mathsf{T}}$: First notice that in this case,

$$\tilde{\sigma}_m \equiv \forall v_0 \cdots \forall v_{n-1} \exists v_n \sigma_{i,n}(v_0, \ldots, v_n)\,.$$

Since $\tilde{\sigma}_m \in \bar{\mathsf{T}}$ and $t_0, \ldots, t_{n-1}$ are closed terms, by L_{10} we get $\bar{\mathsf{T}} \vdash \exists v_n \tilde{\sigma}(v_n)$. Hence, by the CLAIM, $\neg \operatorname{Con}(\bar{\mathsf{T}} + \Sigma)$. This shows that we do not need $\sigma_{i_k,n_k}[c_{j_k}]$ to derive a contradiction from

$$\bar{\mathsf{T}} + \big\{\sigma_{i_1,n_1}[c_{j_1}], \ldots, \sigma_{i_k,n_k}[c_{j_k}]\big\}\,,$$

which is a contradiction to the minimality of the sum $n_1 + \ldots n_k + k$.

Case 2. $\exists v_{n-m-1}$ appears in σ_i: Note that since $c_j \equiv (i, t_0, \ldots, t_{n-1}, n)$ witnesses σ_i,

$$t_{n-m-1} \equiv (i, t_0, \ldots, t_{n-m-2}, n-m-1)$$

witnesses σ_i, too. Similar as above, by L_{10} we get

$$\bar{\mathsf{T}} + \sigma_{i,n-m-1}[t_{n-m-1}] \vdash \exists v_n \tilde{\sigma}(v_n)\,,$$

and with the DEDUCTION THEOREM we obtain

$$\bar{\mathsf{T}} \vdash \sigma_{i,n-m-1}[t_{n-m-1}] \to \exists v_n \tilde{\sigma}(v_n)\,.$$

This shows that if we derive a contradiction from

$$\bar{\mathsf{T}} + \Sigma + \exists v_n \tilde{\sigma}(v_n)\,,$$

then we also derive a contradiction from

$$\bar{\mathsf{T}} + \Sigma + \sigma_{i,n-m-1}[t_{n-m-1}]\,,$$

which is again a contradiction to the minimality of the sum $n_1 + \ldots n_k + k$.

Therefore, $\bar{\mathsf{T}} + \{\sigma_{i_1,n_1}[c_{j_1}], \ldots, \sigma_{i_k,n_k}[c_{j_k}]\}$ is consistent, and since the finitely many $\mathscr{L}_c$-sentences $\sigma_{i_1,n_1}[c_{j_1}], \ldots, \sigma_{i_k,n_k}[c_{j_k}]$ were arbitrary, we obtain that $\bar{\mathsf{T}}_c$ is consistent, which completes the proof. ⊣

The Completeness Theorem for Countable Signatures

In this section, we shall construct a model of the $\mathscr{L}_c$-theory $\bar{\mathsf{T}}_c$, which is of course also a model of the $\mathscr{L}$-theory $\mathsf{T} + \neg\sigma_0$. However, since we extended the signature $\mathscr{L}$, we first have to extend the binary relation $=$, as well as relation symbols in $\mathscr{L}$, to the new closed $\mathscr{L}_c$-terms.

LEMMA 5.2. *The list $\bar{\mathsf{T}}_c$ can be extended to a consistent list $\tilde{\mathsf{T}}$ of $\mathscr{L}_c$-sentences, such that the additional $\mathscr{L}_c$-sentences are variable-free (i.e., they contain neither quantifiers nor free variables) and for each variable-free $\mathscr{L}_c$-sentence σ we have either $\sigma \in \tilde{\mathsf{T}}$ or $\neg\sigma \in \tilde{\mathsf{T}}$.*

Proof. Like in the proof of LINDENBAUM'S LEMMA 4.5, we go through the list of all variable-free $\mathscr{L}_c$-sentences and successively extend the list $\bar{\mathsf{T}}_c$ to a consistent list $\tilde{\mathsf{T}}$. ⊣

Now we are ready to construct the domain of a model of $\tilde{\mathsf{T}}$, which shall be a list of lists. For this, let

$$\Lambda_{\mathcal{T}} = [t_0, t_1, \ldots, t_n, \ldots]$$

be the list of all term-constants (ordered with respect to their codes). We go through the list $\Lambda_{\mathcal{T}}$ and construct step by step a list of lists: First, we set $A_0 := [\,]$. Now, assume that A_n is already defined and consider the $\mathscr{L}_c$-sentences

$$t_n = t_0, \quad t_n = t_1, \quad \ldots \quad t_n = t_{n-1}\,.$$

If, for some m with $0 \leq m < n$, the sentence $t_n = t_m$ belongs to $\tilde{\mathsf{T}}$, then we append t_n to that list in A_n which contains t_m; the resulting list is A_{n+1}. If none of the sentences $t_n = t_m$ (for $0 \leq m < n$) belongs to $\tilde{\mathsf{T}}$ (e.g., if $n = 0$), then $A_{n+1} := A_n + \big[[t_n]\big]$. Finally, let $A = \big[[t_{n_0}, \ldots], [t_{n_1}, \ldots] \ldots\big]$ be the resulting list. Then, A is a finite or potentially infinite list of potentially infinite lists.

The lists in the list A are the objects of the domain of our model $\mathbf{M} \vDash \tilde{\mathsf{T}}$. In order to simplify the notation, for term-constants τ let $\tilde{\tau}$ be the unique list of A which contains τ.

In order to get an $\mathscr{L}_c$-structure $\mathbf{M}$ with domain A, we have to define a mapping which assigns to each constant symbol $c \in \mathscr{L}_c$ an element $c^{\mathbf{M}} \in A$, to each n-ary function symbol $F \in \mathscr{L}$ a function $F^{\mathbf{M}} : A^n \to A$, and to each n-ary relation symbol $R \in \mathscr{L}$ a set $R^{\mathbf{M}} \subseteq A^n$:

- If $c \in \mathscr{L}_c$ is a constant symbol of $\mathscr{L}$ or a special constant, then let
$$c^{\mathbf{M}} := \tilde{c}\,.$$
- If $F \in \mathscr{L}$ is an n-ary function symbol and $\tilde{t}_1, \ldots, \tilde{t}_n$ are elements of A, then let
$$F^{\mathbf{M}}\tilde{t}_1 \cdots \tilde{t}_n := \widetilde{Ft_1 \cdots t_n}\,.$$
- If $R \in \mathscr{L}$ is an n-ary relation symbol and $\tilde{t}_1, \ldots, \tilde{t}_n$ are elements of A, then we define
$$\langle \tilde{t}_1, \ldots, \tilde{t}_n \rangle \in R^{\mathbf{M}} \quad :\Longleftrightarrow \quad Rt_1 \cdots t_n \in \tilde{\mathsf{T}}\,.$$

FACT 5.3. *The definitions above, which rely on representatives of the lists in A, are well-defined.*

Proof. This follows easily by L_{14}, L_{15}, and L_{16}, and the construction of $\tilde{\mathsf{T}}$; the details are left as an exercise to the reader. ⊣

THEOREM 5.4. *The $\mathscr{L}_c$-structure $\mathbf{M}$ is a model of $\tilde{\mathsf{T}}$, and therefore also of the $\mathscr{L}$-theory $\mathsf{T} + \neg\sigma_0$.*

Proof. We have to show that for each $\mathscr{L}_c$-sentence σ, if $\sigma \in \tilde{\mathsf{T}}$ then $\mathbf{M} \vDash \sigma$. We show slightly more, namely that for each $\mathscr{L}_c$-sentence σ we have
$$\tilde{\mathsf{T}} \vdash \sigma \implies \mathbf{M} \vDash \sigma\,,$$
or equivalently, $\mathbf{M} \nvDash \sigma \implies \tilde{\mathsf{T}} \nvdash \sigma$. First, we consider the case when σ is variable-free: The proof is by induction on the number of logical operators. By LEMMA 5.2 we know that for each variable-free $\mathscr{L}_c$-sentence σ we have either $\sigma \in \tilde{\mathsf{T}}$ or $\neg\sigma \in \tilde{\mathsf{T}}$. Hence, we must show that for each variable-free $\mathscr{L}_c$-sentences σ we have $\sigma \in \tilde{\mathsf{T}}$ if and only if $\mathbf{M} \vDash \sigma$.

If σ is variable-free and does not contain logical operators, then σ is atomic. In this case, we have either $\sigma \equiv t_1 = t_2$ (for some term-constants t_1 and t_2) or $\sigma \equiv Rt_1 \cdots t_n$ (for an n-ary relation symbol $R \in \mathscr{L}$ and term-constants $t_1, \ldots, t_n$), and by construction of $\mathbf{M}$, we get $\sigma \in \tilde{\mathsf{T}}$ if and only if $\mathbf{M} \vDash \sigma$ in both cases.

Before we consider the case when σ is variable-free and contains logical operators, recall that for any $\mathscr{L}_c$-sentence $\tilde{\sigma}$ with $\sigma \Leftrightarrow \tilde{\sigma}$, by the SOUNDNESS THEOREM 3.8 we get $\mathbf{M} \vDash \sigma$ if and only if $\mathbf{M} \vDash \tilde{\sigma}$. Therefore, by the THREE-SYMBOLS THEOREM 1.7, we may assume that σ is either of the form $\neg\sigma'$ or of the form $\wedge\sigma_1\sigma_2$. Now, let σ be a non-atomic, variable-free $\mathscr{L}_c$-sentence, and assume that for each variable-free $\mathscr{L}_c$-sentence σ' which contains fewer logical operators than σ, we have $\sigma' \in \tilde{\mathsf{T}}$ if and only if $\mathbf{M} \vDash \sigma'$. By our former assumption, we just have to consider the following two cases:

Case 1. $\sigma \equiv \neg\sigma'$: Since σ' has fewer logical operators than σ, we have $\sigma' \in \tilde{\mathsf{T}}$ if and only if $\mathbf{M} \vDash \sigma'$. This shows that

$$\neg\sigma' \notin \tilde{\mathsf{T}} \iff \mathbf{M} \nvDash \neg\sigma',$$

or equivalently, $\sigma \in \tilde{\mathsf{T}} \iff \mathbf{M} \vDash \neg\sigma$.

Case 2. $\sigma \equiv \wedge\sigma_1\sigma_2$: Since each of σ_1 and σ_2 has fewer logical operators than $\tilde{\sigma}$, we have $\sigma_1 \in \tilde{\mathsf{T}}$ if and only if $\mathbf{M} \vDash \sigma_1$, and $\sigma_2 \in \tilde{\mathsf{T}}$ if and only if $\mathbf{M} \vDash \sigma_2$. Hence, we obtain

$$\wedge\sigma_1\sigma_2 \in \tilde{\mathsf{T}} \iff \sigma_1 \in \tilde{\mathsf{T}} \text{ AND } \sigma_2 \in \tilde{\mathsf{T}} \iff$$
$$\mathbf{M} \vDash \sigma_1 \text{ AND } \mathbf{M} \vDash \sigma_2 \iff \mathbf{M} \vDash \wedge\sigma_1\sigma_2$$

which shows that $\sigma \in \tilde{\mathsf{T}} \iff \mathbf{M} \vDash \sigma$.

We now consider the case that the $\mathscr{L}_c$-sentence $\sigma \in \tilde{\mathsf{T}}$ contains variables: The proof is by induction on the number of different variables which appear in σ. If $\sigma \in \tilde{\mathsf{T}}$ is an $\mathscr{L}_c$-sentence which contains variables, then, by construction of $\bar{\mathsf{T}}_c$, there is a formula in $\bar{\mathsf{T}}_c$ which is in sPNF, say

$$\forall\!\!\!\exists_0 v_0 \cdots \forall\!\!\!\exists_n v_n \sigma_{i,n}(v_0, \ldots, v_n), \quad \text{where } \sigma_{i,n} \text{ is quantifier free,}$$

such that for some natural numbers i, k, n with $k \leq n$ and some term-constants $t_0, \ldots, t_{k-1}$ we have

$$\sigma \equiv \forall\!\!\!\exists_k v_k \cdots \forall\!\!\!\exists_n v_n \sigma_{i,n}(v_0/t_0, \ldots, v_{k-1}/t_{k-1}, v_k, \ldots, v_n).$$

Now, let σ be an $\mathscr{L}_c$-sentence of the above form and assume that for each $\mathscr{L}_c$-sentence σ' which contains fewer variables than σ we have

$$\tilde{\mathsf{T}} \vdash \sigma' \implies \mathbf{M} \vDash \sigma',$$

or equivalently, $\mathbf{M} \nvDash \sigma' \implies \tilde{\mathsf{T}} \nvdash \sigma'$. We are in exactly one of the following two cases:

Case 1. $\forall\!\!\!\exists_k$ is the quantifier $\exists$: Suppose that $\tilde{\mathsf{T}} \vdash \sigma$. Then we have $\sigma \in \bar{\mathsf{T}}_c$, and for the special constant

$$t_k \equiv (i, t_0, \ldots, t_{k-1}, k)$$

and the $\mathscr{L}_c$-sentence

$$\sigma' \equiv \forall\!\!\!\exists_{k+1} v_{k+1} \cdots \forall\!\!\!\exists_n v_n \sigma_{i,n}(v_0/t_0, \ldots, v_k/t_k, v_{k+1}, \ldots),$$

we have $\sigma' \in \bar{\mathsf{T}}_c$, and consequently $\sigma' \in \tilde{\mathsf{T}}$. Now, since σ' has fewer variables than σ, by our assumption we conclude that $\mathbf{M} \vDash \sigma'$, and therefore, by L_{11} and the SOUNDNESS THEOREM 3.8, we obtain $\mathbf{M} \vDash \sigma$. Hence,

$$\tilde{\mathsf{T}} \vdash \sigma \implies \mathbf{M} \vDash \sigma .$$

Case 2. $\mathbb{Y}_k$ is the quantifier $\forall$: Assume that $\mathbf{M} \nvDash \sigma$ and therefore $\mathbf{M} \vDash \neg\sigma$. Now, for the $\mathscr{L}_c$-sentence

$$\tilde{\sigma} \equiv \exists v_k \bar{\mathbb{Y}}_{k+1} v_{k+1} \cdots \bar{\mathbb{Y}}_n v_n \neg\sigma_{i,n}(v_0/t_0, \dots, v_{k-1}/t_{k-1}, v_k, \dots, v_n),$$

where, for $k < i \leq n$, the quantifier $\bar{\mathbb{Y}}_i$ is $\exists$ if $\mathbb{Y}_i$ is $\forall$, and vice versa, we have $\tilde{\sigma} \Leftrightarrow \neg\sigma$ and therefore $\mathbf{M} \vDash \tilde{\sigma}$. For the special constant

$$t_k \equiv (i, t_0, \dots, t_{k-1}, k)$$

and the $\mathscr{L}_c$-sentence

$$\bar{\sigma} \equiv \bar{\mathbb{Y}}_{k+1} v_{k+1} \cdots \bar{\mathbb{Y}}_n v_n \neg\sigma_{i,n}(v_0/t_0, \dots, v_k/t_k, v_{k+1}, \dots, v_n),$$

we obtain $\mathbf{M} \vDash \bar{\sigma}$, i.e., $\mathbf{M} \nvDash \neg\bar{\sigma}$. Now, since $\neg\bar{\sigma}$ has fewer variables than σ, by our assumption we have $\tilde{\mathsf{T}} \nvdash \neg\bar{\sigma}$, i.e.,

$$\tilde{\mathsf{T}} \nvdash \mathbb{Y}_{k+1} v_{k+1} \cdots \mathbb{Y}_n v_n \sigma_{i,n}(v_0/t_0, \dots, v_k/t_k, v_{k+1}, \dots, v_n).$$

Therefore, by L_{10}, we conclude that $\tilde{\mathsf{T}} \nvdash \sigma$, and hence

$$\mathbf{M} \nvDash \sigma \implies \tilde{\mathsf{T}} \nvdash \sigma .$$

So, for each $\mathscr{L}_c$-sentence σ we have that $\tilde{\mathsf{T}} \vdash \sigma$ implies $\mathbf{M} \vDash \sigma$. This shows that $\mathbf{M} \vDash \tilde{\mathsf{T}}$, and $\mathbf{M} \vDash \mathsf{T} + \neg\sigma_0$ in particular. ⊣

The following theorem just summarises what we have achieved so far:

THEOREM 5.5 (GÖDEL'S COMPLETENESS THEOREM). *If $\mathscr{L}$ is a countable signature and T is a consistent set of $\mathscr{L}$-sentences, then T has a model. Moreover, if $\mathsf{T} \nvdash \sigma_0$ (for some $\mathscr{L}$-sentence σ_0), then $\mathsf{T} + \neg\sigma_0$ has a model.*

In our construction, it was essential that the signature $\mathscr{L}$ was countable, so that the symbols in $\mathscr{L}$ could be encoded by finite strings. However, in the more formal setting of axiomatic Set Theory, we can also prove the COMPLETENESS THEOREM for arbitrarily large signatures (see Chapter 15).

Some Consequences and Equivalents

We conclude this chapter by discussing some consequences and equivalent formulations of GÖDEL'S COMPLETENESS THEOREM 5.5 which follow directly or in combination with the COMPACTNESS THEOREM 2.15.

Let $\mathscr{L}$ be a countable signature, T a set of $\mathscr{L}$-sentences, and σ_0 an $\mathscr{L}$-sentence.

- If $\mathsf{T} \nvdash \sigma_0$, then there is an $\mathscr{L}$-structure $\mathbf{M}$ such that $\mathbf{M} \vDash \mathsf{T} + \neg\sigma_0$:

  $$\mathsf{T} \nvdash \sigma_0 \implies \exists \mathbf{M}\,(\mathbf{M} \vDash \mathsf{T} + \neg\sigma_0)$$

  This is just a reformulation of GÖDEL'S COMPLETENESS THEOREM 5.5.

- If T is consistent, then T has a model:

  $$\mathrm{Con}(\mathsf{T}) \implies \exists \mathbf{M}\,(\mathbf{M} \vDash \mathsf{T})$$

  This follows from the fact that $\mathrm{Con}(\mathsf{T})$ is equivalent to the existence of an $\mathscr{L}$-sentence σ_0 such that $\mathsf{T} \nvdash \sigma_0$.

- If each model of T is also a model of σ_0, then $\mathsf{T} \vdash \sigma_0$:

  $$\forall \mathbf{M}\,(\mathbf{M} \vDash \mathsf{T} \implies \mathbf{M} \vDash \sigma_0) \implies \mathsf{T} \vdash \sigma_0$$

  This follows by contraposition: If $\mathsf{T} \nvdash \sigma_0$, then, by GÖDEL'S COMPLETENESS THEOREM 5.5, there is a model $\mathbf{M} \vDash \mathsf{T} + \neg\sigma_0$.

- In combination with the COMPACTNESS THEOREM 2.15, we obtain the following implication:

  If every finite subset T' of T has a model, then T has a model.

  If every finite subset T' of T has a model, then every finite subset T' of T is consistent, and therefore, by the COMPACTNESS THEOREM 2.15, T is consistent. Thus, T has a model.

The most important consequence of GÖDEL'S COMPLETENESS THEOREM 5.5 and the SOUNDNESS THEOREM 3.8 is the following equivalence:

$$\underbrace{\forall \mathbf{M}\,(\mathbf{M} \vDash \mathsf{T} \implies \mathbf{M} \vDash \sigma_0)}_{\text{denoted by } \mathsf{T} \vDash \sigma_0} \iff \mathsf{T} \vdash \sigma_0$$

This equivalence allows us to replace *formal proofs* by *mathematical proofs*. For example, instead of proving formally the uniqueness of the neutral element in groups from the axioms of Group Theory GT, we just show that in every model of GT (i.e., in every group), the neutral element is unique. So, instead of $\mathsf{GT} \vdash \sigma_0$, we just show $\mathsf{GT} \vDash \sigma_0$.

As a last consequence, we would like to mention the so-called *Skolem's Paradox*, which is in fact just the countable version of the DOWNWARD LÖWENHEIM–SKOLEM THEOREM 15.9.

THEOREM 5.6 (SKOLEM'S PARADOX). *If $\mathscr{L}$ is a countable signature and* T *is a consistent set of $\mathscr{L}$-sentences, then* T *has a countable model.*

Proof. In the previous chapter, when we have constructed the universal list of $\mathscr{L}$-sentences, we began with a countable signature $\mathscr{L}$ and a consistent set of $\mathscr{L}$-sentences T, and at the end, we obtained a model of T whose domain was a finite or potentially infinite list of lists. So, the model of T which we constructed is countable. ⊣

What is paradoxical about this statement? For example, the signature of the axioms of Zermelo-Fraenkel Set Theory ZFC only consists of the membership relation $\in$. Hence, if ZFC is consistent, it has a countable model. However, it is easy to prove from the axiom system ZFC that there exist uncountable sets. Nevertheless, this is no contradiction, since countability in the formal theory and on the metalevel simply do not coincide.

NOTES

The COMPLETENESS THEOREM for countable signatures was first proved by Gödel [14, 15]. Later, a modified proof was given by Henkin [22] (see also [24]). The proof given here is essentially Henkin's proof, but in contrast to Henkin's proof, our construction does not rely on the assumption that an *actually infinite* set exists.

EXERCISES

5.0 Let $\mathscr{L}$ be a countable signature and let T be a consistent set of $\mathscr{L}$-sentences. For each subset $\Phi \subseteq \mathsf{T}$, let $\mathbf{M}_\Phi$ be a model for Φ and $\Sigma := \{\mathbf{M}_\Phi : \Phi \subseteq \mathsf{T} \text{ and } \mathrm{Con}(\Phi)\}$, and for each $\mathscr{L}$-sentence $\varphi \in \mathsf{T}$, let $X_\varphi := \{\mathbf{M} \in \Sigma : \mathbf{M} \vDash \varphi\}$.

(a) Show that the set $\{X_\varphi : \varphi \text{ is an } \mathscr{L}\text{-sentence}\}$ is a basis for a topology on Σ.

(b) Show that X_φ is closed for each $\varphi \in \mathsf{T}$.

(c) Use the topological compactness theorem to show that each open covering of Σ contains a finite sub-covering, i.e., the topological space Σ is compact.

5.1 Let DLO be the — assumingly consistent — theory of dense linearly ordered sets without endpoints (see EXERCISE 3.5).

(a) Show that the theory DLO is complete, i.e., for all $\mathscr{L}_{\mathsf{DLO}}$-sentences σ we have *either* $\mathsf{DLO} \vdash \sigma$ *or* $\mathsf{DLO} \vdash \neg\sigma$.

Hint: Assume towards a contradiction that there exists an $\mathscr{L}_{\mathsf{DLO}}$-sentence σ, such that $\mathsf{DLO} \nvdash \neg\sigma$ and $\mathsf{DLO} \nvdash \sigma$. Then $\mathsf{DLO} + \sigma$ and $\mathsf{DLO} + \neg\sigma$ are both consistent, and therefore, by SKOLEM'S PARADOX 5.6, there are countable models $\mathbf{M}$ and $\mathbf{N}$ such that $\mathbf{M} \vDash \mathsf{DLO} + \sigma$ and $\mathbf{N} \vDash \mathsf{DLO} + \neg\sigma$, which contradicts the fact that any two countable models of DLO are isomorphic (see EXERCISE 3.5).

(b) Show that the converse of EXERCISE 3.4 does not hold.

Hint: Let $\mathbb{Q}$ be the set of rational numbers, let $\mathbb{R}$ be the set of real numbers, and let $<$ be the natural ordering on $\mathbb{Q}$ and $\mathbb{R}$, respectively. Then the two non-isomorphic $\mathscr{L}_{\mathsf{DLO}}$-structures $(\mathbb{Q}, <)$ and $(\mathbb{R}, <)$ are both models of DLO.

5.2 Let $\mathscr{L}$ be a countable signature and let T be a consistent set of $\mathscr{L}$-sentences such that T has arbitrarily large finite models.

(a) Show that T has an infinite model.

Hint: Use the COMPACTNESS THEOREM 2.15.

(b) Show that the notion of FINITENESS cannot be formalised in First-Order Logic.

Chapter 6
Language Extensions by Definitions

Sometimes it is convenient to extend a given signature $\mathscr{L}$ by adding new non-logical symbols which have to be properly defined within the language $\mathscr{L}$ or with respect to a given $\mathscr{L}$-theory T. Let the extended signature be $\mathscr{L}^*$ and let the corresponding extended $\mathscr{L}^*$-theory be T^*. Since T is an $\mathscr{L}$-theory, we can only prove $\mathscr{L}$-sentences from T but no $\mathscr{L}^*$-sentences which contain symbols from $\mathscr{L}^* \setminus \mathscr{L}$. However, this does not imply that we can prove substantially more from T^* than from T: It might be that for each $\mathscr{L}^*$-sentence σ^* which is provable from T^* there is an $\mathscr{L}$-sentence $\tilde{\sigma}$ such that $\mathsf{T}^* \vdash \sigma^* \leftrightarrow \tilde{\sigma}$ and $\mathsf{T} \vdash \tilde{\sigma}$ which is indeed the case as we shall see below.

Defining new Relation Symbols

Let us first consider an example from Peano Arithmetic: Extend the signature $\mathscr{L}_{\mathsf{PA}}$ of Peano Arithmetic by adding the binary relation symbol $<$ and denote the extended signature by $\mathscr{L}^*_{\mathsf{PA}} := \mathscr{L}_{\mathsf{PA}} \cup \{<\}$. In order to define the binary relation $<$, we give an $\mathscr{L}_{\mathsf{PA}}$-formula $\psi_<$ with two free variables (e.g., x and y) and say that the relation $x < y$ holds if and only if $\psi_<(x, y)$ holds. In our case, $\psi_<(x, y) \equiv \exists z(x + \mathsf{s}z = y)$. Therefore, we would define the symbol $<$ by stipulating:

$$x < y :\Longleftrightarrow \exists z(x + \mathsf{s}z = y)$$

The problem is now to find for each $\mathscr{L}^*_{\mathsf{PA}}$-sentence σ^* an $\mathscr{L}_{\mathsf{PA}}$-sentence $\tilde{\sigma}$ and an extension PA^* of PA, such that $\mathsf{PA}^* \vdash \sigma^* \leftrightarrow \tilde{\sigma}$ and $\mathsf{PA} \vdash \tilde{\sigma}$ whenever $\mathsf{PA}^* \vdash \sigma^*$.

The following result provides an algorithm which transforms sentences σ^* in the extended language into equivalent sentences $\tilde{\sigma}$ in the original language. In order to prove that the algorithm works, we will make use of GÖDEL'S COMPLETENESS THEOREM 5.5.

L. Halbeisen, R. Krapf, *Gödel's Theorems and Zermelo's Axioms*,
https://doi.org/10.1007/978-3-030-52279-7_6

THEOREM 6.1. *Let $\mathscr{L}$ be a countable signature, let R be an n-ary relation symbol which does not belong to $\mathscr{L}$, and let $\mathscr{L}^* := \mathscr{L} \cup \{R\}$. Furthermore, let $\psi_R(v_1, \ldots, v_n)$ be an $\mathscr{L}$-formula with* $\text{free}(\psi_R) = \{v_1, \ldots, v_n\}$ *and let*

$$\vartheta_R \equiv \forall v_1 \cdots \forall v_n \big(R v_1 \cdots v_n \leftrightarrow \psi_R(v_1, \ldots, v_n) \big) .$$

Finally, let T be a consistent $\mathscr{L}$-theory and let $\mathsf{T}^* := \mathsf{T} + \vartheta_R$. Then there exists an effective algorithm which transforms each $\mathscr{L}^*$-formula φ^* into an $\mathscr{L}$-formula $\tilde{\varphi}$ such that:

(a) *If R does not appear in φ^*, then $\tilde{\varphi} \equiv \varphi^*$.*

(b) $\widetilde{\neg\varphi} \equiv \neg\tilde{\varphi}$ *(for $\varphi^* \equiv \neg\varphi$)*

(c) $\widetilde{\wedge\varphi_1\varphi_2} \equiv \wedge\tilde{\varphi}_1\tilde{\varphi}_2$ *(for $\varphi^* \equiv \wedge\varphi_1\varphi_2$)*

(d) $\widetilde{\exists\nu\varphi} \equiv \exists\nu\tilde{\varphi}$ *(for $\varphi^* \equiv \exists\nu\varphi$)*

(e) $\mathsf{T}^* \vdash \varphi^* \leftrightarrow \tilde{\varphi}$

(f) *If $\mathsf{T}^* \vdash \varphi^*$, then $\mathsf{T} \vdash \tilde{\varphi}$.*

Proof. Let φ^* be an arbitrary $\mathscr{L}^*$-formula. In φ^*, we replace each occurrence of $R(v_1/\tau_1, \ldots, v_n/\tau_n)$ (where $\tau_1, \ldots, \tau_n$ are $\mathscr{L}$-terms) by a particular $\mathscr{L}^*$-formula $\psi'_R(v_1/\tau_1, \ldots, v_n/\tau_n)$ such that

$$\psi'_R(v_1, \ldots, v_n) \Leftrightarrow_{\mathsf{T}} \psi_R(v_1, \ldots, v_n)$$

and none of the bound variables in ψ'_R is among $v_1, \ldots, v_n$ or appears in one of the $\mathscr{L}$-terms $\tau_1, \ldots, \tau_n$. In fact, in order to obtain ψ'_R we just have to rename the bound variables in ψ_R using the VARIABLE SUBSTITUTION THEOREM. For the resulting $\mathscr{L}$-formula $\tilde{\varphi}$, (a)–(d) are obviously satisfied.

We prove (e) and (f) on the semantic level: For this, we first show how we can extend a model $\mathbf{M} \vDash \mathsf{T}$ to a model $\mathbf{M}^* \vDash \mathsf{T}^*$. Let $\mathbf{M}$ be an $\mathscr{L}$-structure with domain A such that for each assignment j we have $(\mathbf{M}, j) \vDash \mathsf{T}$ (i.e., $\mathbf{M} \vDash \mathsf{T}$). We extend $\mathbf{M}$ to an $\mathscr{L}^*$-structure $\mathbf{M}^*$ with the same domain A by stipulating $\mathbf{M}^*|_{\mathscr{L}} := \mathbf{M}$, and for any $a_1, \ldots, a_n \in A$:

$$R^{\mathbf{M}^*}(a_1, \ldots, a_n) \ :\Longleftrightarrow \ \big(\mathbf{M}, j\tfrac{a_1}{v_1} \cdots \tfrac{a_n}{v_n}\big) \vDash \psi_R(v_1, \ldots, v_n) .$$

Then $\mathbf{M}^*$ is an $\mathscr{L}^*$-structure and for every assignment j we have

$$(\mathbf{M}^*, j) \vDash \mathsf{T} \quad \text{and} \quad (\mathbf{M}^*, j) \vDash \vartheta_R .$$

Therefore, we obtain

$$\mathbf{M}^* \vDash \mathsf{T}^* .$$

In order to prove (e), by GÖDEL'S COMPLETENESS THEOREM 5.5 it is enough to show that $\varphi^* \leftrightarrow \tilde{\varphi}$ holds in every model $\mathbf{M}^*$ of T^*. So, let $\mathbf{M}^*$ be an arbitrary model of T^*. In particular, $\mathbf{M}^* \vDash \vartheta_R$. If φ^* does not contain R, then we are done. Otherwise, if φ^* is atomic, then $\varphi^* \equiv R t_1 \cdots t_n$ for some

$\mathscr{L}$-terms $t_1, \ldots, t_n$. Since $\mathbf{M}^* \vDash \vartheta_R$, we get:

$$\mathbf{M}^* \vDash Rt_1 \cdots t_n \leftrightarrow \psi'_R(t_1, \ldots, t_n)$$

This shows $\mathbf{M}^* \vDash \varphi^* \leftrightarrow \tilde{\varphi}$ for atomic formulae, and by (b)–(d) we get the result for arbitrary formulae.

For (f), we first extend an arbitrary model $\mathbf{M} \vDash \mathsf{T}$ to a model $\mathbf{M}^* \vDash \mathsf{T}^*$. By (e), for each $\mathscr{L}^*$-formula φ^* we have:

$$\mathbf{M}^* \vDash \varphi^* \quad \Longleftrightarrow \quad \mathbf{M}^* \vDash \tilde{\varphi}$$

Now, if $\mathsf{T}^* \vdash \varphi^*$, then $\mathbf{M}^* \vDash \varphi^*$, which implies that $\mathbf{M}^* \vDash \tilde{\varphi}$. Since $\tilde{\varphi}$ is an $\mathscr{L}$-formula, we get $\mathbf{M} \vDash \tilde{\varphi}$, and since the model $\mathbf{M}$ of T was arbitrary, by GÖDEL'S COMPLETENESS THEOREM 5.5 we get $\mathsf{T} \vdash \tilde{\varphi}$. ⊣

Defining new Function Symbols

If we define new functions, the situation is slightly more subtle. However, there is also an algorithm which transforms sentences σ^* in the extended language into equivalent sentences $\tilde{\sigma}$ in the original language:

THEOREM 6.2. *Let $\mathscr{L}$ be a countable signature, let f be an n-ary function symbol which does not belong to $\mathscr{L}$, let $\mathscr{L}^* := \mathscr{L} \cup \{f\}$ and let T be a consistent $\mathscr{L}$-theory. Furthermore, let $\psi_f(v_1, \ldots, v_n, y)$ be an $\mathscr{L}$-formula with* $\text{free}(\psi_f) = \{v_1, \ldots, v_n, y\}$ *such that*

$$\mathsf{T} \vdash \forall v_1 \cdots \forall v_n \exists! y\, \psi_f(v_1, \ldots, v_n, y)\,.$$

Finally, let

$$\vartheta_f \equiv \forall v_1 \cdots \forall v_n \forall y \big(fv_1 \cdots v_n = y \leftrightarrow \psi_f(v_1, \ldots, v_n, y)\big)$$

and let $\mathsf{T}^* := \mathsf{T} + \vartheta_f$. Then there exists an effective algorithm which transforms each $\mathscr{L}^*$-formula φ^* into an $\mathscr{L}$-formula $\tilde{\varphi}$ such that:

(a) *If f does not appear in φ^*, then $\tilde{\varphi} \equiv \varphi^*$.*

(b) $\widetilde{\neg\varphi} \equiv \neg\tilde{\varphi}$ *(for $\varphi^* \equiv \neg\varphi$)*

(c) $\widetilde{\wedge\varphi_1\varphi_2} \equiv \wedge\tilde{\varphi}_1\tilde{\varphi}_2$ *(for $\varphi^* \equiv \wedge\varphi_1\varphi_2$)*

(d) $\widetilde{\exists\nu\varphi} \equiv \exists\nu\tilde{\varphi}$ *(for $\varphi^* \equiv \exists\nu\varphi$)*

(e) $\mathsf{T}^* \vdash \varphi^* \leftrightarrow \tilde{\varphi}$

(f) *If $\mathsf{T}^* \vdash \varphi^*$, then $\mathsf{T} \vdash \tilde{\varphi}$.*

Proof. By an *elementary f-term* we mean an $\mathscr{L}^*$-term of the form $ft_1 \cdots t_n$, where $t_1, \ldots, t_n$ are $\mathscr{L}^*$-terms which do not contain the symbol f. We first prove the theorem for atomic $\mathscr{L}^*$-formulae φ^* (i.e., for formulae which are free of quantifiers and logical operators). Let $\varphi^*(f|w)$ be the result of replacing the leftmost occurence of an elementary f-term in φ^* by a new symbol w, which stands for a new variable. Then, the formula

$$\exists w\big(\psi_f(t_1,\dots,t_n,w)\wedge\varphi^*(f|w)\big)$$

is called the *f-transform of* φ^*. If φ^* does not contain f, then let φ^* be its own f-transform. Before we procceed, let us prove the following

Claim. $\mathsf{T}^*\vdash\exists w\big(\psi_f(t_1,\dots,t_n,w)\wedge\varphi^*(f|w)\big)\leftrightarrow\varphi^*$

Proof of Claim. Let $\mathbf{M}^*$ be a model of T^* with domain A, let j be an arbitrary assignment which assigns an element of A to w, and let $\mathbf{M}^*_j := (\mathbf{M}^*, j)$ be the corresponding $\mathscr{L}^*$-interpretation. Assume that

$$\mathbf{M}^*_j\vDash\exists w\big(\psi_f(t_1,\dots,t_n,w)\wedge\varphi^*(f|w)\big)\,.$$

Then, since $\mathsf{T}^*\vdash\forall v_1\cdots\forall v_n\exists!y\,\psi_f(v_1,\dots,v_n,y)$, there exists a unique $b\in A$ such that

$$\mathbf{M}^*_{j\frac{b}{w}}\vDash\psi_f(t_1,\dots,t_n,w)\wedge\varphi^*(f|w)\,,$$

which is the same as saying that

$$\mathbf{M}^*_j\vDash\psi_f(t_1,\dots,t_n,b)\wedge\varphi^*(f|b)\,.$$

Now, since $\mathbf{M}^*_j\vDash\vartheta_f$, b is the same object as $f^{\mathbf{M}^*_j}t_1^{\mathbf{M}^*_j}\cdots t_n^{\mathbf{M}^*_j}$. This implies

$$\mathbf{M}^*_j\vDash ft_1\cdots t_n=b\,,$$

which shows that

$$\mathbf{M}^*_j\vDash\varphi^*\,.$$

For the reverse implication, assume that $\mathbf{M}^*_j\vDash\varphi^*$ and let b be the same object as $f^{\mathbf{M}^*_j}t_1^{\mathbf{M}^*_j}\cdots t_n^{\mathbf{M}^*_j}$. Then $\mathbf{M}^*_j\vDash\varphi^*(f|b)$ and, since $\mathbf{M}^*_j\vDash\vartheta_f$,

$$\mathbf{M}^*_j\vDash\psi_f(t_1,\dots,t_n,w)\leftrightarrow ft_1\cdots t_n=w\,.$$

In particular, we get

$$\mathbf{M}^*_{j\frac{b}{w}}\vDash\psi_f(t_1,\dots,t_n,b)\leftrightarrow ft_1\cdots t_n=b\,,$$

and because $f^{\mathbf{M}^*_j}t_1^{\mathbf{M}^*_j}\cdots t_n^{\mathbf{M}^*_j}$ is the same object as b, we get $\mathbf{M}^*_j\vDash\psi_f(t_1,\dots,t_n,b)$. Since we already know $\mathbf{M}^*_j\vDash\varphi^*(f|b)$, we have:

$$\mathbf{M}^*_j\vDash\psi_f(t_1,\dots,t_n,b)\wedge\varphi^*(f|b)$$

So, there exists a b in A such that

$$\mathbf{M}^*_{j\frac{b}{w}} \vDash \psi_f(t_1,\ldots,t_n,w) \wedge \varphi^*(f\,|\,w)\,,$$

which is the same as saying that

$$\mathbf{M}^*_j \vDash \exists w\big(\psi_f(t_1,\ldots,t_n,w) \wedge \varphi^*(f\,|\,w)\big)\,.$$

Since the model $\mathbf{M}^*$ of T^* was arbitrary, by GÖDEL'S COMPLETENESS THEOREM 5.5 we get $\mathsf{T}^* \vdash \exists w\big(\psi_f(t_1,\ldots,t_n,w) \wedge \varphi^*(f\,|\,w)\big) \leftrightarrow \varphi^*$. $\dashv_{\text{Claim}}$

Since the f-transform $\exists w\big(\psi_f(t_1,\ldots,t_n,w) \wedge \varphi^*(f\,|\,w)\big)$ of φ^* contains one f less than φ^*, if we take successive f-transforms (always introducing new variables), we obtain eventually an atomic $\mathscr{L}$-formula $\tilde{\varphi}$ (i.e., a formula which does not contain f) such that $\mathsf{T}^* \vdash \varphi^* \leftrightarrow \tilde{\varphi}$. We call $\tilde{\varphi}$ the *f-less transform of φ^**.

In order to get f-less transforms of non-atomic $\mathscr{L}^*$-formulae φ^*, we just extend the definition by letting $\widetilde{\neg\varphi}$ be $\neg\tilde{\varphi}$, $\widetilde{\wedge\varphi_1\varphi_2}$ be $\wedge\tilde{\varphi}_1\tilde{\varphi}_2$, and $\widetilde{\exists\nu\varphi}$ be $\exists\nu\tilde{\varphi}$; properties (a)–(e) are then obvious.

It remains to prove property (f). For this, let $\mathbf{M}$ be an abitrary model of T with domain A. Then, since $\mathsf{T} \vdash \forall v_1 \cdots \forall v_n \exists! y\, \psi_f(v_1,\ldots,v_n,y)$, for all $a_1,\ldots,a_n$ in A there exists a unique b in A such that

$$\mathbf{M} \vDash \psi_f(a_1,\ldots,a_n,b),$$

and we define the n-ary function f^* on A by stipulating:

$$f^*(a_1,\ldots,a_n) := b$$

With this definition, we can extend the $\mathscr{L}$-structure $\mathbf{M}$ to an $\mathscr{L}^*$-structure $\mathbf{M}^*$, where we still have $\mathbf{M}^* \vDash \mathsf{T}$. With the definition of f^*, we additionally get $\mathbf{M}^* \vDash \vartheta_f$, which implies $\mathbf{M}^* \vDash \mathsf{T}^*$. If we have $\mathsf{T}^* \vdash \varphi^*$ for some $\mathscr{L}^*$-formula φ^*, then there exists an $\mathscr{L}$-formula $\tilde{\varphi}$ such that $\mathsf{T}^* \vdash \varphi^* \leftrightarrow \tilde{\varphi}$, i.e., $\mathsf{T}^* \vdash \tilde{\varphi}$. Since $\mathsf{T}^* \vdash \tilde{\varphi}$ implies $\mathbf{M}^* \vDash \tilde{\varphi}$, and because $\tilde{\varphi}$ is an $\mathscr{L}$-formula, we have $\mathbf{M} \vDash \tilde{\varphi}$. Now, since the model $\mathbf{M}$ of T was arbitrary, by GÖDEL'S COMPLETENESS THEOREM 5.5 we get $\mathsf{T} \vdash \tilde{\varphi}$. $\dashv$

Defining new Constant Symbols

Constant symbols can be handled like 0-ary function symbols:

FACT 6.3. *Let $\mathscr{L}$ be a countable signature, let c be a constant symbol which does not belong to $\mathscr{L}$, let $\mathscr{L}^* := \mathscr{L} \cup \{c\}$ and let T be a consistent $\mathscr{L}$-theory. Furthermore, let $\psi_c(y)$ be an $\mathscr{L}$-formula with* free$(\psi_c) = \{y\}$ *such that $\mathsf{T} \vdash \exists! y\, \psi_c(y)$. Finally, let*

$$\vartheta_c \equiv \forall y\big(c = y \leftrightarrow \psi_c(y)\big)$$

and let $\mathsf{T}^* := \mathsf{T} + \vartheta_c$. *Then there exists an effective algorithm which transforms each* $\mathscr{L}^*$*-formula* φ^* *into an* $\mathscr{L}$*-formula* $\tilde{\varphi}$ *such that:*

(a) *If* f *does not appear in* φ^*, *then* $\tilde{\varphi} \equiv \varphi^*$.

(b) $\widetilde{\neg\varphi} \equiv \neg\tilde{\varphi}$ *(for* $\varphi^* \equiv \neg\varphi$*)*

(c) $\widetilde{\wedge\varphi_1\varphi_2} \equiv \wedge\tilde{\varphi}_1\tilde{\varphi}_2$ *(for* $\varphi^* \equiv \wedge\varphi_1\varphi_2$*)*

(d) $\widetilde{\exists\nu\varphi} \equiv \exists\nu\tilde{\varphi}$ *(for* $\varphi^* \equiv \exists\nu\varphi$*)*

(e) $\mathsf{T}^* \vdash \varphi^* \leftrightarrow \tilde{\varphi}$

(f) *If* $\mathsf{T}^* \vdash \varphi^*$, *then* $\mathsf{T} \vdash \tilde{\varphi}$.

Proof. The algorithm is constructed in exactly the same way as in the proof of THEOREM 6.2. ⊣

NOTES

In the proof of THEOREM 6.2, we essentially followed the proof of Proposition 2.28 of Mendelson [34].

EXERCISES

6.0 (a) Write the axioms of Group Theory in the language $\mathscr{L}_{\mathsf{GT}^*} = \{\circ\}$, where $\circ$ is a binary function symbol.

(b) Extend the language $\mathscr{L}_{\mathsf{GT}^*}$ with the constant symbol e for the neutral element.

(c) Extend the language $\mathscr{L}_{\mathsf{GT}^*} \cup \{\mathsf{e}\}$ with the unary function symbol $\mathrm{inv}(\,\cdot\,)$, where $\mathrm{inv}(x)$ is the inverse of x with respect to $\circ$.

6.1 Show that in a signature $\mathscr{L}$, constant symbols and function symbols are dispensable (i.e., we only need relation symbols as non-logical symbols).

Hint: Notice that n-ary function symbols can be replaced by $(n+1)$-ary relation symbols, and that constant symbols can be replaced by unary relation symbols.

6.2 (a) Write the axioms of Group Theory in the language $\mathscr{L} = \{R\}$, where R is a ternary relation symbol.

(b) Extend the language $\mathscr{L}$ with the unary relation symbol R_{e} for the neutral element.

(c) Extend the language $\mathscr{L} \cup \{R_{\mathsf{e}}\}$ with the binary relation symbol R_{inv}, where $R_{\mathrm{inv}}(x, y)$ holds if and only if y is the inverse of x.

Hint: Use EXERCISE 6.1.

Part III
Gödel's Incompleteness Theorems

On the syntactical level, an $\mathscr{L}$-theory T is complete if for every $\mathscr{L}$-sentence σ, either $\mathsf{T} \vdash \sigma$ or $\mathsf{T} \vdash \neg\sigma$. On the semantical level, a consistent $\mathscr{L}$-theory is T complete if any two models of T are elementary equivalent.

In this part of the book we shall first provide a few models of Peano Arithmetic PA, where we assume that PA is consistent. Then, we shall prove GÖDEL'S FIRST INCOMPLETENESS THEOREM, which states that Peano Arithmetic PA is not complete, i.e., there is a $\mathscr{L}_{\mathsf{PA}}$-sentence σ, such that neither $\mathsf{PA} \vdash \sigma$ nor $\mathsf{PA} \vdash \neg\sigma$. In a second step we shall prove GÖDEL'S SECOND INCOMPLETENESS THEOREM.

Chapter 7
Countable Models of Peano Arithmetic

By GÖDEL'S COMPLETENESS THEOREM 5.5 we know that every consistent theory T has a model, and if T has an infinite model, then it also has arbitrarily large models. Therefore, if we assume that Peano Arithmetic PA is consistent—which seems sensible—then there exists a model of PA, and because this model is infinite, PA must have arbitrarily large models as well.

In this chapter, we provide a few models of PA. First, we construct the so-called *standard model*, and then we extend this model to countable *non-standard models*.

The Standard Model

For the sake of completeness, let us first recall the language and the seven axioms of Peano Arithmetic PA. The language of PA is $\mathscr{L}_{\mathsf{PA}} = \{\mathsf{0}, \mathsf{s}, +, \cdot\}$, where $\mathsf{0}$ is a constant symbol, s is a unary function symbol, and $+$ and $\cdot$ are binary function symbols.

PA_0: $\neg\exists x(\mathsf{s}x = \mathsf{0})$

PA_1: $\forall x \forall y(\mathsf{s}x = \mathsf{s}y \to x = y)$,

PA_2: $\forall x(x + \mathsf{0} = x)$

PA_3: $\forall x \forall y(x + \mathsf{s}y = \mathsf{s}(x + y))$

PA_4: $\forall x(x \cdot \mathsf{0} = \mathsf{0})$

PA_5: $\forall x \forall y(x \cdot \mathsf{s}y = (x \cdot y) + x)$

If φ is any $\mathscr{L}_{\mathsf{PA}}$-formula with $x \in \text{free}(\varphi)$, then:

PA_6: $\big(\varphi(\mathsf{0}) \land \forall x(\varphi(x) \to \varphi(\mathsf{s}(x)))\big) \to \forall x \varphi(x)$

The domain $\mathbb{N}$ of our standard model consists of the elements in the list of natural numbers as introduced in the introductory chapter. So, each natural number in the set $\mathbb{N}$ is either $\mathbf{0}$ or of the form $\mathbf{s} \cdots \mathbf{s0}$ for some F I N I T E string $\mathbf{s} \cdots \mathbf{s}$. Notice the difference between s (which is an unary function

L. Halbeisen, R. Krapf, *Gödel's Theorems and Zermelo's Axioms*,
https://doi.org/10.1007/978-3-030-52279-7_7

symbol) and $\mathbf{s}$ (which is a symbol which we use to build the elements of the set $\mathbb{N}$, i.e., the objects in the domain of our standard model of Peano Arithmetic). In order to write this more formally, we extend the signature $\mathscr{L}_{\mathsf{PA}}$ by the unary relation symbol $\mathtt{N}$ and add the following statement as a kind of meta-axiom to PA:

$$\Phi \equiv \forall x \left(\{\mathtt{N(0)}, \forall \mathtt{z}(\mathtt{N(z)} \to \mathtt{N(sz)})\} \vdash \mathtt{N(x)} \right)$$

Notice that this statement is *not* a statement in first-order logic since it involves the symbol $\vdash$, which implicitly incorporates the metamathematical notion of F I N I T E N E S S. However, the statement Φ makes sure that every model of $\mathsf{PA} + \Phi$ is isomorphic to the standard model.

Now, we are going to define the standard model of PA with domain $\mathbb{N}$. For this, we first have to define first an $\mathscr{L}_{\mathsf{PA}}$-structure $\mathbb{N}$. If σ and τ are both (possibly empty) finite strings of the form $\mathbf{s} \cdots \mathbf{s}$, then we can interpret the non-logical symbols in $\mathscr{L}_{\mathsf{PA}}$ as follows:

$$\mathsf{0}^{\mathbb{N}} := \mathbf{0}$$

$$\begin{aligned} \mathsf{s}^{\mathbb{N}} :\ & \mathbb{N} \to \mathbb{N} \\ & \sigma\mathbf{0} \mapsto \mathbf{s}\sigma\mathbf{0} \end{aligned}$$

$$\begin{aligned} +^{\mathbb{N}} :\ & \mathbb{N} \times \mathbb{N} \to \mathbb{N} \\ & \langle \sigma\mathbf{0}, \tau\mathbf{0} \rangle \mapsto \sigma\tau\mathbf{0} \end{aligned}$$

$$\begin{aligned} \cdot^{\mathbb{N}} :\ & \mathbb{N} \times \mathbb{N} \to \mathbb{N} \\ & \langle \sigma\mathbf{0}, \tau\mathbf{0} \rangle \mapsto \begin{array}[t]{c} \sigma\sigma \cdots \sigma\mathbf{0} \\ \uparrow\uparrow \cdots \uparrow \\ \underbrace{\mathbf{s}\mathbf{s} \cdots \mathbf{s}}_{\tau} \end{array} \end{aligned}$$

Note that if either σ or τ is the empty string, then $\sigma\mathbf{0} \cdot^{\mathbb{N}} \tau\mathbf{0}$ is $\mathbf{0}$. The main feature of the $\mathscr{L}_{\mathsf{PA}}$-structure $\mathbb{N}$ is that every element of $\mathbb{N}$ corresponds to a certain $\mathscr{L}_{\mathsf{PA}}$-term. In order to prove this, we introduce the following notion: To each finite string $\sigma \equiv \mathbf{s} \cdots \mathbf{s}$ we assign a F I N I T E string $\underline{\sigma} \equiv \mathsf{s} \cdots \mathsf{s}$ such that $\underline{\sigma}$ is obtained from σ by replacing each occurrence of $\mathbf{s}$ by s. As a consequence of this definition, we get the following

FACT 7.1. *For all* F I N I T E *strings σ and τ of the form $\mathbf{s} \cdots \mathbf{s}$, we have:*

(a) *If σ is not the empty string, then* $\mathsf{PA} \vdash \underline{\sigma}\mathsf{0} \neq \mathsf{0}$.

(b) $\mathsf{PA} \vdash \underline{\sigma}\mathsf{0} = \underline{\tau}\mathsf{0} \iff \sigma\mathbf{0} \equiv \tau\mathbf{0}$.

Proof. (a) follows from PA_0, and (b) follows from PA_1 and L_{14}. ⊣

LEMMA 7.2. *Every element of* $\mathbb{N}$ *corresponds to a unique* F I N I T E *application of the function* s *to* $\mathsf{0}$*, or in other words, every element of* $\mathbb{N}$ *is equal to a unique* F I N I T E *application of the function* $\mathsf{s}^{\mathbb{N}}$ *to* $\mathsf{0}^{\mathbb{N}}$*. More formally, for every element* $\sigma\mathbf{0}$ *of* $\mathbb{N}$ *there is a unique* $\mathscr{L}_{\mathsf{PA}}$*-term* $\underline{\sigma}\mathsf{0}$ *such that*

$$(\underline{\sigma}\mathsf{0})^{\mathbb{N}} \quad \text{IS THE SAME OBJECT AS} \quad \sigma\mathbf{0}\,,$$

or equivalently,

$$(\underline{\sigma}\mathsf{0})^{\mathbb{N}} \equiv \sigma\mathbf{0}\,.$$

Proof. By definition of $\mathsf{s}^{\mathbb{N}}$, for every F I N I T E string $\tau \equiv \mathbf{s}\cdots\mathbf{s}$ we get that $\mathsf{s}^{\mathbb{N}}(\tau\mathbf{0})$ is the same element of $\mathbb{N}$ as $\mathbf{s}\tau\mathbf{0}$, and after applying this fact F I N I T E L Y many times we get:

$$\overbrace{\begin{array}{cccc} \mathsf{s}^{\mathbb{N}}\ \mathsf{s}^{\mathbb{N}} & \cdots & \mathsf{s}^{\mathbb{N}} & \mathsf{0}^{\mathbb{N}} \end{array}}^{(\underline{\sigma}\mathsf{0})^{\mathbb{N}}}$$
$$\updownarrow\ \updownarrow\ \cdots\ \updownarrow\ \updownarrow$$
$$\underbrace{\begin{array}{cccc} \mathbf{s}\ \mathbf{s} & \cdots & \mathbf{s} & \mathbf{0} \end{array}}_{\sigma\mathbf{0}}$$

The uniqueness of $\underline{\sigma}\mathsf{0}$ follows from FACT 7.1. ⊣

Now, we are ready to prove that the $\mathscr{L}_{\mathsf{PA}}$-structure $\mathbb{N}$, which is called the **standard model** of Peano Arithmetic, is indeed a model of PA.

THEOREM 7.3. $\mathbb{N} \vDash \mathsf{PA}$.

Proof. By definition of $\mathsf{s}^{\mathbb{N}}$ we get $\mathbb{N} \vDash \mathsf{PA}_0$ and by FACT 7.1 we also have $\mathbb{N} \vDash \mathsf{PA}_1$. Further, by definition of $+^{\mathbb{N}}$ and $\cdot^{\mathbb{N}}$ we get $\mathbb{N} \vDash \mathsf{PA}_2$ and $\mathbb{N} \vDash \mathsf{PA}_4$ respectively. For PA_3, let σ and τ be (possibly empty) finite strings of the form $\mathbf{s}\cdots\mathbf{s}$. Then

$$\sigma\mathbf{0} +^{\mathbb{N}} \mathsf{s}^{\mathbb{N}}\tau\mathbf{0} \equiv \sigma\mathbf{s}\tau\mathbf{0} \equiv \mathbf{s}\sigma\tau\mathbf{0} \equiv \mathsf{s}^{\mathbb{N}}(\sigma\mathbf{0} +^{\mathbb{N}} \tau\mathbf{0}).$$

Similarly, we can show $\mathbb{N} \vDash \mathsf{PA}_5$ (see EXERCISE 7.0). In order to show that $\mathbb{N} \vDash \mathsf{PA}_6$, let $\varphi(x)$ be an $\mathscr{L}_{\mathsf{PA}}$-formula and let us assume that

$$\mathbb{N} \vDash \varphi(\mathsf{0}) \wedge \forall x\big(\varphi(x) \rightarrow \varphi(\mathsf{s}x)\big)\,. \tag{$*$}$$

We have to show that $\mathbb{N} \vDash \forall x\varphi(x)$. By definition of models we get that $\varphi(\mathbf{0})$ holds in $\mathbb{N}$ and for all $n \in \mathbb{N}$: If $\varphi(n)$ holds in $\mathbb{N}$, then also $\varphi(\mathsf{s}^{\mathbb{N}}n)$ holds in $\mathbb{N}$. Let $\sigma\mathbf{0}$ be an arbitrary element of $\mathbb{N}$. Since σ is a F I N I T E string, by $(*)$, the logical axiom L_{10}, and by applying F I N I T E L Y many times Modus Ponens, we get $\mathbb{N} \vDash \varphi(\sigma\mathbf{0})$. Hence, since $\sigma\mathbf{0}$ was arbitrary, $\varphi(n)$ holds in $\mathbb{N}$ for every string $n \in \mathbb{N}$, and therefore, $\mathbb{N} \vDash \forall x\varphi(x)$. ⊣

As a matter of fact, we would like to mention that from a metamathematical point of view, every model of PA must contain an isomorphic copy of the standard model $\mathbb{N}$. Therefore, it would also make sense to call $\mathbb{N}$ the **minimal model** of Peano Arithmetic.

One might be tempted to think that $\mathbb{N}$ is essentially the only model of PA, but this is not the case, as we shall now see.

Countable Non-Standard Models

The previous section shows that every natural number in the standard model $\mathbb{N}$ corresponds to a unique $\mathscr{L}_{\mathsf{PA}}$-term; more precisely, every element $\sigma\mathbf{0}$ of $\mathbb{N}$ is the same object as the term $\underline{\sigma}\mathsf{0}$. In order to simplify notations, we will from now on use variables such as $n, m, \ldots$ to denote elements of $\mathbb{N}$ and $\underline{n}, \underline{m}, \ldots$ their counterpart in the formal language $\mathscr{L}_{\mathsf{PA}}$, i.e., if n stands for $\sigma\mathbf{0}$, then $\underline{n}$ denotes $\underline{\sigma}\mathsf{0}$.

Since every model $\mathbf{M}$ of PA contains $\underline{n}^{\mathbf{M}}$, the **standard natural numbers**, for every $n \in \mathbb{N}$, it is clear that $\mathbf{M}$ contains a copy of the standard model. However, $\mathbf{M}$ can also have **non-standard natural numbers**, i.e., elements which are not interpretations of terms of the form $\underline{n}$. In the following, we present the simplest way to construct such non-standard models.

Let $\mathscr{L}_{\mathsf{PA}^+}$ be the language $\mathscr{L}_{\mathsf{PA}}$ augmented by an additional constant symbol c, which is different from $\mathsf{0}$. Note that by setting

$$x < y :\Longleftrightarrow \exists r(x + \mathsf{s}r = y)$$

one can introduce an ordering in PA, which in the standard model corresponds to the usual ordering of natural numbers (for further details see Chapters 8 and 9). Let PA^+ be the theory whose axioms are PA_0–PA_6 together with the axioms

$$\mathsf{c} > \mathsf{0}$$

$$\mathsf{c} > \mathsf{s0}$$

$$\mathsf{c} > \mathsf{ss0}$$

$$\mathsf{c} > \mathsf{sss0}$$

$$\vdots$$

Hence, PA^+ is $\mathsf{PA} \cup \{\mathsf{c} > \underline{n} : n \in \mathbb{N}\}$.

LEMMA 7.4. Con(PA^+), *i.e., the theory* PA^+ *is consistent.*

Proof. By the COMPACTNESS THEOREM it suffices to prove that every FINITE subset of PA^+ is consistent. Let T be a FINITE subset of PA^+. Now let $n \in \mathbb{N}$ be maximal such that the formula $\mathsf{c} > \underline{n}$ belongs to T. Notice that such n exists, since T is finite. Then we can define a model $\mathbf{M}$ of T with domain $\mathbb{N}$ by interpreting the constant and function symbols by $\mathsf{0}^{\mathbf{M}} \equiv \mathbf{0}$, $\mathsf{s}^{\mathbf{M}} \equiv \mathbf{s}$, $+^{\mathbf{M}} \equiv +^{\mathbb{N}}$, $\cdot^{\mathbf{M}} \equiv \cdot^{\mathbb{N}}$ and $\mathsf{c}^{\mathbf{M}} \equiv \mathbf{s}n$. Since $\mathbb{N} \vDash \mathsf{PA}$, we get that $\mathbf{M} \vDash \mathsf{PA}$ and by construction $\mathbf{M} \vDash \mathsf{c} > \underline{m}$ for every $m \leq n$, and hence $\mathbf{M} \vDash \mathsf{T}$. ⊣

Now, since $\mathscr{L}_{\mathsf{PA}+}$ is a countable signature, by THEOREM 5.6 it follows that PA^+ has a countable model $\mathbf{M}$ which is also a **non-standard model** of PA, i.e., a model which is not isomorphic to the standard model $\mathbb{N}$. What does the order structure of $\mathbf{M}$ look like?

Note that c has a successor $\mathsf{s}c = c + 1$, and $c + 1$ in turn has a successor, and so on. Furthermore, since

$$\mathsf{PA} \vdash \forall x\big(x = \mathsf{0} \vee \exists y(x = \mathsf{s}y)\big)$$

(see LEMMA 8.4), c also has a predecessor, i.e., there exists $c - 1$ with the property that $\mathsf{s}(c - 1) = c$, and the same argument yields that c in fact has infinitely many predecessors, which are all non-standard. Hence, the order structure of c and its predecessors and successors corresponds to $(\mathbb{Z}, <)$, so there are infinitely many such $\mathbb{Z}$-chains. Moreover, each multiple of c yields a further copy of a $\mathbb{Z}$-chain. Now, one can easily prove in PA that every number is even or odd (see EXERCISE 8.1), and hence there is d such that $2d = c$ or $2d = c + 1$. We denote d by $\frac{c}{2}$. This shows that between the copy of the standard model and the $\mathbb{Z}$-chain given by c, there is a further $\mathbb{Z}$-chain given by $\frac{c}{2}$ and its predecessors and successors. In fact, the $\mathbb{Z}$-chains are ordered like $(\mathbb{Q}, <)$ (see EXERCISE 7.6).

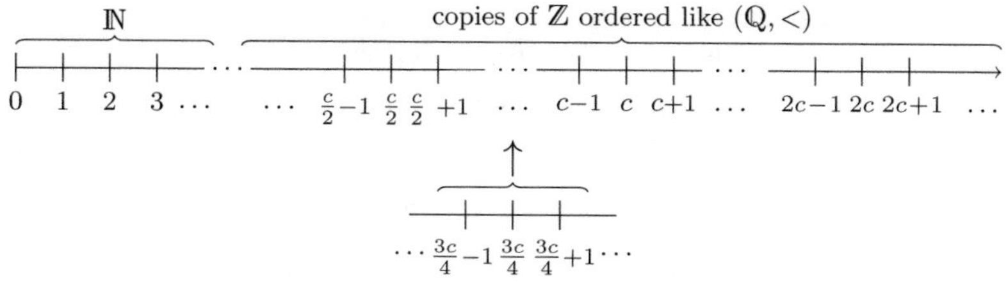

Note that the proof of LEMMA 7.4 implies that there are non-standard models of PA which are elementarily equivalent to $\mathbb{N}$. To see this, let $\mathbf{Th}(\mathbb{N})$ denote the theory of all $\mathscr{L}_{\mathsf{PA}}$-sentences which are true in $\mathbb{N}$. Then one could simply replace PA by $\mathbf{Th}(\mathbb{N})$ in LEMMA 7.4 and thus obtain a model of $\mathbf{Th}(\mathbb{N})$ augmented by all formulae of the form $\mathsf{c} > \underline{n}$ for $n \in \mathbb{N}$. By construction, this model is elementarily equivalent to $\mathbb{N}$. For a more general result see EXERCISE 7.2.

NOTES

An early attempt at formalising arithmetic was given by Grassmann [20] in 1861, who defined addition and multiplication and proved elementary results such as the associative and commutative laws using induction. Dedekind [6] also identified induction as a key principle in 1888, as well as the first two axioms of Peano Arithmetic; however, he introduced them as a definition rather than as axioms. Peano [41] presented his five axioms in 1889, where he only introduces zero and the successor function axiomatically, and the induction axiom is given in second-order logic in the following form: Every set of natural numbers which contains 0 and is closed under the successor function is the set of all natural numbers. The version of Peano's Axioms formalised in first-order logic — where the induction axiom is replaced by an axiom schema, and the axioms defining addition and multipliation are included — goes back to the advent of first-order logic in the 1920's. The first explicit construction of a non-standard model of arithmetic was given by Skolem in [51]. For further reading on non-standard models consult [28].

Exercises

7.0 Prove that $\mathbb{N} \vDash \mathsf{PA}_5$.

7.1 Prove that PA_0 and PA_1 are independent of the other axioms of PA.

7.2 Show that there are uncountably many countable models of PA which are all elementarily equivalent and pairwise non-isomorphic.

Hint: Let $\mathbb{P}$ be the set of prime numbers and let c be a constant symbol which is different from $\mathsf{0}$. For any distinct prime numbers p and q, let $\varphi_{p,q}$ be the formula

$$p \mid \mathsf{c} \wedge q \nmid \mathsf{c}.$$

For every subset $S \subseteq \mathbb{P}$, let Φ_S be the collection of all formulae $\varphi_{p,q}$ such that $p \in S$ and $q \notin S$. Now, for each $S \subseteq \mathbb{P}$, $\mathbb{N}$ is a model for every finite subset of $\mathsf{T}(\mathbb{N}) + \Phi_S$, and hence, for every $S \subseteq \mathbb{P}$, $\mathsf{T}(\mathbb{N}) + \Phi_S$ has a countable model, say $\mathbf{N}_S$. Notice that for all these models $\mathbf{N}_S$ we have $\mathbf{N} \vDash \mathsf{T}(\mathbb{N})$, and that for each model $\mathbf{N}_S$, there are only countably many subsets $S \subseteq \mathbb{P}$ such that $\mathbf{N}_S \vDash \Phi_S$. Since by Cantor's Theorem 13.6 the set of all subsets $S \subseteq \mathbb{P}$ is uncountable, we obtain uncountably many countable models $\mathbf{N}_S$ of PA which are pairwise non-isomorphic.

7.3 Prove the following so-called *Overspill Principle*: If $\mathbf{M}$ is a non-standard model of PA with domain M, φ is a formula with $n+1$ free variables and $b_1, \ldots, b_n \in M$, then

$$\mathbf{M} \vDash \varphi(\underline{n}, b_1, \ldots, b_n) \quad \text{for all } n \in \mathbb{N}$$

implies that there is a non-standard element $a \in M$ such that

$$\mathbf{M} \vDash \forall x(x < a \rightarrow \varphi(x, b_1, \ldots, b_n)).$$

7.4 Show that it is not possible to introduce a relation standard(x) by a language extension of $\mathscr{L}_{\mathsf{PA}}$ such that for every model $\mathbf{M}$ of PA with domain M and for every $a \in M$, we have $\mathbf{M} \vDash \text{standard}(a)$ if and only if $a = \underline{n}^{\mathbf{M}}$ for some $n \in \mathbb{N}$.

7.5 Let $\mathbf{M}$ be a non-standard model of PA with domain M. Show that there is an $a \in M$ such that every standard prime number divides a.

7.6 Let $\mathbf{M}$ be a countable non-standard model of PA with domain M. For every non-standard element c of M, let

$$\mathbb{Z}_c :\equiv \{d \in M : \text{there exists } n \in \mathbb{N} \text{ such that } d + n = c \text{ or } c + n = d\},$$

and let $\mathbb{Z}_c < \mathbb{Z}_d$, if $c + n < d$ for all $n \in \mathbb{N}$. Show that $\{\mathbb{Z}_c : c \in M \text{ is non-standard}\}$ is a dense linearly ordered set (see Exercise 3.5) and use Exercise 5.1 to conclude that the order structure of $\mathbf{M}$ corresponds to the disjoint union of $\mathbb{N}$ and $\mathbb{Q} \times \mathbb{Z}$.

Chapter 8
Arithmetic in Peano Arithmetic

In this chapter, we take a closer look at Peano Arithmetic (PA) which we have defined in Chapter 1. In particular, we prove within PA some basic arithmetical results, starting with the commutativity and associativity of addition and multiplication, culminating in some results about coprimality. This paves the way for the coding of finite sequences of numbers, which will be covered in the next chapter. Furthermore, we introduce some alternative formulations of the Induction Schema PA_6.

Addition & Multiplication

In this section, we verify the basic computation rules of PA involving addition and multiplication. Since the complete proofs are very long and tedious, we will only show the commutativity of $+$ in an elaborate way. Subsequently, we will use semi-formal proofs as described in Chapter 1 which include enough details to allow the reader to reconstruct a corresponding formal proof.

LEMMA 8.1. $\mathsf{PA} \vdash \forall x \forall y(x + y = y + x)$

Proof. We proceed by induction on x. Thus, we have to show

(a) $\mathsf{PA} \vdash \forall y(\mathsf{0} + y = y + \mathsf{0})$, and

(b) $\mathsf{PA} \vdash \forall y(x + y = y + x) \rightarrow \forall y(\mathsf{s}x + y = y + \mathsf{s}x)$.

For (a), we first prove

$$\mathsf{PA} \vdash \forall y(\mathsf{0} + y = y)$$

by induction on y. The base case $\mathsf{0} + \mathsf{0} = \mathsf{0}$ is clearly an instance of PA_2, and for the induction step, we assume $\mathsf{0}+y = y$ for some y. Then $\mathsf{0}+\mathsf{s}y = \mathsf{s}(\mathsf{0}+y)$ by PA_3 and $\mathsf{s}(\mathsf{0}+y) = \mathsf{s}y$ by assumption. In order to keep the notation short, we just write $\mathsf{0}+\mathsf{s}y = \mathsf{s}(\mathsf{0}+y) = \mathsf{s}y$ instead of $\mathsf{0}+\mathsf{s}y = \mathsf{s}(\mathsf{0}+y) \wedge \mathsf{s}(\mathsf{0}+y) = \mathsf{s}y$. So, by PA_6 we obtain $\forall y(\mathsf{0} + y = y)$, and since by PA_2 we have $\forall y(y + \mathsf{0} = y)$, by symmetry and transitivity of $=$ we have $\forall y(\mathsf{0} + y = y + \mathsf{0})$.

L. Halbeisen, R. Krapf, *Gödel's Theorems and Zermelo's Axioms*,
https://doi.org/10.1007/978-3-030-52279-7_8

As a prerequisite for (b) we need

$$\mathsf{PA} \vdash \forall y\big(\mathtt{s}x + y = \mathtt{s}(x + y)\big)$$

which is again verified by induction on y: If $y = \mathsf{0}$, note that by PA_2 we have $\mathtt{s}x + \mathsf{0} = \mathtt{s}x = \mathtt{s}(x + \mathsf{0})$. For the induction step, assume $\mathtt{s}x + y = \mathtt{s}(x + y)$. Then, by PA_3, we have $\mathtt{s}x + \mathtt{s}y = \mathtt{s}(\mathtt{s}x + y) = \mathtt{s}(\mathtt{s}(x + y)) = \mathtt{s}(x + \mathtt{s}y)$.

Now, we are ready to prove (b): Assume that $x + y = y + x$ for some x and for all y. Then $\mathtt{s}x + y = \mathtt{s}(x + y) = \mathtt{s}(y + x) = y + \mathtt{s}x$ by our computation above and PA_3, which, by PA_6, shows (b). ⊣

In a similar manner, we can derive other basic calculation rules whose proofs are left as an exercise for the reader.

LEMMA 8.2.

(a) $\mathsf{PA} \vdash \forall x \forall y \forall z\big((x + y) + z = x + (y + z)\big)$

(b) $\mathsf{PA} \vdash \forall x \forall y \forall z\big((x \cdot y) \cdot z = x \cdot (y \cdot z)\big)$

(c) $\mathsf{PA} \vdash \forall x \forall y(x \cdot y = y \cdot x)$

(d) $\mathsf{PA} \vdash \forall x \forall y \forall z\big(x \cdot (y + z) = (x \cdot y) + (x \cdot z)\big)$

From now on, we will make use of these rules without explicitly mentioning them anymore. The next lemma shows injectivity of left — and by commutativity also right — addition.

LEMMA 8.3. $\mathsf{PA} \vdash \forall x \forall y \forall z(x + y = x + z \to y = z)$

Proof. The proof is by induction on x. The base case follows from the proof of LEMMA 8.1. For the induction step, assume

$$\forall y \forall z(x + y = x + z \to y = z)$$

and let $\mathtt{s}x + y = \mathtt{s}x + z$. Then $\mathtt{s}(x + y) = \mathtt{s}x + y = \mathtt{s}x + z = \mathtt{s}(x + z)$, where the first and the third equality again follow from LEMMA 8.1 and PA_3. Then by PA_2 we obtain $x + y = x + z$ and in particular $y = z$. ⊣

The next result is crucial, because — as we will see in Chapter 10 — it is the only application of PA_6 which is indispensable for the proof of the FIRST INCOMPLETENESS THEOREM 10.6.

LEMMA 8.4. $\mathsf{PA} \vdash \forall x\big(x = \mathsf{0} \lor \exists y(x = \mathtt{s}y)\big)$

Proof. We proceed by induction on x. The base case is trivial and the induction step follows from the fact that x witnesses $\exists y(\mathtt{s}x = \mathtt{s}y)$. ⊣

From now on, we will use the convention that $\cdot$ binds stronger than $+$ and omit the multiplication sign, e.g., the term $xy + z$ stands for $(x \cdot y) + z$. Furthermore, by associativity of $+$ and $\cdot$ we may omit parentheses whenever we have pure products of pure sums of terms.

In order to keep the notation short, for $\mathscr{L}_{\mathsf{PA}}$-formulae φ we define

$$\forall x \neq 0\,(\varphi(x)) :\Longleftrightarrow \forall x\big(x \neq 0 \to \varphi(x)\big).$$

The next result shows a property of multiplication which is similar to the one given in LEMMA 8.3 for addition.

LEMMA 8.5.

(a) $\mathsf{PA} \vdash \forall x \forall y\big(xy = 0 \leftrightarrow (x = 0 \vee y = 0)\big)$

(b) $\mathsf{PA} \vdash \forall x \neq 0\, \forall y \forall z(xy = xz \to y = z)$

Proof. For (a), let $xy = 0$ and suppose towards a contradiction that $x, y \neq 0$. Then by LEMMA 8.4 there are x', y' such that $x = \mathsf{s}x'$ and $y = \mathsf{s}y'$. By PA_5 and PA_3, we obtain

$$0 = xy = \mathsf{s}x' \cdot \mathsf{s}y' = \mathsf{s}x' \cdot y' + \mathsf{s}x' = \mathsf{s}(\mathsf{s}x' \cdot y' + x')\,,$$

which contradicts PA_0.

For (b), suppose that $x \neq 0$. We proceed by induction on y. If $y = 0$, then $xy = 0$. So, $xy = xz$ implies $xz = 0$ and by (a) we obtain $z = 0$ and consequently $y = z$. Now assume that

$$\forall z(xy = xz \to y = z).$$

Let z be arbitrary such that $x \cdot \mathsf{s}y = xz$. By (a), we can rule out the possibility that $z = 0$. Hence, by LEMMA 8.4, there is a z' such that $z = \mathsf{s}z'$. Therefore, by PA_5,

$$xy + x = x \cdot \mathsf{s}y = xz = x \cdot \mathsf{s}z' = xz' + x\,.$$

Using LEMMATA 8.1 and 8.3 we obtain that $xy = xz'$ and thus the induction hypothesis implies $y = z'$. Therefore, we finally get $\mathsf{s}y = \mathsf{s}z' = z$ as desired. ⊣

The Natural Ordering on Natural Numbers

In Chapter 6, we have seen how to extend languages by incorporating new symbols for relations, functions or constants. In this sense, we can now introduce the binary relations $\leq$ and $<$ in PA by stipulating

$$x \leq y :\Longleftrightarrow \exists r(x + r = y)\,,$$

$$x < y :\Longleftrightarrow x \leq y \wedge x \neq y\,.$$

An alternative definition of $x < y$ is given by

$$x < y :\Longleftrightarrow \exists r \neq 0\,(x + r = y)\,.$$

Furthermore, we define

$$x \geq y :\Longleftrightarrow y \leq x$$
$$x > y :\Longleftrightarrow y < x.$$

Now, we define **bounded quantification** by stipulating

$$\exists x \lhd y\, \varphi(x) :\Longleftrightarrow \exists x\big(x \lhd y \wedge \varphi(x)\big),$$

$$\forall x \lhd y\, \varphi(x) :\Longleftrightarrow \forall x\big(x \lhd y \rightarrow \varphi(x)\big),$$

where $\lhd$ stands either for $<$ or for $\leq$. The next result shows some properties of $<$ and $\leq$.

LEMMA 8.6.

(a) $\mathsf{PA} \vdash \forall x \forall y(x < \mathrm{s}y \leftrightarrow x \leq y)$
(b) $\mathsf{PA} \vdash \forall x \forall y(x < y \leftrightarrow \mathrm{s}x \leq y)$

Proof. We only consider (a) and leave (b) as an excercise. Fix x and y. Firstly, assume that $x < \mathrm{s}y$ and take $r \neq 0$ such that $x + r = \mathrm{s}y$. By LEMMA 8.4 we find an r' such that $r = \mathrm{s}r'$. Then $\mathrm{s}(x + r') = x + \mathrm{s}r' = x + r = \mathrm{s}y$ by PA_3, and by PA_2 we obtain $x + r' = y$, which shows that $x \leq y$.

Conversely, let $x \leq y$ and take r such that $x + r = y$. Then $x + \mathrm{s}r = \mathrm{s}(x + r) = \mathrm{s}y$, which shows that $x < \mathrm{s}y$. ⊣

The next result implies that $\leq$ defines a total ordering on the natural numbers.

LEMMA 8.7.

(a) $\mathsf{PA} \vdash \forall x(x \leq x)$
(b) $\mathsf{PA} \vdash \forall x \forall y(x \leq y \wedge y \leq x \rightarrow x = y)$
(c) $\mathsf{PA} \vdash \forall x \forall y \forall z(x \leq y \wedge y \leq z \rightarrow x \leq z)$
(d) $\mathsf{PA} \vdash \forall x \forall y(x < y \vee x = y \vee x > y)$

Proof. Condition (a) is a trivial consequence of PA_2.

For (b), assume that $x \leq y$ and $y \leq x$. Then there are r, s such that $x + r = y$ and $y + s = x$. We obtain that

$$y + (s + r) = (y + s) + r = x + r = y = y + 0.$$

By LEMMA 8.3, this implies $s + r = 0$ and hence, by PA_0, $s = 0 = r$, which shows that $x = y$.

For (c), let $x \leq y$ and $y \leq z$ and take witnesses r, s satisfying $x + r = y$ and $y + s = z$, respectively. Then $x + (r + s) = (x + r) + s = y + s = z$ and thus $x \leq z$.

We show (d) by induction on x. If $x = 0$, we can make a case distinction according to LEMMA 8.4: If $y = 0$ then $x = y$ and otherwise $x < y$. For the induction step, fix y and assume that $x < y \vee x = y \vee x > y$. Now, we make a case distinction, where in the case of $x < y$, LEMMA 8.6 implies that $\mathsf{s}x \leq y$ and thus either $\mathsf{s}x < y$ or $\mathsf{s}x = y$. Secondly, if $x = y$ then

$$\mathsf{s}x = \mathsf{s}y = \mathsf{s}(y + 0) = y + \mathsf{s}0,$$

which shows that $\mathsf{s}x > y$. The case of $x > y$ is similar. ⊣

Finally, one can show that addition and multiplication with non-zero numbers preserve the natural odering (the proof is left as an excercise to the reader):

LEMMA 8.8.

(a) $\mathsf{PA} \vdash \forall x \forall y \forall z \big(x \leq y \leftrightarrow (x + z \leq y + z)\big)$

(b) $\mathsf{PA} \vdash \forall x \forall y \forall z \neq 0 \big(x \leq y \leftrightarrow (x \cdot z \leq y \cdot z)\big)$

Subtraction & Divisibility

With the help of the ordering that we have introduced in the previous section, we are ready to define a version of subtraction which rounds up to 0 in order to preserve non-negativity. For this, we first show the following

LEMMA 8.9. $\mathsf{PA} \vdash \forall x \forall y \big(x \leq y \rightarrow \exists! r(x + r = y)\big)$

Proof. Assume that $x \leq y$. The existence of r follows directly from the definition of $\leq$, and the uniqueness of r is a consequence of LEMMA 8.3. ⊣

Therefore, we can define within PA the binary function $-$, called **bounded subtraction**, by stipulating

$$x - y = z :\Longleftrightarrow (y \leq x \wedge y + z = x) \vee (x < y \wedge z = 0).$$

Observe that $\mathsf{PA} \vdash \forall x \forall y \leq x((x - y) + y = x)$, from which we can easily derive computation rules for bounded subtraction such as

$$\mathsf{PA} \vdash \forall x \forall y \forall z \big(x(y - z) = xy - xz\big), \text{ or}$$

$$\mathsf{PA} \vdash \forall x \forall y \forall z \big(x \leq y \rightarrow (x - z \leq y - z)\big).$$

Let us now turn to divisibility, which can easily be formalised by stipulating

$$x \mid y :\Longleftrightarrow \exists r(rx = y).$$

If the binary divisibility relation $\mid$ holds for the ordered pair (x, y), then we say that x *divides* y. Without much effort, one can verify that the divisibility relation is reflexive, antisymmetric, and transitive. For this reason, we will omit the proof of the next result.

LEMMA 8.10.

(a) $\mathsf{PA} \vdash \forall x(x \mid x)$
(b) $\mathsf{PA} \vdash \forall x \forall y(x \mid y \wedge y \mid x \rightarrow x = y)$
(c) $\mathsf{PA} \vdash \forall x \forall y \forall z(x \mid y \wedge y \mid z \rightarrow x \mid z)$

Also without much effort, we can prove the following

LEMMA 8.11.

(a) $\mathsf{PA} \vdash \forall x \forall y \forall z(x \mid y \wedge x \mid z \rightarrow x \mid y \pm z)$, *where the symbol* $\pm$ *stands for either* $+$ *or* $-$.
(b) $\mathsf{PA} \vdash \forall x \forall y \forall z(x \mid y \rightarrow x \mid yz)$

Proof. For (a), assume that x divides y and z. Then there are r, s such that $y = rx$ and $z = sx$. Then $y \pm z = rx \pm sx = (r \pm s)x$, thus x divides $y \pm z$. Condition (b) is obvious. ⊣

In most textbooks, one defines two numbers to be *coprime* (or *relatively prime*), if they have no common divisor. Nevertheless, for our purpose it is more convenient to use the following equivalent definition:

$$\operatorname{coprime}(x, y) \;:\Longleftrightarrow\; x \neq \mathsf{0} \wedge y \neq \mathsf{0} \wedge \forall z\big(x \mid yz \rightarrow x \mid z\big)$$

Since we will be working with this somewhat unusual definition of relative primality, we first check that it is a symmetric relation.

LEMMA 8.12. $\mathsf{PA} \vdash \forall x \forall y\big(\operatorname{coprime}(x, y) \leftrightarrow \operatorname{coprime}(y, x)\big)$

Proof. Assume coprime(x, y). We have to show that for every z we have that $y \mid xz$ implies $y \mid z$. So, let z be such that $y \mid xz$. Since $y \mid xz$, there is an r with $yr = xz$. Furthermore, since $x \mid xz$ and $xz = yr$, we get $x \mid yr$, and by coprime(x, y) we have $x \mid r$. Thus, there is an s such that $xs = r$, and hence, $xsy = ry = yr = xz$. Now, by LEMMA 8.5 we obtain $sy = z$, and therefore $y \mid z$ as desired. ⊣

If the binary relation coprime holds for x and y, then we say that *x and y are coprime*.

LEMMA 8.13. $\mathsf{PA} \vdash \forall x \forall y \forall k\big(k \mid x \wedge \operatorname{coprime}(x, y) \rightarrow \operatorname{coprime}(k, y)\big)$.

Proof. Assume that x and y are coprime. Let k be a divisor of x and let r be such that $rk = x$. Assume $y \mid kz$ for some arbitrary z. We have to show that $y \mid z$. First, notice that by LEMMA 8.11 (b) we have $y \mid rkz$, and since $rkz = xz$, we have $y \mid xz$. Now, since coprime(x, y), we obtain $y \mid z$ as desired. ⊣

The following result is crucial for the construction of *Gödel's β-function* (see THEOREM 9.10), which will be the key to the FIRST INCOMPLETENESS THEOREM 10.6.

LEMMA 8.14. $\mathsf{PA} \vdash \forall k \forall x \neq 0 \forall j \Big(k \mid x \to \text{coprime}\big(1 + (j+k)x, 1 + jx\big)\Big)$

Proof. We first show $\mathsf{PA} \vdash \forall x \neq 0 \forall j \big(\text{coprime}(x, 1 + jx)\big)$, i.e., we show that for all z,

$$x \mid (1 + jx)z \ \to \ x \mid z\,.$$

For this, suppose $x \mid (1+jx)z$ for some arbitrary z. Since $(1+jx)z = z+jxz$, by LEMMA 8.11 (b) we have $x \mid jxz$, and as a consequence of LEMMA 8.11 (a) we obtain $x \mid z$.

Now, let k and $x \neq 0$ be such that $k \mid x$. Notice that since $x \neq 0$, this implies that $k \neq 0$. Furthermore, let j be arbitrary but fixed. We have to show

$$\text{coprime}\big(1 + jx, 1 + (j+k)x\big)\,,$$

i.e., we have to show that for all z,

$$(1 + jx) \mid \big(1 + (j+k)x\big)z \ \to \ (1 + jx) \mid z\,.$$

First, notice that

$$\big(1 + (j+k)x\big)z = (1 + jx + kx)z = (1 + jx)z + kxz\,.$$

Assume now that for some z,

$$(1 + jx) \mid (1 + jx)z + kxz\,.$$

By LEMMA 8.11 (b) we have $(1+jx) \mid (1+jx)z$, and by LEMMA 8.11 (a) this implies $(1+jx) \mid kxz$. Now, since $\text{coprime}(x, 1+jx)$, as shown above, we get

$$(1 + jx) \mid x(kz) \to (1 + jx) \mid kz\,.$$

Finally, since by assumption $k \mid x$, by LEMMA 8.13 and $\text{coprime}(x, 1+jx)$ we get $\text{coprime}(k, 1 + jx)$. Hence, we obtain $(1 + jx) \mid z$ as desired. ⊣

Alternative Induction Schemata

A fundamental principle in elementary number theory states that if there is a natural number fulfilling some property Ψ, then there must be a least natural number satisfying Ψ. This principle can be shown in PA; actually, every instance of this principle (i.e., by considering Ψ to be some $\mathscr{L}_{\mathsf{PA}}$-formula) is equivalent to the corresponding instance of the Induction Schema PA_6. In order to prove this, we need another induction principle which will turn out to be quite useful for further proofs in this book.

PROPOSITION 8.15 (STRONG INDUCTION PRINCIPLE). *Let $\varphi(x)$ be an $\mathscr{L}_{\mathsf{PA}}$-formula. Then in* PA, *$\varphi$ satisfies the following* ***principle of strong induction****:*

$$\mathsf{PA} \vdash \forall x\big(\forall y < x\, \varphi(y) \to \varphi(x)\big) \to \forall x \varphi(x)$$

Proof. Suppose $\forall x\big(\forall y < x\, \varphi(y) \to \varphi(x)\big)$. Using PA_6, we first show $\forall x \psi(x)$ for

$$\psi :\equiv \forall y < x\, \varphi(y).$$

Notice that $\psi(0)$ vacuously holds, since there is no $y < 0$ with $\neg\varphi(y)$. Now, if $\psi(x)$ holds, then by our assumption we have $\varphi(x)$. So, we have $\psi(x)$ and $\varphi(x)$, which is the same as $\psi(\mathrm{s}x)$. Therefore, by PA_6 we obtain $\forall x \psi(x)$. Now, because for every x, $\psi(\mathrm{s}x)$ implies $\varphi(x)$, we finally obtain $\forall x \varphi(x)$. ⊣

PROPOSITION 8.16 (LEAST NUMBER PRINCIPLE). *Let $\varphi(x)$ be an $\mathscr{L}_{\mathsf{PA}}$-formula. Then*

$$\mathsf{PA} \vdash \exists x \varphi(x) \to \exists x\big(\varphi(x) \land \forall y < x\, \neg\varphi(y)\big).$$

Informally, the LEAST NUMBER PRINCIPLE states that if there is a witness to an arithmetic statement, then there is always a least witness. This principle is often used in the following equivalent form: If a universally quantified formula does not hold, then there is a least counterexample.

Proof of Proposition 8.16. By TAUTOLOGY (K) and the 3-SYMBOLS THEOREM 1.7, we have

$$\exists x \varphi(x) \to \exists x\big(\varphi(x) \land \forall y < x\, \neg\varphi(y)\big) \;\Leftrightarrow\; \forall x\, \neg\varphi(x) \lor \exists x\big(\varphi(x) \land \forall y < x\, \neg\varphi(y)\big),$$

where the latter statement is equivalent to the implication

$$\forall x\big(\neg\varphi(x) \lor \neg\forall y < x\, \neg\varphi(y)\big) \to \forall x\, \neg\varphi(x).$$

Now, by TAUTOLOGY (K) this implication is equivalent to

$$\forall x\big(\forall y < x\, \neg\varphi(y) \to \neg\varphi(x)\big) \to \forall x\, \neg\varphi(x),$$

which is the STRONG INDUCTION PRINCIPLE 8.15 applied to the formula $\neg\varphi(x)$. Consequently, we have $\mathsf{PA} \vdash \exists x \varphi(x) \to \exists x\big(\varphi(x) \land \forall y < x\, \neg\varphi(y)\big)$. ⊣

Relative Primality Revisited

We conclude this chapter by providing an alternative definition of relative primality, which shall be useful in the next chapter. First, we introduce the *Principle of Division with Remainder*:

PROPOSITION 8.17 (PRINCIPLE OF DIVISION WITH REMAINDER).

$$\mathsf{PA} \vdash \forall x \, \forall y > 0 \, \exists q \, \exists r \big(x = qy + r \wedge r < y \big).$$

Proof. Let $\varphi(x) \equiv \forall y > 0 \, \exists q \, \exists r \big(x = qy + r \wedge r < y \big)$. The proof is by induction on x. Obviously, we have $\varphi(0)$. Now, assume that we have $\varphi(x)$ for some x, i.e., for each $y > 0$ there are q, r such that

$$x = qy + r \wedge r < y .$$

If we replace x by $\mathrm{s}x$, then for each $y > 0$ there are q, r such that

$$\mathrm{s}x = qy + \mathrm{s}r \wedge \mathrm{s}r \leq y .$$

If $\mathrm{s}r < y$, let $r' := \mathrm{s}r$ and $q' := q$, and if $\mathrm{s}r = y$, let $r' := 0$ and $q' := \mathrm{s}q$. Now, in both cases we obtain

$$\mathrm{s}x = q'y + r' \wedge r' < y ,$$

which shows $\varphi(\mathrm{s}x)$. ⊣

The following result gives a connection between the PRINCIPLE OF DIVISION WITH REMAINDER and the relatively prime numbers:

LEMMA 8.18. *For any $x, y > 0$ with $x = qy + r$ and $r < y$ we have*

$$\mathsf{PA} \vdash \operatorname{coprime}(y, x) \leftrightarrow \operatorname{coprime}(y, r) .$$

Proof. By definition we have $\operatorname{coprime}(y, x) \leftrightarrow \forall z (y \mid xz \to y \mid z)$, and since $x = qy + r$, we obtain

$$\operatorname{coprime}(y, x) \leftrightarrow \forall z (y \mid yqz + rz \to y \mid z) .$$

Now, by LEMMA 8.11 we have $(y \mid yqz + rz) \leftrightarrow (y \mid rz)$, and therefore we obtain

$$\operatorname{coprime}(y, x) \;\leftrightarrow\; \forall z (y \mid rz \to y \mid z) \;\leftrightarrow\; \operatorname{coprime}(y, r) .$$

⊣

Now we are ready to give the promised alternative definition of relative primality.

PROPOSITION 8.19.

$$\mathsf{PA} \vdash \forall x \forall y \Big(\operatorname{coprime}(x, y) \;\leftrightarrow\; x \neq 0 \wedge y \neq 0 \wedge \forall z \big((z \mid x \wedge z \mid y) \to z = 1 \big) \Big).$$

Proof. The statement is obvious for $x = y$, or if at least one of x and y is equal to 1. Therefore, without loss of generality, let us assume that $x > y > 1$.

($\rightarrow$) The proof is by contraposition. Assume that there is a z such that $z \mid x$, $z \mid y$, and $z > 1$. Then, there is a $u < x$ such that $uz = x$. Now, since $z \mid y$, we obtain $x \mid yu$, and since $u < x$, we have $x \nmid u$, which implies $\neg\,\mathrm{coprime}(x, y)$.

($\leftarrow$) Assume towards a contradiction that there is a pair of numbers (x, y) with $x > y > 0$ such that for all z we have

$$(z \mid x \wedge z \mid y) \rightarrow z = 1\,,$$

but $\neg\,\mathrm{coprime}(x, y)$. By the LEAST NUMBER PRINCIPLE, let (x_0, y_0) be such a pair of numbers where x_0 is minimal. Let q and r be such that $x_0 = qy_0 + r$. Since $\neg\,\mathrm{coprime}(x_0, y_0)$, by LEMMA 8.18 we have $\neg\,\mathrm{coprime}(y_0, r)$. On the other hand, if there is a $z_0 > 1$ with $z_0 \mid y_0$ and $z_0 \mid r$, then this would imply that

$$z_0 \mid qy_0 + r,$$

i.e., $z_0 \mid x_0$. But since $z_0 > 1$, this contradicts the fact that $(z_0 \mid x_0 \wedge z_0 \mid y_0) \rightarrow z_0 = 1$. Therefore, for the pair (y_0, r) we have $\neg\,\mathrm{coprime}(y_0, r)$, for all z we have

$$(z \mid y_0 \wedge z \mid r) \rightarrow z = 1\,,$$

and in addition we have $y_0 < x_0$, which is a contradiction to the minimality of x_0. ⊣

As an immediate consequence of PROPOSITION 8.19, we get the following

COROLLARY 8.20. *For all x and y, the following statement is provable in* PA:

$$\mathrm{coprime}(x, y) \;\leftrightarrow\; x \neq 0 \wedge y \neq 0 \wedge \forall z < (x + y)\big((z \mid x \wedge z \mid y) \rightarrow z = 1\big)$$

EXERCISES

8.0 Prove that addition is associtative, i.e., $\mathsf{PA} \vdash \forall x \forall y \forall z (x + (y + z) = (x + y) + z)$.

8.1 Introduce the unary relations $\mathrm{even}(x)$ and $\mathrm{odd}(x)$ formalising evenness and oddness, and show that

$$\mathsf{PA} \vdash \forall x\big(\mathrm{even}(x) \vee \mathrm{odd}(x)\big).$$

8.2 Show that BÉZOUT'S LEMMA is provable in PA, i.e., show that

$$\mathsf{PA} \vdash \forall x \forall y\Big(\mathrm{coprime}(x, y) \leftrightarrow \big(x \neq 0 \wedge y \neq 0 \wedge \exists a < y\, \exists b < x\, (ax + 1 = by)\big)\Big).$$

Hint: First, show $\exists a \exists b (ax + 1 = by) \leftrightarrow \exists a' \exists b' (a'x = b'y + 1)$ (e.g., let $a' := y - a$ and $b' := x - b$). Then use the PRINCIPLE OF DIVISION WITH REMAINDER and the LEAST NUMBER PRINCIPLE.

8.3 Prove PA_6 from PA_0–PA_5 and the LEAST NUMBER PRINCIPLE.

8.4 Prove the following alternative induction principle:

$$\mathsf{PA} \vdash \big(\varphi(1) \wedge \forall x(\varphi(x) \rightarrow \varphi(2x) \wedge \varphi(x - 1))\big) \rightarrow \forall x \varphi(x)$$

Chapter 9
Gödelisation of Peano Arithmetic

The key ingredient for Gödel's Incompleteness Theorems is the so-called Gödelisation process which allows us to code terms, formulae and even proofs within PA. In order to achieve this, we introduce Gödel's β-function, with the help of which one can encode any F I N I T E sequence of natural numbers by a single natural number.

Natural Numbers in Peano Arithmetic

As we have already seen in Chapter 7, every standard natural number corresponds to a unique $\mathscr{L}_{\mathsf{PA}}$-term. More precisely, every element $\sigma\mathbf{0}$ of $\mathbb{N}$ corresponds to a term $\underline{\sigma}0$. In order to simplify notations, from now on we will use variables such as $n, m, \ldots$ to denote elements of $\mathbb{N}$ and $\underline{n}, \underline{m}, \ldots$ their counterpart in the formal language $\mathscr{L}_{\mathsf{PA}}$, i.e., if n stands for $\sigma\mathbf{0}$, then $\underline{n}$ denotes $\underline{\sigma}0$. Then FACT 7.1 yields

$$n \equiv m \iff \mathsf{PA} \vdash \underline{n} = \underline{m}.$$

Moreover, by definition of $\underline{n}$ for $n \in \mathbb{N}$ we have:

$$\underline{\mathbf{0}} \equiv 0$$
$$\underline{\mathsf{s}n} \equiv \mathsf{s}\underline{n}$$

Furthermore, we define

$$\bigvee_{k=\mathbf{0}}^{n-1} x = \underline{k} \;:\equiv\; x = \underline{\mathbf{0}} \vee x = \underline{\mathsf{s}\mathbf{0}} \vee \cdots \vee x = \underline{n-1}\,.$$

L. Halbeisen, R. Krapf, *Gödel's Theorems and Zermelo's Axioms*,
https://doi.org/10.1007/978-3-030-52279-7_9

PROPOSITION 9.1. *Any two natural numbers $n, m \in \mathbb{N}$ satisfy the following properties:*

N_0: $\mathsf{PA} \vdash \mathsf{s}\underline{n} = \underline{\mathsf{s}n}$

N_1: $\mathsf{PA} \vdash \underline{m} + \underline{n} = \underline{m +^{\mathbb{N}} n}$

N_2: $\mathsf{PA} \vdash \underline{m} \cdot \underline{n} = \underline{m \cdot^{\mathbb{N}} n}$

N_3: *If $m \equiv n$ then $\mathsf{PA} \vdash \underline{m} = \underline{n}$, and if $m \not\equiv n$ then $\mathsf{PA} \vdash \underline{m} \neq \underline{n}$.*

N_4: *If $m < n$ then $\mathsf{PA} \vdash \underline{m} < \underline{n}$, and if $m \not< n$ then $\mathsf{PA} \vdash \underline{m} \not< \underline{n}$.*

N_5: $\mathsf{PA} \vdash \forall x \big(x < \underline{n} \leftrightarrow \bigvee_{k=\mathbf{0}}^{n-1} x = \underline{k} \big)$

Before we give a proof of PROPOSITION 9.1, let us recall the Induction Principle that we have introduced in the introductory chapter: *If a statement A holds for $\mathbf{0}$ and if whenever A holds for a natural number n in $\mathbb{N}$ then it also holds for $n+1$, then the statement A holds for all natural numbers n in $\mathbb{N}$.* Note that this Induction Principle is more general than what we obtain from the induction axiom PA_6 in the standard model $\mathbb{N}$. The reason is that PA_6 is restricted to properties which can be described by an $\mathscr{L}_{\mathsf{PA}}$-formula, whereas the Induction Principle applies to any statement about standard natural numbers. In order to distinguish between the Induction Principle for standard natural numbers and induction within PA using PA_6, we shall call the former *metainduction*.

Proof of Proposition 9.1. N_0 follows directly from the definition of $\underline{n}$ for natural numbers $n \in \mathbb{N}$.

We prove N_1 by metainduction on n. The case $n \equiv \mathbf{0}$ is obviously true, since $\underline{\mathbf{0}}$ is 0. For the induction step, let us assume $\mathsf{PA} \vdash \underline{m} + \underline{n} = \underline{m +^{\mathbb{N}} n}$. Using N_0 and PA_3 both within PA and in $\mathbb{N}$ we obtain

$$\mathsf{PA} \vdash \underline{m} + \underline{\mathbf{s}n} = \underline{m} + \mathsf{s}\underline{n} = \mathsf{s}(\underline{m} + \underline{n}) = \mathsf{s}(\underline{m +^{\mathbb{N}} n}) = \underline{\mathbf{s}(m +^{\mathbb{N}} n)} = \underline{m +^{\mathbb{N}} \mathbf{s}n}.$$

The proof of N_2 is similar and is left as an exercise to the reader.

The first part of N_3 follows from FACT 7.1, and the second part is a consequence of N_4, since whenever $m \not\equiv n$, then either $m < n$ or $n < m$.

Let us now turn to N_4. If $m < n$, then there is $k \in \mathbb{N}$ such that $m +^{\mathbb{N}} k \equiv n$ and $k \not\equiv \mathbf{0}$. By N_3 and N_1 we get $\mathsf{PA} \vdash \underline{m} + \underline{k} = \underline{m +^{\mathbb{N}} k} = \underline{n}$. It remains to show that $\mathsf{PA} \vdash \underline{k} \neq 0$. Since $k \not\equiv \mathbf{0}$, it is of the form $\mathbf{s}k'$ for some $k' \in \mathbb{N}$. Thus by N_0 and PA_0, $\mathsf{PA} \vdash \underline{k} = \underline{\mathbf{s}k'} = \mathsf{s}\underline{k'} \neq 0$. The second statement of N_4 follows from the first one and N_3 by observing that if $m \not< n$, then either $m \equiv n$ or $n < m$.

In order to prove N_5, we proceed by metainduction on n. The case $n \equiv \mathbf{0}$ is trivially satisfied. Now, assume that N_5 holds for some n and let $x < \underline{\mathbf{s}n} = \mathsf{s}\underline{n}$. Then, by LEMMA 8.6 we get $x \leq \underline{n}$, i.e., either $x < \underline{n}$ or $x = \underline{n}$. Since the first case is equivalent to $\bigvee_{k=\mathbf{0}}^{n-1} x = \underline{k}$ by assumption, we obtain $\bigvee_{k=\mathbf{0}}^{n} x = \underline{k}$ as desired. The converse is a consequence of N_4. ⊣

On the one hand, by the SOUNDNESS THEOREM 3.8 we know that every statement which is provable within PA holds in every model of PA, in particular in the standard model $\mathbb{N}$. On the other hand, not every statement which is true in $\mathbb{N}$ must be provable within PA. In this respect, PROPOSITION 9.1 gives us a few statements which are true in the standard model $\mathbb{N}$ and which are provable within PA. In order to obtain more such statements, we shall introduce the notion of $\mathbb{N}$-*conformity*; but before doing so, we have to give a few preliminary notions.

We call an $\mathscr{L}_{\mathsf{PA}}$-formula φ a **strict ∃-formula** if it is built up from atomic formulae and negated atomic formulae using $\wedge$, $\vee$, existential quantification $\exists\nu$ and bounded universal quantification, i.e., "$\forall\nu < \tau$" for some term τ. Furthermore, φ is said to be an **∃-formula** if there is a strict ∃-formula ψ such that $\varphi \Leftrightarrow_{\mathsf{PA}} \psi$. By exchanging the role of universal and existential quantification in the above definition, we can analogously define (**strict**) **∀-formulae**. Furthermore, if a formula is both an ∃- and a ∀-formula, then we call it a **Δ-formula**. In particular, every formula which contains only bounded quantifiers is a Δ-formula.

Example 9.2. The formulae "$x < y$" and "$x \mid y$" are Δ-formulae:

$$x \leq y \Leftrightarrow_{\mathsf{PA}} \exists r < y\,(x + r = y) \qquad \text{and} \qquad x \mid y \Leftrightarrow_{\mathsf{PA}} \exists r < \mathsf{s}y\,(rx = y)$$

PROPOSITION 9.3. *Let $\varphi(x_1, \ldots, x_n)$ be a formula whose free variables are among $x_1, \ldots, x_n$, and let $a_1, \ldots, a_n \in \mathbb{N}$.*

(a) *If φ is an ∃-formula and $\mathbb{N} \vDash \varphi(a_1, \ldots, a_n)$, then $\mathsf{PA} \vdash \varphi(\underline{a_1}, \ldots, \underline{a_n})$.*
(b) *If φ is a ∀-formula and $\mathbb{N} \vDash \neg\varphi(a_1, \ldots, a_n)$, then $\mathsf{PA} \vdash \neg\varphi(\underline{a_1}, \ldots, \underline{a_n})$.*

Proof. Observe first that (b) follows from (a), since the negation of a ∀-formula is an ∃-formula. Furthermore, note that it is enough to prove (a) for strict ∃-formulae. We proceed by induction on the construction of φ.

- If φ is an atomic formula, then it is of the form $\tau_0(x_1, \ldots, x_n) = \tau_1(x_1, \ldots, x_n)$ for some terms τ_0, τ_1 whose variables are among $x_1, \ldots, x_n$. We show by induction on the construction of terms that for every term $\tau(x_1, \ldots, x_n)$ and for all standard natural numbers $a_1, \ldots, a_n \in \mathbb{N}$, we have

$$\mathsf{PA} \vdash \underline{\tau^{\mathbb{N}}(a_1, \ldots, a_n)} = \tau(\underline{a_1}, \ldots, \underline{a_n}). \qquad (*)$$

The statement is clear for terms τ of the form $\tau \equiv \nu$ (for a variable ν) or $\tau \equiv 0$. If τ is of the form $\mathsf{s}\tau'$ for some term τ', then by the induction hypothesis we have $\mathsf{PA} \vdash \underline{a} = \tau'(\underline{a_1}, \ldots, \underline{a_n})$, where $a = \tau'^{\mathbb{N}}(a_1, \ldots, a_n) \in \mathbb{N}$. Therefore, $\tau^{\mathbb{N}}(a_1, \ldots, a_n)$ is $\mathsf{s}a$. Then by N_0 we have

$$\mathsf{PA} \vdash \underline{\mathsf{s}a} = \mathsf{s}\underline{a} = \mathsf{s}\tau'(\underline{a_1}, \ldots, \underline{a_n}) = \tau(\underline{a_1}, \ldots, \underline{a_n})$$

as desired. The proofs for terms of the form $\tau_0 + \tau_1$ or $\tau_0 \cdot \tau_1$ are similar using N_1 and N_2, respectively. This shows $(*)$.

Now assume $\mathbb{N} \models \tau_0(a_1,\dots,a_n) = \tau_1(a_1,\dots,a_n)$ and put $a \equiv \tau_0(a_1,\dots,a_n)$ and $b \equiv \tau_1(a_1,\dots,a_n)$. Then by (∗) and N_3 we get

$$\mathsf{PA} \vdash \tau_0(\underline{a_1},\dots,\underline{a_n}) = \underline{a} = \underline{b} = \tau_1(\underline{a_1},\dots,\underline{a_n}).$$

- Suppose that $\varphi(x_1,\dots,x_n) \equiv \varphi_0(x_1,\dots,x_n) \wedge \varphi_1(x_1,\dots,x_n)$ and $\mathbb{N} \models \varphi(a_1,\dots,a_n)$. Then $\mathbb{N} \models \varphi_0(a_1,\dots,a_n)$ and $\mathbb{N} \models \varphi_1(a_1,\dots,a_n)$. By induction hypothesis, $\mathsf{PA} \vdash \varphi_0(\underline{a_1},\dots\underline{a_n})$ and $\mathsf{PA} \vdash \varphi_1(\underline{a_1},\dots,\underline{a_n})$. Using $(\mathsf{I}\wedge)$ this shows that $\mathsf{PA} \vdash \varphi(\underline{a_1},\dots,\underline{a_n})$. The disjunctive case is similar.
- Let now $\varphi(x_1,\dots,x_n) \equiv \forall y < \tau(x_1,\dots,x_n)\, \psi(x_1,\dots,x_n,y)$ and suppose that $\mathbb{N} \models \varphi(a_1,\dots,a_n)$. Let $a \equiv \tau^{\mathbb{N}}(a_1,\dots,a_n)$. Then for every $b < a$ we have $\mathbb{N} \models \psi(a_1,\dots,a_n,b)$. Hence, by induction hypothesis, for every $b < a$ we have $\mathsf{PA} \vdash \psi(\underline{a_1},\dots,\underline{a_n},\underline{b})$, and by (∗), $\mathsf{PA} \vdash \underline{a} = \tau(\underline{a_1},\dots,\underline{a_n})$. Now using N_5 we obtain

$$\mathsf{PA} \vdash \ \varphi(\underline{a_1},\dots,\underline{a_n}) \leftrightarrow \forall y \Big(\bigvee_{b=0}^{a-1} y = \underline{b} \rightarrow \psi(\underline{a_1},\dots,\underline{a_n},y) \Big).$$

  The right-hand side can clearly be derived in PA.
- Finally, let $\varphi(x_1,\dots,x_n) \equiv \exists y \psi(x_1,\dots,x_n,y)$. Then $\mathbb{N} \models \varphi(a_1,\dots,a_n)$ implies that there exists $b \in \mathbb{N}$ such that $\mathbb{N} \models \psi(a_1,\dots,a_n,b)$. Inductively, we get $\mathbb{N} \models \psi(\underline{a_1},\dots,\underline{a_n},\underline{b})$, which completes the proof.

⊣

Note that any constants, relations, and functions that one can define in PA in the sense of Chapter 6 can be interpreted in the standard model $\mathbb{N}$.

A relation $R(x_1,\dots,x_n)$ defined by

$$R(x_1,\dots,x_n) :\Longleftrightarrow \psi_R(x_1,\dots,x_n)$$

is said to be $\mathbb{N}$-**conform** if for all $a_1,\dots,a_n \in \mathbb{N}$ the following two properties are satisfied:

(a) If $\mathbb{N} \models \psi_R(a_1,\dots,a_n)$, then $\mathsf{PA} \vdash \psi_R(\underline{a_1},\dots,\underline{a_n})$.

(b) If $\mathbb{N} \models \neg\psi_R(a_1,\dots,a_n)$, then $\mathsf{PA} \vdash \neg\psi_R(\underline{a_1},\dots,\underline{a_n})$.

For the sake of simplicity, the formula ψ_R is also called $\mathbb{N}$-conform.

Now, let f be a function symbol whose defining formula is $\psi_f(x_1,\cdots,x_n,y)$, i.e., $\mathsf{PA} \vdash \forall x_1 \dots \forall x_n \exists! y\, \psi_f(x_1,\dots,x_n,y)$ and

$$f x_0 \cdots x_n = y :\Longleftrightarrow \psi_f(x_1,\dots,x_n,y).$$

Then we say that f is $\mathbb{N}$-**conform** if its defining formula ψ_f is $\mathbb{N}$-conform. Let $f^{\mathbb{N}}$ be the interpretation of f in $\mathbb{N}$. If f is $\mathbb{N}$-conform, then for all $a_1,\dots,a_n \in \mathbb{N}$

$$\mathsf{PA} \vdash \psi_f(\underline{a_1},\dots,\underline{a_n},\underline{f^{\mathbb{N}}(a_1,\dots,a_n)})\,,$$

and hence

$$\mathsf{PA} \vdash f(\underline{a_1}, \dots, \underline{a_n}) = \underline{f^{\mathbb{N}}(a_1, \dots, a_n)}.$$

To see this, suppose that f is $\mathbb{N}$-conform. For the sake of simplicity, suppose that $n \equiv 1$ and let $a \in \mathbb{N}$. Then $\mathbb{N} \vDash \psi_f(a, f^{\mathbb{N}}(a))$, hence by $\mathbb{N}$-conformity we get $\mathsf{PA} \vdash \psi_f(\underline{a}, \underline{f^{\mathbb{N}}(a)})$. On the other hand, we have $\mathsf{PA} \vdash \psi_f(\underline{a}, f(\underline{a}))$ and hence by functionality of ψ_f we get $\mathsf{PA} \vdash f(\underline{a}) = \underline{f^{\mathbb{N}}(a)}$.

COROLLARY 9.4.

(a) *Every relation which is defined by a Δ-formula is $\mathbb{N}$-conform.*

(b) *Every function which is defined by an $\exists$-formula is $\mathbb{N}$-conform.*

Proof. Condition (a) follows directly from PROPOSITION 9.3. For (b), it suffices to prove that every function whose defining formula is an $\exists$-formula is already a Δ-formula. Suppose that f is defined by the $\exists$-formula ψ_f, i.e.,

$$f(x_1, \dots, x_n) = y :\Longleftrightarrow \psi_f(x_1, \dots, x_n, y).$$

Now note that by functionality of ψ_f we have

$$\psi_f(x_1, \dots, x_n, y) \Leftrightarrow_{\mathsf{PA}} \forall z(\psi_f(x_1, \dots, x_n, z) \to z = y).$$

Moreover, TAUTOLOGY (K) yields

$$\forall z(\psi_f(x_1, \dots, x_n, z) \to z = y) \Leftrightarrow \forall z(\neg\psi_f(x_1, \dots, x_n, z) \vee z = y),$$

which is a $\forall$-formula. ⊣

Example 9.5. The binary coprimality relation "coprime" is $\mathbb{N}$-conform. To see this, first notice that by the previous example, the defining formula of the divisibility relation is a Δ-formula, and therefore, by COROLLARY 9.4, the symbol $\mid$ is $\mathbb{N}$-conform. Furthermore, by COROLLARY 8.20 the defining formula of "coprime" is equivalent to a Δ-formula, and therefore, by COROLLARY 9.4 the binary relation "coprime" is $\mathbb{N}$-conform.

Gödel's β-Function

The main goal of this section is to define a binary function (the so-called **β-function** introduced by Kurt Gödel) which encodes a F I N I T E sequence of natural numbers $c_0, \cdots, c_{n-1}$ in the standard model by a single number c such that for all $i < n$,

$$\mathsf{PA} \vdash \beta(\underline{c}, \underline{i}) = \underline{c_i}.$$

In fact, one can even do better than that and introduce a function β such that for every unary function f definable in Peano Arithmetic,

$$\mathsf{PA} \vdash \forall k \exists c \forall i < k\big(\beta(c, i) = f(i)\big).$$

The first step is to encode *ordered pairs* of numbers by introducing a binary pairing function op. We define

$$\mathrm{op}(x, y) = z :\Longleftrightarrow (x + y) \cdot (x + y) + x + 1 = z\,.$$

Furthermore, we define the unary relation *not an ordered pair* "nop" and the two binary functions *first element* "fst" and *second element* "snd" by stipulating

$$\begin{aligned} \mathrm{nop}(c) &:\Longleftrightarrow \neg\exists x \exists y \big(\mathrm{op}(x, y) = c\big)\,, \\ \mathrm{fst}(c) = x &:\Longleftrightarrow \exists y \big(\mathrm{op}(x, y) = c\big) \vee \big(\mathrm{nop}(c) \wedge x = 0\big)\,, \\ \mathrm{snd}(c) = y &:\Longleftrightarrow \exists x \big(\mathrm{op}(x, y) = c\big) \vee \big(\mathrm{nop}(c) \wedge y = 0\big)\,. \end{aligned}$$

In particular, whenever $\mathrm{op}(x, y) = c$, then

$$c = \mathrm{op}\big(\mathrm{fst}(c), \mathrm{snd}(c)\big)\,.$$

Until now, we did not show that the above definitions are well-defined. This, however, follows from the following

LEMMA 9.6. $\mathsf{PA} \vdash \mathrm{op}(x, y) = \mathrm{op}(x', y') \to x = x' \wedge y = y'$.

Proof. Assume that $\mathrm{op}(x, y) = \mathrm{op}(x', y')$. We first show that this implies $x + y = x' + y'$: Suppose towards a contradiction that $x + y < x' + y'$. Then, by PA_3 and LEMMA 8.6, we obtain $\mathtt{s}(x + y) = x + \mathtt{s}y \leq x' + y'$. Therefore,

$$\begin{aligned} \mathrm{op}(x', y') = \mathrm{op}(x, y) &= (x + y) \cdot (x + y) + x + 1 \\ &\leq (x + \mathtt{s}y) \cdot (x + y) + (x + \mathtt{s}y) \\ &= (x + \mathtt{s}y) \cdot (x + \mathtt{s}y) \\ &= \mathtt{s}(x + y) \cdot \mathtt{s}(x + y) \\ &\leq (x' + y') \cdot (x' + y') \\ &< \mathrm{op}(x', y'), \end{aligned}$$

which is obviously a contradiction. By symmetry, the relation $x' + y' < x + y$ can also be ruled out, and therefore we have that $\mathrm{op}(x, y) = \mathrm{op}(x', y')$ implies $x + y = x' + y'$. Now, if $x + y = x' + y'$, then $(x + y) \cdot (x + y) = (x' + y') \cdot (x' + y')$, and since $\mathrm{op}(x, y) = \mathrm{op}(x', y')$, by LEMMA 8.3 we obtain $x + 1 = x' + 1$, which implies $x = x'$ and also $y = y'$. ⊣

Now we are ready to define the β-function. Let

$$\gamma(a, i, y, x) :\equiv \big(1 + (\mathrm{op}(a, i) + 1) \cdot y\big) \mid x$$

and define

$$\begin{aligned}\beta(c,i) = a \;:\Longleftrightarrow\; & \big(\mathrm{nop}(c) \wedge a = 0\big) \vee \\ & \exists x \exists y \Big(c = \mathrm{op}(x,y) \wedge \Big(\big(\neg\exists b\, \gamma(b,i,y,x) \wedge a = 0\big) \vee \\ & \qquad \big(\gamma(a,i,y,x) \wedge \neg\exists b < a\, \gamma(b,i,y,x)\big)\Big)\Big).\end{aligned}$$

Slightly less formal, we can define $\beta(c,i)$ by stipulating

$$\beta(c,i) = \begin{cases} 0 & \text{if } \mathrm{nop}(c), \\ 0 & \text{if } c = \mathrm{op}(x,y) \wedge \neg\exists b\, \gamma(b,i,y,x), \\ a & \text{if } c = \mathrm{op}(x,y) \wedge a = \min\big\{b : \gamma(b,i,y,x)\big\}. \end{cases}$$

Observe that as a consequence of the LEAST NUMBER PRINCIPLE, β is a binary function.

Before we can encode finite sequences with the β-function, we have to prove a few auxiliary results. The first one states that for every m there exists a y which is a multiple of $\mathrm{lcm}(1,\ldots,m)$.

LEMMA 9.7. $\mathsf{PA} \vdash \forall m \exists y \forall k\big((k \neq 0 \wedge k \leq m) \to k \mid y\big)$.

Proof. We proceed by induction on m. The case when $m = 0$ is clear. Assume that there is a y such that for every k with $0 < k \leq m$ we have $k \mid y$. Let $y' = y \cdot \mathsf{s}m$. Then, by LEMMA 8.11, every k with $0 < k \leq \mathsf{s}m$ divides y'. ⊣

As described in Chapter 6, for any $\mathscr{L}_{\mathsf{PA}}$-formula φ which is *functional*, i.e., $\mathsf{PA} \vdash \forall x \exists! y \varphi(x,y)$, we can introduce a function symbol F_φ by stipulating

$$F_\varphi(x) = y \;:\Longleftrightarrow\; \varphi(x,y)\,.$$

If F is defined by some functional $\mathscr{L}_{\mathsf{PA}}$-formula, then we say that F is **definable** in PA.

The next result shows that for every function F which is definable in PA and for every $k > 0$, we can define $\max\big\{F(0),\ldots,F(k-1)\big\}$.

LEMMA 9.8. *Let F be a function which is definable in* PA. *Then*

$$\mathsf{PA} \vdash \forall k > 0\, \exists! x\big(\exists i < k(F(i) = x) \wedge \forall i < k\big(F(i) \leq x\big)\big).$$

Proof. We prove the statement by induction on k starting with 1. For $k = 1$, one can clearly take $x = F(0)$. Assume that there is a unique x and there is $i_0 < k$ such that $F(i_0) = x$ and for all $i < k$, $F(i) \leq x$. Now if $F(k) \leq x$, then set $x' = x$; otherwise let $x' = F(k)$. Then for every $i < \mathsf{s}k$, we have $F(i) \leq x$ and the first condition is also satisfied since x' is either $F(i_0)$ or $F(k)$; uniqueness is trivial. ⊣

This leads to the following definition:

$$\max\nolimits_{i<k} F(i) = x \;:\Longleftrightarrow\; \exists i < k\big(F(i) = x\big) \wedge \forall i < k\big(F(i) \leq x\big)$$

The next result plays an important role in the coding of finite sequences.

LEMMA 9.9. *Let G be an unary, strictly increasing function which is definable in* PA *and let $\varphi(\nu)$ be an $\mathscr{L}_{\mathsf{PA}}$-formula. Then*

$$\mathsf{PA} \vdash \forall m\Big(\forall j < m \forall j' < m(j \neq j' \to \text{coprime}(G(j), G(j'))) \\ \to \exists x \forall j < m\,\big(G(j) \mid x \leftrightarrow \varphi(j)\big)\Big).$$

Proof. We proceed by induction on m starting with $m = 1$. If $\varphi(0)$ holds, let $x := G(0)$, otherwise let $x := 1$. For the induction step, assume that for all distinct $j, j' \leq \mathsf{s}m$, $G(j)$ and $G(j')$ are coprime and that there is an x such that for all $j < m$, $G(j) \mid x \leftrightarrow \varphi(j)$. By the LEAST NUMBER PRINCIPLE, let x_0 be the least such x. Now we consider the following two cases: If $\varphi(m)$ holds, let $x_1 := G(m) \cdot x_0$, otherwise, let $x_1 := x_0$. It remains to show that for all $j \leq m$ we have $G(j) \mid x_1 \leftrightarrow \varphi(j)$.

If $\varphi(m)$ fails (i.e., $x_1 = x_0$), then, by induction hypothesis, for all $j < m$ we have $G(j) \mid x_0 \leftrightarrow \varphi(j)$ and $\text{coprime}(G(j), G(m))$, where the latter implies by the choice of x_0 that $G(m) \nmid x_0$. To see this, assume that $G(m) \mid x_0$. Then there is an r such that $G(m) \cdot r = x_0$, and since $m \geq 1$, G is strictly increasing and $G(0) \neq 0$ by $\text{coprime}(G(0), G(1))$, we get that $G(m) > 1$ and consequently $r < x_0$. Moreover, since for all $j < m$ we have $\text{coprime}(G(j), G(m))$, this implies

$$\forall j < m\big(G(j) \mid \underbrace{G(m) \cdot r}_{= x_0} \;\leftrightarrow\; G(j) \mid r\big),$$

which contradicts the minimality of x_0.

If $\varphi(m)$ holds (i.e., $x_1 = G(m) \cdot x_0$), then, since $\text{coprime}(G(j), G(m))$ for all $j < m$ we have

$$G(j) \mid \underbrace{G(m) \cdot x_0}_{= x_1} \;\leftrightarrow\; G(j) \mid x_0 .$$

Furthermore, we obviously have $G(m) \mid x_1$, which completes the proof. ⊣

The following theorem states how the β-function can be used to code finite sequences.

THEOREM 9.10. *Let F be a function which is definable in* PA. *Then*

$$\mathsf{PA} \vdash \forall k\, \exists c\, \forall i < k\, \big(\beta(c, i) = F(i)\big).$$

Proof. Fix an arbitrary number k. Let $F'(i) := \text{op}\big(F(i), i\big) + 1$ and let

$$m := \max_{i<k} F'(i)\,.$$

By LEMMA 9.7 there is a y such that every $j \leq m$ divides y. Furthermore, by LEMMA 8.14 we have for all u with $u \mid y$ (i.e., $1 \leq u \leq m$) and for all w

$$\text{coprime}\big(1 + wy, 1 + (w + u)y\big)\,.$$

In particular, if $i < j < m$, then for $w := i + 1$ and $u := j - i$, we obtain

$$\text{coprime}\big(1 + (i+1)y, 1 + (j+1)y\big).$$

Finally, define the unary function G by

$$G(j) = z \ :\Longleftrightarrow\ z = 1 + (j+1)y,$$

and let

$$\varphi_0(z) \ :\equiv\ \exists i < k\big(z = \text{op}(F(i), i)\big).$$

Then G is a strictly increasing function and we can apply LEMMA 9.9 in order to find a number x such that for all $j < m$, where $m \geq \text{op}\big(F(i), i\big)$ (for all $i < k$), we have

$$G(j) \mid x \ \leftrightarrow\ \varphi_0(j),$$

in other words,

$$\forall j < m\Big(1 + (j+1)y \mid x \ \leftrightarrow\ \exists i < k\,\big(j = \text{op}(F(i), i)\big)\Big).$$

It remains to show that for $c = \text{op}(x, y)$ we have $\beta(c, i) = F(i)$ for all $i < k$. By our assumption on x, we have $1 + (\text{op}(F(i), i) + 1)y \mid x$ and $\gamma\big(F(i), i, y, x\big)$. Therefore, it is enough to check that $F(i)$ is minimal with this property. Assume towards a contradiction that there is an $a < F(i)$ with $\gamma(a, i, y, x)$, i.e.,

$$1 + (\text{op}(a, i) + 1)y \mid x.$$

Then, by the formula φ_0, there is a j with $j = \text{op}(a, i) = \text{op}(F(i'), i')$ for some $i' < k$. Thus, by LEMMA 9.6, we have $i = i'$ and $a = F(i') = F(i)$, which is a contradiction to the assumption $a < F(i)$. ⊣

Note that all functions — in particular the β-function — which we have introduced in this section can be defined by an $\exists$-formula and are therefore $\mathbb{N}$-conform.

Encoding Finite Sequences

This section aims at showing how the β-function can be used to encode a finite sequence of numbers; but what is meant by the words "finite" and "number"? In the standard model, this coincides with FINITENESS and the usual natural numbers. In general, however, this means that the sequence has a limited length k for some k, i.e., in non-standard models its length can actually be a non-standard number.

In a naive way, sequences of natural numbers can be viewed as functions from some $\{0, \cdots, n\}$ to the natural numbers, where $\{0, \cdots, n\}$ is the domain of the function. In PA, however, we cannot specify the domain of a definable function, which is why we will use $\beta(\,\cdot\,, 0)$ to encode the length of a sequence.

Concretely, we will encode $\langle F(i) \mid i < n\rangle$ using some c (whose existence is guaranteed by THEOREM 9.10) such that

$$\begin{aligned} \beta(c, 0) &= n \\ \forall i < n(\beta(c, i+1) &= F(i)). \end{aligned}$$

This motivates us to introduce the functions

$$\begin{aligned} \mathrm{lh}(c) &:\equiv \beta(c, 0) \\ c_i &:\equiv \beta(c, i+1). \end{aligned}$$

We will also call $\mathrm{lh}(c)$ the *length* of c. Furthermore, we define s to be a *sequence*, (denoted $\mathrm{seq}(s)$), if s is the smallest code for $\langle s_i \mid i < \mathrm{lh}(s)\rangle$:

$$\mathrm{seq}(s) :\Longleftrightarrow \forall t < s\big(\mathrm{lh}(t) = \mathrm{lh}(s) \to \exists i < \mathrm{lh}(s)(t_i \neq s_i)\big).$$

Note that the definition of seq assures that codes for finite sequences are unique, i.e.,

$$\mathsf{PA} \vdash \big(\mathrm{seq}(s) \wedge \mathrm{seq}(s') \wedge \mathrm{lh}(s) = \mathrm{lh}(s') \wedge \forall i < \mathrm{lh}(s)(s_i = s'_i)\big) \to s = s'.$$

Example 9.11. The simplest example is the empty sequence $\langle\,\rangle$ which is defined by $\langle\,\rangle = s :\Longleftrightarrow \mathrm{seq}(s) \wedge \mathrm{lh}(s) = 0$. By taking a closer look at the definition of the β-function, one can easily see that $\langle\,\rangle$ is actually 0, since it is the smallest code s with $\beta(s, 0) = 0$.

Secondly, we consider one-element sequences: The sequence just consisting of x is given by

$$\langle x\rangle = s :\Longleftrightarrow \mathrm{seq}(s) \wedge \mathrm{lh}(s) = 1 \wedge s_0 = x.$$

In the same way, one can define two-element sequences as

$$\langle x, y\rangle = s :\Longleftrightarrow \mathrm{seq}(s) \wedge \mathrm{lh}(s) = 2 \wedge s_0 = x \wedge s_1 = y.$$

More generally, if F is definable in PA, then one can define

$$\langle F(i) \mid i < k\rangle = s :\Longleftrightarrow \mathrm{seq}(s) \wedge \mathrm{lh}(s) = k \wedge \forall i < k\big(s_i = F(i)\big).$$

THEOREM 9.10 assures that such a number s always exists and since it is the least code it is unique.

The functions $c, i \mapsto c_i$, lh and $\langle\cdot\rangle$ are all defined by $\exists$-formulae and are thus $\mathbb{N}$-conform as a consequence of COROLLARY 9.4. We will use the same notation for the corresponding function in $\mathbb{N}$, for example we write $\langle n, m\rangle$ for $\langle n, m\rangle^{\mathbb{N}}$.

Next, we show how finite sequences can be concatenated.

PROPOSITION 9.12.

$$\mathsf{PA} \vdash \forall s \forall s' \exists t \Big(\mathrm{seq}(t) \wedge \mathrm{lh}(t) = \mathrm{lh}(s) + \mathrm{lh}(s') \wedge \forall i < \mathrm{lh}(s)(t_i = s_i) \wedge \forall i < \mathrm{lh}(s')(t_{\mathrm{lh}(s)+i} = s'_i) \Big)$$

Proof. Put $F(0) = \mathrm{lh}(s) + \mathrm{lh}(t)$, $F(i) = \beta(s,i)$ for $0 < i < \mathrm{lh}(s) + 1$, and $F(i) = \beta(t, i - \mathrm{lh}(s))$ for $i \geq \mathrm{lh}(s) + 1$. This clearly defines a function, so we can apply THEOREM 9.10 and obtain a code t such that

$$\text{for all } i < \mathrm{lh}(s) + \mathrm{lh}(t) + 1 \text{ we have } \beta(t,i) = F(i)\,.$$

In particular, this means that $\mathrm{lh}(t) = \mathrm{lh}(s) + \mathrm{lh}(s')$, $(t)_i = \beta(t, i+1) = \beta(s, i+1) = s_i$ for $i < \mathrm{lh}(s)$. Similarly, we get $t_{\mathrm{lh}(s)+i} = s'_i$ for $i < \mathrm{lh}(s')$. The LEAST NUMBER PRINCIPLE then enables us to choose t minimal with the properties from above, i.e., such that $\mathrm{seq}(t)$. ⊣

With PROPOSITION 9.12, we can define

$$s^\frown s' = t :\iff \mathrm{seq}(t) \wedge \mathrm{lh}(t) = \mathrm{lh}(s) + \mathrm{lh}(s') \wedge \forall i < \mathrm{lh}(s)(t_i = s_i) \wedge \forall i < \mathrm{lh}(s')(t_{\mathrm{lh}(s)+i} = s'_i)\,.$$

Note that by PROPOSITION 9.12, $s^\frown s'$ is functional. Moreover, it is easy to check that concatenation is associative, i.e.,

$$\mathsf{PA} \vdash (s^\frown s')^\frown s'' = s^\frown(s'^\frown s'').$$

Therefore, we can omit the brackets and write $s^\frown s'^\frown s''$ instead of $s^\frown(s'^\frown s'')$.

Encoding Power Functions

In the previous paragraphs, we have seen how the β-function allows us to encode finite sequences. Now we will use these insights to show how recursive functions can be defined in PA. We will not do this in general, since the only crucial function we need is the power function. Further examples of recursive functions can be found in the exercises.

The definability of the power function is remarkable, since it means that we can define exponentiation from addition and multiplication; however, as we will see in Chapter 12, multiplication cannot be defined from addition. The idea is to interpret the power x^k as the sequence $\langle 1, x, \cdots, x^{k-1}, x^k \rangle$ of length $k+1$.

We introduce the function x^k by stipulating

$$x^k = y :\iff \exists t \big(\mathrm{seq}(t) \wedge \mathrm{lh}(t) = \mathsf{s}k \wedge t_0 = 1 \wedge \forall i < k (t_{\mathsf{s}i} = x \cdot t_i) \wedge t_k = y \big).$$

Why is x^k functional? Clearly, the function x^k has (if defined) a unique value. In order to see that it is always defined, we can use induction: For $k = 0$ it is clear. Now assume that there is a sequence s of length $k+1$ such that $s_0 = 1$ and for all $i < k$ we have $s_{\mathsf{s}i} = x \cdot s_i$. Consider $t = s^\frown\langle x \cdot s_k\rangle$. Then t is a sequence of length $k+2$ which satisfies the desired properties.

Note that the power function is defined by an $\exists$-formula and therefore $\mathbb{N}$-conform by COROLLARY 9.4. Furthermore, observe that the power function fulfils the usual recursive definition, i.e.,

$$\mathsf{PA} \vdash \forall x(x^0 = 1)$$
$$\mathsf{PA} \vdash \forall x \forall k(x^{\mathsf{s}k} = x \cdot x^k).$$

Our next aim is to encode terms, formulae and proofs by making use of unique prime decomposition. This can be shown in PA. However, for us it suffices to show that the function mapping x, y, z to $2^x \cdot 3^y \cdot 5^z$ is injective. The general result is left as an excercise to the interested reader. We define primality by

$$\mathrm{prime}(x) :\Longleftrightarrow x > 1 \wedge \forall z\big(z \mid x \to (z = x \vee z = 1)\big).$$

If $\mathrm{prime}(x)$, we say that x is *prime*. Note that prime can be defined by an $\exists$-formula, since $\forall z$ can be replaced by $\forall z \leq x$. By using the fact that $2, 3$ and 5 are prime in the standard model $\mathbb{N}$ and by PROPOSITION 9.3, we obtain

$$\mathsf{PA} \vdash \mathrm{prime}(2) \wedge \mathrm{prime}(3) \wedge \mathrm{prime}(5).$$

Moreover, prime decomposition up to 5 is easily seen to be unique: One just has to show that

$$\mathsf{PA} \vdash 2^x \cdot 3^y \cdot 5^z = 2^{x'} \cdot 3^{y'} \cdot 5^{z'} \to x = x' \wedge y = y' \wedge z = z'.$$

This is usually proved by induction on $x + y + z$. Note that the simplest way to achieve this is to use the following characterisation of primality (see EXERCISE 9.1):

$$\mathsf{PA} \vdash \mathrm{prime}(x) \leftrightarrow \forall y \forall z(x \mid yz \to x \mid y \vee x \mid z)$$

Encoding Terms and Formulae

In a first step, every logical and every non-logical symbol ζ of Peano Arithmetic is assigned a natural number $\#\zeta$ in $\mathbb{N}$, called **Gödel number** of ζ. Since from now on, we will often switch between the meta-level and the formal level, we will always explicitly mention whenever we are reasoning formally, i.e., within PA. Otherwise the proofs will be on the meta-level.

Symbol ζ	Gödel number $\#\zeta$
0	$\mathbf{0}$
s	$\mathbf{2}$
$+$	$\mathbf{4}$
$\cdot$	$\mathbf{6}$
$=$	$\mathbf{8}$
$\neg$	$\mathbf{10}$
$\wedge$	$\mathbf{12}$
$\vee$	$\mathbf{14}$
$\rightarrow$	$\mathbf{16}$
$\exists$	$\mathbf{18}$
$\forall$	$\mathbf{20}$
v_0	$\mathbf{1}$
v_1	$\mathbf{3}$
$\vdots$	$\vdots$
v_n	$\mathbf{2} \cdot n + \mathbf{1}$

In the previous section, we introduced power functions in PA. Since $\mathbb{N} \vDash \mathsf{PA}$, such functions also exist in $\mathbb{N}$, and we will use the same notation n^k as in PA. By $\mathbb{N}$-conformity we have $\mathsf{PA} \vdash \underline{n}^{\underline{k}} = \underline{n^k}$ for all $n, k \in \mathbb{N}$. Note that by THEOREM 1.7 it would already suffice to just gödelize the logical operators $\neg, \wedge$ and $\exists$. Next we encode terms and formulae.

Term τ	Gödel number $\#\tau$	Formula φ	Gödel number $\#\varphi$
0	$\mathbf{0}$	$\tau_0 = \tau_1$	$\mathbf{2}^{\# =} \cdot \mathbf{3}^{\# \tau_0} \cdot \mathbf{5}^{\# \tau_1}$
v_n	$\mathbf{2} \cdot n + \mathbf{1}$	$\neg\psi$	$\mathbf{2}^{\# \neg} \cdot \mathbf{3}^{\# \psi}$
$\mathsf{s}t$	$\mathbf{2}^{\# \mathsf{s}} \cdot \mathbf{3}^{\# t}$	$\psi_0 \wedge \psi_1$	$\mathbf{2}^{\# \wedge} \cdot \mathbf{3}^{\# \psi_1} \cdot \mathbf{5}^{\# \psi_1}$
$t_0 + t_1$	$\mathbf{2}^{\# +} \cdot \mathbf{3}^{\# t_0} \cdot \mathbf{5}^{\# t_1}$	$\psi_0 \vee \psi_1$	$\mathbf{2}^{\# \vee} \cdot \mathbf{3}^{\# \psi_0} \cdot \mathbf{5}^{\# \psi_1}$
$t_0 \cdot t_1$	$\mathbf{2}^{\# \cdot} \cdot \mathbf{3}^{\# t_0} \cdot \mathbf{5}^{\# t_1}$	$\psi_0 \rightarrow \psi_1$	$\mathbf{2}^{\# \rightarrow} \cdot \mathbf{3}^{\# \psi_0} \cdot \mathbf{5}^{\# \psi_1}$
		$\exists x \psi$	$\mathbf{2}^{\# \exists} \cdot \mathbf{3}^{\# x} \cdot \mathbf{5}^{\# \psi}$
		$\forall x \psi$	$\mathbf{2}^{\# \forall} \cdot \mathbf{3}^{\# x} \cdot \mathbf{5}^{\# \psi}$

Observe that by the uniqueness of the prime decomposition up to 5, every natural number encodes at most one variable, term or formula. So far, we have only assigned a natural number in the standard model to each symbol, term, and formula. However, we want to do this within Peano Arithmetic. This can be achieved by stipulating

$$\ulcorner \zeta \urcorner :\equiv \underline{\# \zeta}$$

for an arbitrary symbol, term or formula ζ. This allows us to express in PA that some number is the code of a variable, term or formula. However, we can easily formalize this so-called **Gödelisation** process, where $2 :\equiv \mathsf{ss0}$, $3 :\equiv \mathsf{sss0}$, and $5 :\equiv \mathsf{sssss0}$.

$$\begin{aligned} \mathrm{succ}(n) &:\equiv \mathsf{2}^{\ulcorner \mathsf{s} \urcorner} \cdot \mathsf{3}^n & \mathrm{add}(n,m) &:\equiv \mathsf{2}^{\ulcorner + \urcorner} \cdot \mathsf{3}^n \cdot \mathsf{5}^m \\ \mathrm{mult}(n,m) &:\equiv \mathsf{2}^{\ulcorner \cdot \urcorner} \cdot \mathsf{3}^n \cdot \mathsf{5}^m & \mathrm{eq}(t,t') &:\equiv \mathsf{2}^{\ulcorner = \urcorner} \cdot \mathsf{3}^t \cdot \mathsf{5}^{t'} \\ \mathrm{not}(f) &:\equiv \mathsf{2}^{\ulcorner \neg \urcorner} \cdot \mathsf{3}^f & \mathrm{and}(f,f') &:\equiv \mathsf{2}^{\ulcorner \wedge \urcorner} \cdot \mathsf{3}^f \cdot \mathsf{5}^{f'} \\ \mathrm{or}(f,f') &:\equiv \mathsf{2}^{\ulcorner \vee \urcorner} \cdot \mathsf{3}^f \cdot \mathsf{5}^{f'} & \mathrm{imp}(f,f') &:\equiv \mathsf{2}^{\ulcorner \to \urcorner} \cdot \mathsf{3}^f \cdot \mathsf{5}^{f'} \\ \mathrm{ex}(v,f) &:\equiv \mathsf{2}^{\ulcorner \exists \urcorner} \cdot \mathsf{3}^v \cdot \mathsf{5}^f & \mathrm{all}(v,f) &:\equiv \mathsf{2}^{\ulcorner \forall \urcorner} \cdot \mathsf{3}^v \cdot \mathsf{5}^f \end{aligned}$$

In order to simplify the notation, for terms $\tau, \tau_0, \ldots, \tau_n$, we define

$$\tau \in \{\tau_0, \ldots, \tau_n\} \quad :\equiv \quad \bigvee_{i=0}^{n} \tau = \tau_i \,.$$

Now we are ready to provide a formalised version of construction of terms and formulae:

$$\mathrm{var}(v) :\Longleftrightarrow \exists n(v = \mathsf{2} \cdot n + \mathsf{1})$$

$$\begin{aligned} \mathrm{c\_term}(c,t) :\Longleftrightarrow\ & \mathrm{seq}(c) \wedge c_{\mathrm{lh}(c)-1} = t \wedge \\ & \forall k < \mathrm{lh}(c)\Big(\mathrm{var}(c_k) \vee c_k = \mathsf{0} \vee \\ & \quad \exists i < k\, \exists j < k\big(c_k \in \big\{\, \mathrm{succ}(c_i), \mathrm{add}(c_i,c_j), \mathrm{mult}(c_i,c_j)\big\}\big)\Big) \end{aligned}$$

$$\mathrm{term}(t) :\Longleftrightarrow \exists c\big(\mathrm{c\_term}(c,t)\big)$$

$$\begin{aligned} \mathrm{c\_fml}(c,f) :\Longleftrightarrow\ & \mathrm{seq}(c) \wedge c_{\mathrm{lh}(c)-1} = f \wedge \\ & \forall k < \mathrm{lh}(c)\Big(\exists t\, \exists t'\big(\mathrm{term}(t) \wedge \mathrm{term}(t') \wedge c_k = \mathrm{eq}(t,t')\big) \vee \\ & \quad \exists i,j < k\big(c_k \in \big\{\, \mathrm{not}(c_i), \mathrm{and}(c_i,c_j), \mathrm{or}(c_i,c_j), \mathrm{imp}(c_i,c_j)\big\}\big) \vee \\ & \quad \exists i < k\, \exists v\big(\mathrm{var}(v) \wedge c_k \in \big\{\, \mathrm{ex}(v,c_i), \mathrm{all}(v,c_i)\big\}\big)\Big) \end{aligned}$$

$$\mathrm{fml}(f) :\Longleftrightarrow \exists c\big(\mathrm{c\_fml}(c,f)\big)$$

Note that all the above relations are defined by $\exists$-formulae.

Example 9.13. Let us consider the term $\tau \equiv \mathsf{s}v_n + \mathsf{0}$. In the standard model $\mathbb{N}$, the sequence $c \equiv \langle \#v_n, \#\mathsf{s}v_n, \#\mathsf{0}, \#\mathsf{s}v_n + \mathsf{0}\rangle$ encodes τ, i.e., $\mathbb{N} \vDash \mathrm{c\_term}(c, \#\tau)$. By PROPOSITION 9.3 this implies $\mathsf{PA} \vdash \mathrm{c\_term}(\underline{c}, \ulcorner\tau\urcorner)$.

LEMMA 9.14. *For $n \in \mathbb{N}$ we have*

(a) $\mathbb{N} \vDash \mathrm{var}(n)$ *if and only if* $n \equiv \#\nu$ *for some variable* ν.

(b) $\mathbb{N} \vDash \mathrm{term}(n)$ *if and only if* $n \equiv \#\tau$ *for some* $\mathscr{L}_{\mathsf{PA}}$*-term* τ.

(c) $\mathbb{N} \vDash \mathrm{fml}(n)$ *if and only if* $n \equiv \#\varphi$ *for some* $\mathscr{L}_{\mathsf{PA}}$*-formula* φ.

Proof. Condition (a) is obvious. For (b), we first prove that $\mathbb{N} \vDash \mathrm{term}(\#\tau)$ for every term τ. We proceed by induction on the term construction of τ. If $\tau \equiv \mathsf{0}$ or τ is a variable, then clearly $\mathbb{N} \vDash \mathrm{c\_term}(\langle\#\tau\rangle, \#\tau)$ and hence the claim follows. Now, if $\tau \equiv \mathsf{s}\tau'$ for some term τ' with $\mathbb{N} \vDash \mathrm{term}(\#\tau')$, and

$c \in \mathbb{N}$ is a code with $\mathbb{N} \vDash \text{c\_term}(c, \#\tau')$, then $\mathbb{N} \vDash \text{succ}(\#\tau')$, and hence, $\mathbb{N} \vDash \text{c\_term}(c^\frown\langle\#\tau\rangle, \#\tau)$. The other cases are similar. For the converse, we use the principle of strong induction in $\mathbb{N}$. Suppose that the claim holds for all $m < n$ in $\mathbb{N}$ and let $\mathbb{N} \vDash \text{term}(n)$. If $n \equiv \mathbf{0}$ then $n \equiv \#0$, and if $n \equiv 2m+1$ for some m, then $n \equiv \#v_m$. Let $\mathbb{N} \vDash \text{c\_term}(c, n)$ for some $c \in \mathbb{N}$ with $\text{lh}(c) > 1$. Now in $\mathbb{N}$ we have either $n \equiv \text{succ}^{\mathbb{N}}(c_i)$ for some $i < \text{lh}(c)$, $n \equiv \text{add}^{\mathbb{N}}(c_i, c_j)$ or $n \equiv \text{mult}^{\mathbb{N}}(c_i, c_j)$ for $i, j < \text{lh}(c)$. In the first case, note that $\mathbb{N} \vDash \text{c\_term}(\langle c_k \mid k < \mathsf{s}i\rangle, c_i)$. By our induction hypothesis, we can take a term τ such that $c_i \equiv \#\tau$. But then, by $\mathbb{N}$-conformity, we have $n \equiv \text{succ}^{\mathbb{N}}(c_i) \equiv (2^{\ulcorner \mathsf{s} \urcorner} \cdot 3^{\underline{c_i}})^{\mathbb{N}} \equiv 2^{\#\mathsf{s}} \cdot 3^{c_i} \equiv \#\mathsf{s}\tau$ and hence, n encodes $\mathsf{s}\tau$. The other cases are similar.

The corresponding statement for formulae is proved in the same way and is therefore left as an exercise. ⊣

Note that the relations var, term and formula are $\exists$-formulae. Therefore, by combining LEMMA 9.14 and PROPOSITION 9.1 we obtain: If ν is a variable, τ is a term and φ is a formula, then

$$\begin{aligned} &\mathsf{PA} \vdash \text{var}(\ulcorner \nu \urcorner) \\ &\mathsf{PA} \vdash \text{term}(\ulcorner \tau \urcorner) \\ &\mathsf{PA} \vdash \text{formula}(\ulcorner \varphi \urcorner). \end{aligned}$$

Before we proceed to gödelise logical axioms, the axioms of Peano Arithmetic and formal proofs, we have to deal with substitution: First, we introduce new relations which check wether a code for a variable appears in the code of a term or formula, respectively.

$$\begin{aligned} \text{var\_in\_term}(v, t) \;:\Longleftrightarrow\; &\text{var}(v) \land \exists c\Big(\text{c\_term}(c, t) \land \\ &\forall c' < c \,\neg\, \text{c\_term}(c', t) \land \exists i < \text{lh}(c)(c_i = v)\Big) \end{aligned}$$

Note that the minimality of c is necessary since otherwise, any code for a variable could appear in the sequence of codes of the term construction. The same holds for the following relation var_in_fml:

$$\begin{aligned} \text{var\_in\_fml}(v, f) :\Longleftrightarrow\; &\exists c\Big(\text{c\_fml}(c, f) \land \forall c' < c \,\neg\, \text{c\_fml}(c', f) \land \\ &\exists i < \text{lh}(c)\, \exists t_0\, \exists t_1 \big(\text{term}(t_0) \land \text{term}(t_1) \land c_i = \text{eq}(t_0, t_1) \land \\ &\big(\text{var\_in\_term}(v, t_0) \lor \text{var\_in\_term}(v, t_1)\big)\big)\Big) \end{aligned}$$

$$\begin{aligned} \text{free}(v, f) :\Longleftrightarrow\; &\exists c\Big(\text{c\_fml}(c, f) \land \text{var\_in\_fml}(v, f) \land \\ &\forall i < \text{lh}(c)\, \forall j < i\big(c_i \neq \text{ex}(v, c_j) \land c_i \neq \text{all}(v, c_j)\big)\Big) \end{aligned}$$

For the sake of simplicity, we permit the substitution $\varphi(x/\tau)$ only if it is admissible and x as well as all variables in τ appear only free in φ. This does not impose a restriction, since by renaming of variables, every formula is equivalent to one in which no variable occurs both bound and free. We can thus define

$$\begin{aligned}\text{sb\_adm}(v,t,f) :\Longleftrightarrow\ & \text{var\_in\_fml}(v,f) \wedge \text{free}(v,f) \wedge \\ & \forall v' < t\big(\text{var\_in\_term}(v',t) \rightarrow \text{free}(v',f)\big).\end{aligned}$$

Note that the relations var_in_term, var_in_fml, free, and sb_adm are all $\exists$-formulae: The only unbounded universal quantifier appears in the relation sb_adm, where var_in_term occurs as negated — recall that $\text{var\_in\_term}(v',t) \rightarrow \text{free}(v',f)$ is equivalent to $\neg\,\text{var\_in\_term}(v',t) \vee \text{free}(v',f)$. However, the existential quantifier in the definition of $\text{var\_in\_term}(v',t)$ can be replaced by a bounded one, since the code of t has to be smaller than the code of f.

The next relation expresses that c' encodes the construction of the term obtained from the term with code t by replacing every occurrence of the code v of a variable by the code t_0.

$\text{c\_sb\_term}(c,c',c'',v,t_0,t,t') \ :\Longleftrightarrow$

$\text{var}(v) \wedge \text{c\_term}(c,t) \wedge \text{c\_term}(c',t') \wedge \text{c\_term}(c'',t_0) \ \wedge \text{lh}(c') = \text{lh}(c'') + \text{lh}(c)$,

and for all $k < \text{lh}(c'')$ we have $c'_k = c''_k$, and for all $k < \text{lh}(c)$ we have:

$$\begin{aligned}(\text{var}(c_k) \ \wedge \ c_k \neq v \rightarrow c'_k = c_k) \ \wedge & (c_k = \mathsf{0} \rightarrow c'_{\text{lh}(c'')+k} = \mathsf{0}) \\ & \wedge \ (c_k = v \rightarrow c'_{\text{lh}(c'')+k} = t_0)\end{aligned}$$

$$\begin{aligned}\forall i < k\, \forall j < k\, \big(\ & c_k = \text{succ}(c_i) && \rightarrow c'_{\text{lh}(c'')+k} = \text{succ}(c'_{\text{lh}(c'')+i}) && \wedge \\ & c_k = \text{add}(c_i,c_j) && \rightarrow c'_{\text{lh}(c'')+k} = \text{add}(c'_{\text{lh}(c'')+i}, c'_{\text{lh}(c'')+j}) && \wedge \\ & c_k = \text{mult}(c_i,c_j) && \rightarrow c'_{\text{lh}(c'')+k} = \text{mult}(c'_{\text{lh}(c'')+i}, c'_{\text{lh}(c'')+j}) && \big)\end{aligned}$$

By omitting the codes c, c', c'' we can describe term substitution by

$$\text{sb\_term}(v,t_0,t,t') :\Longleftrightarrow \exists c\, \exists c'\, \exists c'' \big(\text{c\_sb\_term}(c,c',c'',v,t_0,t,t')\big).$$

Informally, if t encodes the term τ, v the variable ν, t_0 the term τ_0, and t' encodes $\tau(\nu/\tau_0)$, then the relation $\text{sb\_term}(v,t_0,t,t')$ holds.

For formulae, we proceed similarly, except that we first have to make sure that the substitution is admissible.

$\text{c\_sb\_fml}(c,c',v,t_0,f,f') \ :\Longleftrightarrow$

$\text{c\_fml}(c,f) \wedge \text{c\_fml}(c',f') \wedge \text{sb\_adm}(v,t_0,f) \wedge \text{lh}(c') = \text{lh}(c)$,

and for all $k < \text{lh}(c)$ we have:

$$\begin{aligned}\forall t\, \forall t'\, \forall s\, \forall s' \Big(\big(c_k = \text{eq}(t,t') \wedge \text{sb\_term}(v,t_0,t,s) \wedge \\ \text{sb\_term}(v,t_0,t',s')\big) \ \rightarrow c'_k = \text{eq}(s,s') \Big)\end{aligned}$$

$$\begin{aligned}\forall i < k\, \forall j < k\, \big(\; c_k &= \mathrm{not}(c_i) &&\to c'_k = \mathrm{not}(c'_i) \;\wedge\\ c_k &= \mathrm{or}(c_i, c_j) &&\to c'_k = \mathrm{or}(c'_i, c'_j) \;\wedge\\ c_k &= \mathrm{and}(c_i, c_j) &&\to c'_k = \mathrm{and}(c'_i, c'_j) \;\wedge\\ c_k &= \mathrm{imp}(c_i, c_j) &&\to c'_k = \mathrm{imp}(c'_i, c'_j) \;\big)\end{aligned}$$

$$\begin{aligned}\forall i < k\, \forall v\, \big(\; c_k &= \mathrm{all}(v, c_i) \to c'_k = \mathrm{ex}(v, c'_i) \;\wedge\\ c_k &= \mathrm{ex}(v, c_i) \to c'_k = \mathrm{ex}(v, c'_i) \;\big)\end{aligned}$$

Again, by leaving out the sequence codes, we define

$$\mathrm{sb\_fml}(v, t_0, f, f') :\Longleftrightarrow \exists c \exists c' (\mathrm{c\_sb\_fml}(c, c', v, t_0, f, f').$$

Informally, if f encodes the formula φ, v the variable ν, t_0 the term τ, and if the substitution $\varphi(\nu/\tau)$ is admissible and f' encodes $\varphi(\tau)$, then the relation $\mathrm{sb\_fml}(v, t_0, f, f')$ holds.

LEMMA 9.15. *Let τ and τ_0 be two terms, φ a formula and ν a variable such that the substitution $\varphi(\nu/\tau_0)$ is admissible. Then we have:*

(a) $\mathsf{PA} \vdash \mathrm{sb\_term}(\ulcorner\nu\urcorner, \ulcorner\tau_0\urcorner, \ulcorner\tau\urcorner, t) \leftrightarrow t = \ulcorner\tau(\nu/\tau_0)\urcorner$

(b) $\mathsf{PA} \vdash \mathrm{sb\_fml}(\ulcorner\nu\urcorner, \ulcorner\tau_0\urcorner, \ulcorner\varphi\urcorner, f) \leftrightarrow f = \ulcorner\varphi(\nu/\tau_0)\urcorner$

Proof. This follows from the definition of the relations sb_term and sb_fml using induction on the term construction of τ and the formula construction of φ. ⊣

Example 9.16. Let $\tau \equiv \mathtt{s}x + y$ and $\tau_0 \equiv \mathtt{0}$. Then the sequence $\langle \ulcorner x\urcorner, \ulcorner \mathtt{s}x\urcorner, \ulcorner y\urcorner, \ulcorner \mathtt{s}x + y\urcorner\rangle$ encodes $\ulcorner\tau\urcorner$ and $\langle\ulcorner\mathtt{0}\urcorner\rangle$ encodes τ_0. Now, when coding $\tau(x/\tau_0)$, we first take the code of τ_0 and then replace every occurrence of x in the code of τ by τ_0. This gives

$$\langle \ulcorner\mathtt{0}\urcorner, \ulcorner\mathtt{s0}\urcorner, \ulcorner y\urcorner, \ulcorner\mathtt{s0} + y\urcorner\rangle$$

which encodes $\tau(x/\tau_0)$.

Encoding Formal Proofs

Finally, we can use the machinery as developed in the previous section to encode axioms and formal proofs. We first show how to achieve this in $\mathbb{N}$. For this, recall that a formal proof of some $\mathscr{L}_{\mathsf{PA}}$-formula φ is a finite sequence $\varphi_0, \ldots, \varphi_n$ with $\varphi_n \equiv \varphi$ such that each φ_i is an instance of a logical axiom, an axiom of PA or is obtained from preceding elements of the sequence by using Modus Ponens or Generalisation. Hence we can code a formal proof of φ by

$\langle \#\varphi_0, \ldots, \#\varphi_n \rangle$. Conversely, from such a code we can recover the sequence $\varphi_0, \ldots, \varphi_n$ and hence reconstruct a formal proof of φ.

As for terms and formulae, we proceed to code formal proofs in PA. The goal is to define a relation prv with the property that $\mathbb{N} \vDash \mathrm{prv}(\ulcorner\varphi\urcorner)$ for some formula φ if and only if there is a formal proof of φ, i.e., $\mathsf{PA} \vdash \varphi$. The following examples illustrate how axioms can be formalised in PA:

$$\mathrm{ax\_L}_1(f) :\Longleftrightarrow \exists f' \exists f'' \big(\mathrm{fml}(f') \land \mathrm{fml}(f'') \land f = \mathrm{imp}(f', \mathrm{imp}(f'', f'))\big)$$
$$\mathrm{ax\_L}_{10}(f) :\Longleftrightarrow \exists f' \exists f'' \exists v \exists t \big(\mathrm{sb\_fml}(v, t, f', f'') \land f = \mathrm{imp}(\mathrm{all}(v, f'), f'')\big)$$
$$\mathrm{ax\_L}_{14}(f) :\Longleftrightarrow \exists t \big(\mathrm{term}(t) \land f = \mathrm{eq}(t, t)\big)$$

PA_0 and the Induction Schema are gödelised as follows:

$$\mathrm{ax\_PA}_0(f) :\Longleftrightarrow f = \ulcorner \forall v_0 \neg(\mathsf{s}v_0 = \mathsf{0}) \urcorner$$
$$\begin{aligned}\mathrm{ax\_PA}_6(f) :\Longleftrightarrow & \exists f' \exists f'' \exists f''' \exists v \exists g \big(\mathrm{free}(v, f') \land \mathrm{sb\_fml}(v, \ulcorner \mathsf{0} \urcorner, f', f'') \\ & \land \mathrm{sb\_fml}(v, \mathrm{succ}(v), f', f''') \land g = \mathrm{all}(v, \mathrm{imp}(f', f''')) \\ & \land f = \mathrm{imp}(\mathrm{and}(f'', g), \mathrm{all}(v, f'))\big)\end{aligned}$$

We leave it to the reader to formalize the other axioms. Similarly, we define axioms:

$$\mathrm{log\_ax}(f) :\Longleftrightarrow \mathrm{ax\_L}_0(f) \lor \cdots \lor \mathrm{ax\_L}_{16}(f)$$
$$\mathrm{peano\_ax}(f) :\Longleftrightarrow \mathrm{ax\_PA}_0(f) \lor \cdots \lor \mathrm{ax\_PA}_6(f)$$
$$\mathrm{ax}(f) :\Longleftrightarrow \mathrm{log\_ax}(f) \lor \mathrm{peano\_ax}(f)$$

Next, we formalize the inference rules Modus Ponens and Generalisation:

$$\mathrm{mp}(f', f'', f) :\Longleftrightarrow \mathrm{fml}(f') \land \mathrm{fml}(f) \land f'' = \mathrm{imp}(f', f)$$
$$\mathrm{gen}(v, f', f) :\Longleftrightarrow \mathrm{var}(v) \land \mathrm{fml}(f') \land f = \mathrm{all}(v, f')$$

Finally, we encode formal proofs as sequences of codes of formulae which are either axioms or produced by one of the inference rules. Therefore, we define the predicates $\mathrm{c\_prv}(c, f)$ in order to specify that c encodes a proof of the formula coded by f and prv to express provability.

$$\begin{aligned}\mathrm{c\_prv}(c, f) :\Longleftrightarrow & \mathrm{seq}(c) \land c_{\mathrm{lh}(c)-1} = f \land \forall k < \mathrm{lh}(c) \big(\mathrm{ax}(c_k) \lor \\ & \exists i < k \exists j < k \big(\mathrm{mp}(c_i, c_j, c_k) \lor \exists v (\mathrm{gen}(v, c_i, c_k))\big)\big) \\ \mathrm{prv}(f) :\Longleftrightarrow & \exists c (\mathrm{prv}(c, f)).\end{aligned}$$

Note that in the standard model $\mathbb{N}$ we have $\mathbb{N} \vDash \mathrm{c\_prv}(c, \#\varphi)$ if and only if c encodes a sequence $\langle \#\varphi_0, \ldots, \#\varphi_n \rangle$, where $\varphi_0, \ldots, \varphi_n$ is a formal proof of φ.

LEMMA 9.17. *Let $n, c \in \mathbb{N}$ be natural numbers. Then $\mathbb{N} \vDash \mathrm{c\_prv}(c, n)$ if and only if c encodes a formal proof of some $\mathscr{L}_{\mathsf{PA}}$-formula φ with $\#\varphi = n$. In particular, $\mathbb{N} \vDash \mathrm{prv}(\#\varphi)$ if and only if $\mathsf{PA} \vdash \varphi$.*

Proof. Note first that $\mathbb{N} \vDash \mathrm{ax}(m)$ for some $m \in \mathbb{N}$ if and only if $m = \#\psi$ encodes an instance of a logical axiom or an axiom of PA. The proof is the same as the proof of LEMMA 9.14, where in the forward direction, one proceeds by induction on $\mathrm{lh}(c)$, and for the converse by induction on the length of the formal proof of φ.

For the second part, suppose that $\mathbb{N} \vDash \mathrm{prv}(\#\varphi)$. Then there is $c \in \mathbb{N}$ which encodes a formal proof of φ, and hence, $\mathsf{PA} \vdash \varphi$. ⊣

Note that LEMMA 9.17 does not hold if we replace $\mathbb{N}$ by a non-standard model $\mathbf{M}$ of PA: If $\mathbf{M} \vDash \mathrm{c\_prv}(c, \ulcorner\varphi\urcorner)$ and $c^{\mathbf{M}}$ is a non-standard number, then the "proof" encoded by $c^{\mathbf{M}}$ is of non-standard length and therefore, we cannot conclude that $\mathsf{PA} \vdash \varphi$.

COROLLARY 9.18. *Let φ be an $\mathscr{L}_{\mathsf{PA}}$-formula. If $\mathsf{PA} \vdash \varphi$ then there is $n \in \mathbb{N}$ such that $\mathsf{PA} \vdash \mathrm{c\_prv}(\underline{n}, \ulcorner\varphi\urcorner)$. In particular, if $\mathsf{PA} \vdash \varphi$ then $\mathsf{PA} \vdash \mathrm{prv}(\ulcorner\varphi\urcorner)$.*

Proof. Suppose that $\mathsf{PA} \vdash \varphi$. Then by LEMMA 9.17 we have $\mathbb{N} \vDash \mathrm{c\_prv}(c, \#\varphi)$ for some $c \in \mathbb{N}$. Observe that the relations c_prv and prv are defined by an $\exists$-formula. Hence, it follows from PROPOSITION 9.3 that $\mathsf{PA} \vdash \mathrm{c\_prv}(\underline{c}, \ulcorner\varphi\urcorner)$, and in particular, $\mathsf{PA} \vdash \mathrm{prv}(\ulcorner\varphi\urcorner)$. ⊣

NOTES

In his proof of the incompleteness theorems [16], Gödel used the β-function and unique prime decomposition in order to encode finite sequences by a single number. There are, however, other ways to achieve this (e.g., the coding provided by Smullyan in [52]). In our presentation, we mainly followed Shoenfield [47].

EXERCISES

9.0 Prove the statement N_2.

9.1 (a) Show that $\mathsf{PA} \vdash \forall x \geq 2\, \exists y(\mathrm{prime}(y) \land y \mid x)$, i.e., every $x \geq 2$ has a prime divisor.

Hint: Use the LEAST NUMBER PRINCIPLE.

(b) Show that $\mathsf{PA} \vdash \forall x(\mathrm{prime}(x) \leftrightarrow \forall y \forall z(x \mid yz \rightarrow x \mid y \lor x \mid z))$.

Hint: Use the PRINCIPLE OF DIVISION WITH REMAINDER.

9.2 Introduce a factorial function ! such that $n! = 1 \cdot \ldots \cdot n$, and show that it is $\mathbb{N}$-conform.

9.3 Introduce a function $\mathrm{lcm}_{i<k} F(i)$ for a function F definable in PA such that $\mathrm{lcm}_{i<k} F(i)$ is the least common multiple of $F(0), \ldots, F(k-1)$ and show that it is $\mathbb{N}$-conform.

9.4 State and prove in PA that every number has a unique prime decomposition, i.e., prove that every number is a product of primes, and show that this is unique up to permutations of the factors.

9.5 Give the Gödel code of the formula $\forall x(0 \neq x)$.

9.6 An alternative way to define Gödel coding is to use the existence and uniqueness of the base b notation for $b \geq 2$.

(a) Show that it suffices to gödelise only b symbols for some $b \in \mathbb{N}$.

(b) Prove (in PA) that for every number n there are $n_0, \ldots, n_k$ such that

$$n = n_k b^k + \ldots + n_1 b + n_0 \equiv: (n_k \ldots n_0)_b.$$

(c) Show that there is a function $*$ definable in PA such that

$$(n_k \ldots n_0)_b * (m_l \ldots m_0)_b = (n_k \ldots n_0 m_l \ldots m_0)_b.$$

(d) Use (a) and (b) to give an alternative of Gödel coding using base b notation rather than the unique prime decomposition.

9.7 Show that there is a function which truncates sequences, i.e., introduce a binary function $\upharpoonright$ such that $\mathsf{PA} \vdash \mathrm{seq}(s) \land k < \mathrm{lh}(s) \rightarrow \mathrm{seq}(s \upharpoonright k) \land \mathrm{lh}(s \upharpoonright k) = k \land \forall i < k((s \upharpoonright k)_i = s_i)$.

9.8 Prove that whenever $\mathbb{N} \vDash \mathrm{term}(n)$ for some $n \in \mathbb{N}$, then $n = \#\tau$ for some term τ. State and prove a similar result for variables and formulae.

9.9 Fill out the details of the proof of LEMMA 9.15.

9.10 Show how the remaining logical axioms and axioms of PA can be gödelised.

Chapter 10
The First Incompleteness Theorem

In 1931, Gödel proved his FIRST INCOMPLETENESS THEOREM which states that if PA is consistent, then it is incomplete, i.e., there is a $\mathscr{L}_{\mathsf{PA}}$-sentence σ such that $\mathsf{PA} \nvdash \sigma$ and $\mathsf{PA} \nvdash \neg\sigma$. In this chapter, we prove the FIRST INCOMPLETENESS THEOREM not only for PA but also for weaker and stronger theories.

The Provability Predicate

In this section, we state some properties of the provability predicate that we introduced in Chapter 9.

LEMMA 10.1.

(a) $\mathsf{PA} \vdash \mathrm{prv}(x) \wedge \mathrm{prv}\big(\mathrm{imp}(x, y)\big) \rightarrow \mathrm{prv}(y)$

(b) $\mathsf{PA} \vdash \mathrm{prv}(x) \wedge \mathrm{prv}(y) \rightarrow \mathrm{prv}\big(\mathrm{and}(x, y)\big)$.

Proof. For (a), note that $\mathrm{prv}(x)$ and $\mathrm{prv}(\mathrm{imp}(x, y))$ imply $\mathrm{mp}(x, \mathrm{imp}(x, y), y)$. Now, if c, c' satisfy $\mathrm{c\_prv}(c, x)$ and $\mathrm{c\_prv}(c', \mathrm{imp}(x, y))$, respectively, then the concatenation of the codes yields $\mathrm{c\_prv}(c^\frown c'^\frown \langle y \rangle, y)$ and hence $\mathrm{prv}(y)$ as desired.

For (b), assume $\mathrm{prv}(x)$ and $\mathrm{prv}(y)$. In particular, this implies $\mathrm{fml}(x)$ and $\mathrm{fml}(y)$. Note that using the formalised version of the axiom L_5, we obtain

$$\mathsf{PA} \vdash \mathrm{prv}\Big(\mathrm{imp}\big(y, \mathrm{imp}\big(x, \mathrm{and}(x, y)\big)\big)\Big).$$

Using $\mathrm{prv}(y)$ and (a), we get $\mathrm{prv}\big(\mathrm{imp}\big(x, \mathrm{and}(x, y)\big)\big)$, and a further application of (a) yields $\mathrm{prv}(\mathrm{and}(x, y))$. ⊣

As an immediate consequence of LEMMA 10.1, we obtain the following

L. Halbeisen, R. Krapf, *Gödel's Theorems and Zermelo's Axioms*,
https://doi.org/10.1007/978-3-030-52279-7_10

COROLLARY 10.2. *Let φ and ψ be $\mathscr{L}_{\mathsf{PA}}$-formulae. Then we have*

(a) $\mathsf{PA} \vdash \operatorname{prv}(\ulcorner \varphi \to \psi \urcorner) \to (\operatorname{prv}(\ulcorner \varphi \urcorner) \to \operatorname{prv}(\ulcorner \psi \urcorner))$,

(b) $\mathsf{PA} \vdash \operatorname{prv}(\ulcorner \varphi \urcorner) \wedge \operatorname{prv}(\ulcorner \psi \urcorner) \to \operatorname{prv}(\ulcorner \varphi \wedge \psi \urcorner)$.

Note that (a) corresponds to a formalised version of the inference rule (MP).

COROLLARY 10.3. *Let φ and ψ be $\mathscr{L}_{\mathsf{PA}}$-formulae. Then the following statements hold:*

(a) *If* $\varphi \Leftrightarrow_{\mathsf{PA}} \psi$, *then* $\operatorname{prv}(\ulcorner \varphi \urcorner) \Leftrightarrow_{\mathsf{PA}} \operatorname{prv}(\ulcorner \psi \urcorner)$.

(b) $\operatorname{prv}(\ulcorner \varphi \urcorner) \wedge \operatorname{prv}(\ulcorner \psi \urcorner) \Leftrightarrow_{\mathsf{PA}} \operatorname{prv}(\ulcorner \varphi \wedge \psi \urcorner)$.

Proof. For (a), assume that $\varphi \Leftrightarrow_{\mathsf{PA}} \psi$. By symmetry, it suffices to verify that $\mathsf{PA} \vdash \operatorname{prv}(\ulcorner \varphi \urcorner) \to \operatorname{prv}(\ulcorner \psi \urcorner)$. Since $\mathsf{PA} \vdash \varphi \to \psi$, COROLLARY 9.18 yields $\mathsf{PA} \vdash \operatorname{prv}(\ulcorner \varphi \to \psi \urcorner)$. The assertion then follows from COROLLARY 10.2 using Modus Ponens.

For (b), note that by COROLLARY 10.2.(b), it suffices to prove $\mathsf{PA} \vdash \operatorname{prv}(\ulcorner \varphi \wedge \psi \urcorner) \to \operatorname{prv}(\ulcorner \varphi \urcorner) \wedge \operatorname{prv}(\ulcorner \psi \urcorner)$. But this is a direct consequence of COROLLARY 10.2.(a) using L_3 and L_4. ⊣

The Diagonalisation Lemma

We already know that every standard natural number is either **0** or the successor of a standard natural number. Hence, we can introduce a binary relation which states that x is the code of a standard natural number:

$$\begin{aligned} \mathrm{c\_nat}(c, n, x) :\Longleftrightarrow \operatorname{seq}(c) \wedge \ \operatorname{lh}(c) = \mathsf{s}n \wedge c_0 = \ulcorner 0 \urcorner \wedge \\ \forall i < n \left(c_{\mathsf{s}i} = \operatorname{succ}(c_i) \wedge c_n = x\right) \end{aligned}$$

$$\operatorname{nat}(n, x) :\Longleftrightarrow \exists c \left(\mathrm{c\_nat}(c, n, x)\right)$$

Clearly, it follows from the definition that

$$\mathsf{PA} \vdash \mathrm{c\_nat}(c, n, x) \to \mathrm{c\_nat}\big(c^\frown \langle \operatorname{succ}(x)\rangle, \mathsf{s}n, \operatorname{succ}(x)\big)\,.$$

LEMMA 10.4. *For any natural number $n \in \mathbb{N}$ we have $\mathsf{PA} \vdash \operatorname{nat}(\underline{n}, \ulcorner \underline{n} \urcorner)$. In particular, if φ is an $\mathscr{L}_{\mathsf{PA}}$-formula, then $\mathsf{PA} \vdash \operatorname{nat}(\ulcorner \varphi \urcorner, \ulcorner \ulcorner \varphi \urcorner \urcorner)$.*

Proof. We proceed by metainduction on n. For $n \equiv \mathbf{0}$, the term $\underline{\mathbf{0}}$ is the same as 0, and clearly, the code c of the singleton sequence $\langle \ulcorner 0 \urcorner \rangle$ witnesses $\mathrm{c\_nat}(c, 0, \ulcorner 0 \urcorner)$. Now, suppose that for some c and $n \in \mathbb{N}$ we have $\mathrm{c\_nat}(c, \underline{n}, \ulcorner \underline{n} \urcorner)$. Let c'' be the code for $\langle \ulcorner \underline{\mathsf{s}n} \urcorner \rangle$ and let $c' := c^\frown c''$. Notice that since $\operatorname{lh}(c) = \mathsf{s}\underline{n}$, we have $\operatorname{lh}(c') = \mathsf{ss}\underline{n}$, and by definition of c' we have $(c')_{\mathsf{s}\underline{n}} = \ulcorner \underline{\mathsf{s}n} \urcorner = \operatorname{succ}(\ulcorner \underline{n} \urcorner)$. Using the induction hypothesis and the above observation, we obtain $\mathrm{c\_nat}(c', \underline{\mathsf{s}n}, \ulcorner \underline{\mathsf{s}n} \urcorner)$ as desired. ⊣

Finally, we define the Gödel number of a number by stipulating

$$\mathrm{gn}(n) = x :\Longleftrightarrow \mathrm{nat}(n, x) \vee \big(\neg\exists y(\mathrm{nat}(n, y)) \wedge x = \mathsf{0}\big).$$

This indeed defines a function, since using the definition of the predicate seq, one can easily show that

$$\mathsf{PA} \vdash \mathrm{nat}(n, x) \wedge \mathrm{nat}(n, y) \to x = y\,.$$

In particular, by LEMMA 10.4 we have

$$\mathsf{PA} \vdash \mathrm{gn}(\ulcorner\varphi\urcorner) = \ulcorner\ulcorner\varphi\urcorner\urcorner. \qquad (*)$$

We have now assembled all the ingredients to prove the DIAGONALISATION LEMMA, which will be crucial in the proof of the FIRST INCOMPLETENESS THEOREM.

THEOREM 10.5 (DIAGONALISATION LEMMA). *Let $\varphi(\nu)$ be an $\mathscr{L}_{\mathsf{PA}}$-formula with one free variable ν which does not occur bound in φ. Then there exists an $\mathscr{L}_{\mathsf{PA}}$-sentence σ_φ such that*

$$\mathsf{PA} \vdash \sigma_\varphi \leftrightarrow \varphi(\nu/\ulcorner\sigma_\varphi\urcorner)\,, \qquad \text{i.e., } \sigma_\varphi \Leftrightarrow_{\mathsf{PA}} \varphi(\ulcorner\sigma_\varphi\urcorner).$$

Proof. We define

$$\psi(v_0) :\equiv \forall v_1\Big(\mathrm{sb\_fml}(\ulcorner v_0\urcorner, \mathrm{gn}(v_0), v_0, v_1) \to \varphi(\nu/v_1)\Big)$$

and

$$\sigma_\varphi :\equiv \psi(v_0/\ulcorner\psi\urcorner).$$

In other words, $\sigma_\varphi \equiv \psi(\ulcorner\psi\urcorner)$ and $\ulcorner\sigma_\varphi\urcorner = \ulcorner\psi(\ulcorner\psi\urcorner)\urcorner$. Since $\ulcorner v_0\urcorner = \mathsf{s0}$, we have

$$\begin{aligned}
\sigma_\varphi &\equiv \forall v_1\big(\mathrm{sb_fml}(\mathsf{s0}, \mathrm{gn}(\ulcorner\psi\urcorner), \ulcorner\psi(v_0)\urcorner, v_1) \to \varphi(\nu/v_1)\big)\\
&\Leftrightarrow_{\mathsf{PA}} \forall v_1\big(\mathrm{sb_fml}(\mathsf{s0}, \ulcorner\ulcorner\psi\urcorner\urcorner, \ulcorner\psi(v_0)\urcorner, v_1) \to \varphi(v_1)\big)\\
&\Leftrightarrow_{\mathsf{PA}} \forall v_1\big(v_1 = \ulcorner\psi(v_0/\ulcorner\psi\urcorner)\urcorner \to \varphi(v_1)\big)\\
&\Leftrightarrow_{\mathsf{PA}} \varphi(\ulcorner\psi(v_0/\ulcorner\psi\urcorner)\urcorner)\\
&\equiv \varphi(\ulcorner\psi(\ulcorner\psi\urcorner)\urcorner)\\
&\equiv \varphi(\ulcorner\sigma_\varphi\urcorner),
\end{aligned}$$

where the first equivalence follows from $(*)$, the second equivalence follows from LEMMA 9.15, and the third equivalence follows from L_{10} and L_{14}. ⊣

The DIAGONALISATION LEMMA is often called FIXPOINT LEMMA, since the sentence σ_φ can be conceived as a fixed point of φ. It is a powerful tool, since it allows us to make self-referential statements, i.e., for a formula φ with one free variable it provides a sentence σ_φ which states "I have the property φ".

The First Incompleteness Theorem

Now we are ready to prove the First Incompleteness Theorem:

Theorem 10.6 (First Incompleteness Theorem for PA). *PA is incomplete.*

Proof. By the Diagonalisation Lemma there is an $\mathscr{L}_{\mathsf{PA}}$-sentence σ such that

$$\sigma \Leftrightarrow_{\mathsf{PA}} \neg\,\mathrm{prv}(\ulcorner\sigma\urcorner).$$

To see this, let $\varphi(v_0) :\equiv \neg\,\mathrm{prv}(v_0)$. Then $\sigma_\varphi \Leftrightarrow_{\mathsf{PA}} \varphi(\ulcorner\sigma_\varphi\urcorner)$ and

$$\varphi(\ulcorner\sigma_\varphi\urcorner) \equiv \varphi(v_0/\ulcorner\sigma_\varphi\urcorner) \equiv \neg\,\mathrm{prv}(v_0/\ulcorner\sigma_\varphi\urcorner) \equiv \neg\,\mathrm{prv}(\ulcorner\sigma_\varphi\urcorner)\,.$$

Assume towards a contradiction that PA is complete. We have to consider the following two cases:

Case 1: $\mathsf{PA} \vdash \sigma$. On the one hand, by Corollary 9.18 we have $\mathsf{PA} \vdash \mathrm{prv}(\ulcorner\sigma\urcorner)$. On the other hand, since $\sigma \Leftrightarrow_{\mathsf{PA}} \neg\,\mathrm{prv}(\ulcorner\sigma\urcorner)$, we have $\mathsf{PA} \vdash \neg\,\mathrm{prv}(\ulcorner\sigma\urcorner)$. Therefore, $\mathsf{PA} \vdash \bot$, i.e., PA is inconsistent.

Case 2: $\mathsf{PA} \vdash \neg\sigma$. From

$$\neg\sigma \Leftrightarrow_{\mathsf{PA}} \neg\neg\,\mathrm{prv}(\ulcorner\sigma\urcorner) \Leftrightarrow_{\mathsf{PA}} \mathrm{prv}(\ulcorner\sigma\urcorner)$$

we obtain $\mathsf{PA} \vdash \mathrm{prv}(\ulcorner\sigma\urcorner)$. In particular, $\mathbb{N} \vDash \mathrm{prv}(\#\sigma)$, so there exists an $n \in \mathbb{N}$ with $\mathbb{N} \vDash \mathrm{c\_prv}(n, \#\sigma)$. Thus, by Lemma 9.17, n encodes a formal proof of σ, which implies $\mathsf{PA} \vdash \sigma$. Therefore, by our assumption, we have $\mathsf{PA} \vdash \bot$, i.e., PA is inconsistent.

Summing up, we have

$$\mathsf{PA} \vdash \sigma \quad \text{or} \quad \mathsf{PA} \vdash \neg\sigma \quad \Longleftrightarrow \quad \neg\,\mathrm{Con}(\mathsf{PA}),$$

which shows that PA is incomplete. ⊣

Remarks

- In the above proof of Theorem 10.6 we proved that a sentence σ with the property $\sigma \Leftrightarrow_{\mathsf{PA}} \neg\,\mathrm{prv}(\ulcorner\sigma\urcorner)$ witnesses the incompleteness of PA. In $\mathbb{N}$, however, σ is true: Note that if $\mathbb{N} \vDash \neg\sigma$, then $\mathbb{N} \vDash \mathrm{prv}(\#\sigma)$. On the other hand, Lemma 9.17 implies $\mathsf{PA} \vdash \sigma$, and hence, $\mathbb{N} \vDash \sigma$, which is a contradiction. Observe that in $\mathbb{N}$ the sentence σ expresses "I am not provable" — where the expression "provable" is meant with respect to prv — which is, of course, true.

- With respect to the model $\mathbb{N}$, we have

$$\mathbb{N} \vDash \operatorname{prv}(\ulcorner\sigma\urcorner) \iff \mathsf{PA} \vdash \sigma$$

for every $\mathscr{L}_{\mathsf{PA}}$-sentence σ. However, by the FIRST INCOMPLETENESS THEOREM we know that this is generally not the case for arbitrary models $\mathbf{M} \vDash \mathsf{PA}$.

- The existence of a sentence σ such that $\mathsf{PA} \nvdash \sigma$ as well as $\mathsf{PA} \nvdash \neg\sigma$ implies that both theories $\mathsf{PA} + \neg\sigma$ and $\mathsf{PA} + \sigma$ are consistent. Hence, by GÖDEL'S COMPLETENESS THEOREM 5.5, there are models $\mathbf{M}_{\neg\sigma}$ and $\mathbf{M}_\sigma$ for $\mathsf{PA} + \neg\sigma$ and $\mathsf{PA} + \sigma$, respectively. Notice that the models $\mathbf{M}_{\neg\sigma}$ and $\mathbf{M}_\sigma$ are not elementarily equivalent, and therefore, they are also not isomorphic.

- Let σ_0 be such that neither $\mathsf{PA} \vdash \sigma_0$ nor $\mathsf{PA} \vdash \neg\sigma_0$. Since $\mathsf{PA} \nvdash \sigma_0$, we find that the $\mathscr{L}_{\mathsf{PA}}$-theory $\mathsf{PA} + \neg\sigma_0$ is consistent. Now, by adding the sentence σ_0 as a new axiom to PA, in the same way as above we can construct an $\mathscr{L}_{\mathsf{PA}}$-sentence σ_1 such that neither $\mathsf{PA} + \neg\sigma_0 \vdash \sigma_1$ nor $\mathsf{PA} + \neg\sigma_0 \vdash \neg\sigma_1$ (see below). Proceeding this way, we see that we cannot complete PA by just adding finitely many axioms (for a stronger result see THEOREM 10.11 & 10.12).

Completeness and Incompleteness of Theories of Arithmetic

A first attempt to deal with the incompleteness phenomenon might be to replace PA with $\mathsf{T} \equiv \mathsf{PA} + \sigma$, since $\mathbb{N} \vDash \mathsf{T}$. Moreover, the gödelisation process could be done in the same way, where one would just need to code an additional axiom, namely σ. This, however, would lead to a modified provability predicate $\operatorname{prv}_{\mathsf{T}}$ which additionally allows formal proofs to be initialised by σ. One could then prove a version of the DIAGONALISATION LEMMA which allows us to define a version σ_{T} of σ with the property

$$\mathsf{T} \vdash \sigma_{\mathsf{T}} \leftrightarrow \neg \operatorname{prv}_{\mathsf{T}}(\ulcorner\sigma_{\mathsf{T}}\urcorner).$$

But then, we obtain a version of the FIRST INCOMPLETENESS THEOREM, since $\mathsf{T} \nvdash \sigma_{\mathsf{T}}$ and $\mathsf{T} \nvdash \neg\sigma_{\mathsf{T}}$. This suggests that THEOREM 10.6 can be generalised. Such a generalisation is the very goal of this section, whereby we consider both theories which are weaker and stronger than PA. We investigate how much of PA is really needed for the proof of the FIRST INCOMPLETENESS THEOREM. As we have seen, exponentiation can be expressed using addition and multiplication. Therefore, one idea might be to leave out multiplication and thus delete PA_4 and PA_5. However, the resulting theory, called **Presburger Arithmetic**, will turn out to be complete (see Chapter 12).

Robinson Arithmetic

The most critical axiom is certainly the Induction Schema PA_6, so we might consider the theory with PA_6 deleted. This is still not strong enough, but we will see that one instance of PA_6 actually suffices: **Robinson Arithmetic** RA is the axiom system consisting of PA_0-PA_5 and the additional axiom

$$\forall x\big(x = \mathsf{0} \vee \exists y(x = \mathsf{s}y)\big).$$

The language of RA is also $\mathscr{L}_{\mathsf{PA}}$, so we can express the same statements as in PA, which implies that RA must be incomplete. On the other hand, we can prove much less in RA than in PA. For example, RA is so weak that we cannot even prove $\forall x(\mathsf{0} + x = x)$.

Example 10.7. We show that $\mathsf{RA} \nvdash \forall x(\mathsf{0} + x = x)$, which implies that we cannot prove within RA that addition is commutative. In order to achieve this, we provide a model $\mathbf{M}$ of RA in which $\forall x(\mathsf{0}+x = x)$ is false. The domain of the model is $M = \mathbb{N} \cup \{a, b\}$, where a and b are two distinct objects which do not belong to $\mathbb{N}$. Furthermore, let $\bar{a} \equiv b$ and $\bar{b} \equiv a$. Then we can interpret $\mathsf{0}^{\mathbf{M}}$ by $\mathbf{0}$ and define the functions $\mathsf{s}^{\mathbf{M}}$, $+^{\mathbf{M}}$, and $\cdot^{\mathbf{M}}$ as follows:

$$\mathsf{s}^{\mathbf{M}}(x) \equiv \begin{cases} \mathsf{s}^{\mathbb{N}}(x) & x \in \mathbb{N} \\ x & x \in \{a, b\} \end{cases}$$

$$x +^{\mathbf{M}} y \equiv \begin{cases} x +^{\mathbb{N}} y & x, y \in \mathbb{N} \\ x & y \in \mathbb{N} \text{ and } x \notin \mathbb{N} \\ \bar{y} & y \notin \mathbb{N} \end{cases}$$

$$x \cdot^{\mathbf{M}} y \equiv \begin{cases} x \cdot^{\mathbb{N}} y & x, y \in \mathbb{N} \\ y & y \in \{0, a, b\} \\ \bar{x} & y \neq 0 \text{ and } x \in \{a, b\} \end{cases}$$

It is easy to check that $\mathbf{M}$ is a model of RA, and that $\mathbf{0}+^{\mathbf{M}}b \equiv a \not\equiv b \equiv b+^{\mathbf{M}}\mathbf{0}$.

Note that N_0–N_5 in Proposition 9.1 are also provable in RA, since the proof uses metainduction rather than induction in PA and the only non-trivial argument uses LEMMA 8.6, which can easily be seen to hold in RA.

$\mathbb{N}$-*Conformity Revisited*

In what follows, we prove that all relations and functions that were introduced in Chapters 8 and 9 are $\mathbb{N}$-conform. For this purpose, we prove that each such relation and function can be defined both by an $\exists$-formula and a $\forall$-formula. The representations with an $\exists$-formula are already given, and by the proof of COROLLARY 9.4.(b), functions defined by an $\exists$-formula always have an equivalent definition by a $\forall$-formula. The only relations whose representation by a $\forall$-formula is non-trivial, are term, fml, as well as all relations used to formalise substitution and formal proofs. Note that if we are able to show that the existential quantifiers in term and fml can be replaced by bounded existential quantifiers, then the same can be achieved for all subsequent relations.

LEMMA 10.8. *If ψ is a formula of the form $\psi \equiv \exists c(\mathrm{seq}(c) \wedge \varphi(c))$ for some Δ-formula φ, and if there is a term τ whose variables are among $\mathrm{free}(\psi)$ such that*

$$\mathsf{PA} \vdash \mathrm{seq}(c) \wedge \varphi(c) \rightarrow \big(\mathrm{lh}(c) < \tau \wedge \forall i < \mathrm{lh}(c)(c_i < \tau)\big),$$

then ψ is a Δ-formula as well.

Proof. We go once more through the proof of THEOREM 9.10 and show that the quantifier $\exists c$ can be replaced by a bounded quantifier. For this purpose, suppose that $F(i)$ is a function defined by a Δ-formula. Let $F'(i) = \mathrm{op}(\tau, i)+1$ and $m = \max_{i<\tau} F'(i)$. Moreover, note that by Exercise 9.2 we can define factorials in PA; so, let $y := m!$. Furthermore, put $G(j) = 1 + (j+1)y$. By LEMMA 8.14, we have that $G(i)$ and $G(j)$ are coprime for all $i, j < m$. Now, LEMMA 9.9 allows us to pick x with $\chi(x)$, where

$$\chi(x) \equiv \forall j < m\Big(G(j) \mid x \leftrightarrow \exists i < \tau\big(j = \mathrm{op}(\tau, i)\big)\Big).$$

We check that if $F(i) < \tau$ for every $i < \tau$ then we can find an upper bound τ' whose variables coincide with the variables of τ such that there is $c < \tau'$ with $\beta(c, i) = F(i)$ for all $i < \tau$. If this can be accomplished, then we have

$$\psi \Leftrightarrow_{\mathsf{PA}} \exists c < \tau'(\mathrm{seq}(c) \wedge \varphi(c)).$$

To see this, suppose that $\mathrm{seq}(c) \wedge \varphi(c)$ with $c \geq \tau'$. Now take $F(i) := \beta(c, i) < \tau$. By our assumption, there is $c' < \tau' \leq c$ with $\beta(c', i) = F(i) = \beta(c, i)$ for all $i < \tau$. Moreover, note that $\mathrm{lh}(c') = \beta(c', 0) = \beta(c, 0) = \mathrm{lh}(c)$ and $\mathrm{lh}(c') = F(0) < \tau$, and hence $c'_i = c_i$ for all $i < \mathrm{lh}(c)$, contradicting $\mathrm{seq}(c)$.

It remains to find τ'. Note that we clearly have $m \leq \tau_1$ with $\tau_1 \equiv \mathrm{op}(\tau, \tau)+1$ and hence $y \leq \tau_1!$. Furthermore, we have $G(j) < 1+(\tau_1+1)!$ for each $j < m$. Therefore, since $G(i)$ and $G(j)$ are coprime for all $i, j < m$, we can find x which satisfies $\chi(x)$ such that $x < \tau_2$ with $\tau_2 \equiv (1+(\tau_1+1)!)^{\tau_1}$. In particular, there is $c = \mathrm{op}(x, y)$ with $\mathrm{seq}(c) \wedge \varphi(c)$ and $c < \mathrm{op}(\tau_1, \tau_2)$. ⊣

LEMMA 10.9. *The relations* term *and* fml *are $\mathbb{N}$-conform.*

Proof. We want to apply LEMMA 10.8 to the defining formulae of term and formula, respectively. Since both cases are similar, we only consider term. We prove that $\exists c\big(\mathrm{c\_term}(c, t)\big)$ is equivalent to the formula

$$\varphi(t) \equiv \exists c(\mathrm{c\_term}(c, t) \wedge \forall i < \mathrm{lh}(c) \forall j < i(c_j < c_i)).$$

Then LEMMA 10.8 for $\tau \equiv t+1$ concludes the proof. We proceed by strong induction on $\mathrm{lh}(c)$. If $\mathrm{lh}(c) = 1$, then there is nothing to prove. Suppose now that $\mathrm{term}(t) \to \varphi(t)$ holds for all $t' < t$ and assume $\mathrm{c\_term}(c, t)$. If $t = 0$ or $\mathrm{var}(t)$, then $\mathrm{c\_term}(\langle t \rangle, t)$ and hence $\varphi(t)$ holds. Therefore, we have either $t = \mathrm{succ}(c_i)$ or $t = \mathrm{add}(c_i, c_j)$ or $t = \mathrm{mult}(c_i, c_j)$ for $i, j < \mathrm{lh}(c)$. We only focus on the first case, since the others can be handled in the same way. Note that by Exercise 9.7 we can restrict c to $\langle c_j \mid j \leq i \rangle$, which we denote by $c \restriction \mathsf{s}c_i$. Clearly, $c_i < c$ and $\mathrm{c\_term}(c \restriction \mathsf{s}i, c_i)$. Hence, by our induction hypothesis, there is d with $\mathrm{c\_term}(d, c_i)$ and $d_k < d_j$ for all $j < \mathrm{lh}(d)$ and $k < j$. But then, $d^\frown \langle t \rangle$ witnesses $\varphi(t)$. ⊣

Let us now turn back to RA: LEMMA 10.9 implies that if $n \in \mathbb{N}$ is a natural number which is not the Gödel number of a term or formula, then

$$\mathsf{RA} \vdash \neg\,\mathrm{term}(\underline{n}) \qquad \text{and} \qquad \mathsf{RA} \vdash \neg\,\mathrm{fml}(\underline{n})\,.$$

Moreover, since the relation c_prv is also a Δ-formula, we have

$$\mathsf{RA} \vdash \neg\,\mathrm{c\_prv}(\underline{n}, \ulcorner \sigma \urcorner)$$

whenever n does not encode a formal proof of σ. However, the existential quantifier in the definition of the provability relation prv cannot be bounded, since otherwise, $\mathsf{RA} \nvdash \sigma$ would imply $\mathsf{RA} \vdash \neg\,\mathrm{prv}(\ulcorner \sigma \urcorner)$, which contradicts the incompleteness of RA.

Generalising the First Incompleteness Theorem

There are two ways to generalise the FIRST INCOMPLETENESS THEOREM: Firstly, one can modify the underlying language, and secondly, one can use a different axiom system. If the language satisfies $\mathscr{L} \supseteq \mathscr{L}_{\mathsf{PA}}$ and we have $\mathbb{N}$-conformity for all relevant relations, then, as we shall see, the proof can easily be transferred to the new setting. However, there are two issues at stake, namely the gödelisation of the language and the gödelisation of the axioms. The coding of terms, formulae and proofs can then be realised in the same way as in Chapter 9.

A language $\mathscr{L} \supseteq \mathscr{L}_{\mathsf{PA}}$ is said to be **gödelisable** if it is countable. Note that if $\mathscr{L}$ is gödelisable, then its constant symbols, relation and function symbols admit a Gödel coding as described in Chapter 9. A theory T in some gödelisable language $\mathscr{L} \supseteq \mathscr{L}_{\mathsf{PA}}$ is **gödelisable**, if there is a Δ-formula ax_{T} in the language $\mathscr{L}_{\mathsf{PA}}$ with the property that

$$\mathbb{N} \vDash \mathrm{ax}_{\mathsf{T}}(\#\varphi) \quad \text{if and only if} \quad \varphi \in \mathsf{T}\,,$$

where $\#\varphi$ is the Gödel code of φ. As in the case of PA, we introduce Gödel codes on the formal level by stipulating $\ulcorner \varphi \urcorner :\equiv \underline{\#\varphi}$. Note that if T is gödelisable and satisfies N_0–N_5, then by COROLLARY 9.4, every Δ-formula

φ in the language $\mathscr{L}_{\mathsf{PA}}$ is $\mathbb{N}$-conform. In particular, by LEMMA 10.9 it is possible to define Δ-formulae $\mathrm{term}_{\mathsf{T}}$ and $\mathrm{fml}_{\mathsf{T}}$ such that

$$\mathbb{N} \vDash \mathrm{term}_{\mathsf{T}}(n) \iff n \equiv \#\tau \text{ for some } \mathscr{L}\text{-term } \tau\,,$$
$$\mathbb{N} \vDash \mathrm{fml}_{\mathsf{T}}(n) \iff n \equiv \#\varphi \text{ for some } \mathscr{L}\text{-formula } \varphi\,.$$

Moreover, by the gödelisability of T, the axioms can be coded by some Δ-formula ax_{T}. One can then proceed to define a Δ-formula $\mathrm{c\_prv}_{\mathsf{T}}$ and an $\exists$-formula $\mathrm{prv}_{\mathsf{T}}$ such that

$$\mathbb{N} \vDash \mathrm{c\_prv}_{\mathsf{T}}(n, \#\varphi) \iff n \text{ codes a formal proof of } \varphi\,,$$
$$\mathbb{N} \vDash \mathrm{prv}_{\mathsf{T}}(\#\varphi) \iff \mathsf{T} \vdash \varphi$$

for every $n \in \mathbb{N}$ and $\mathscr{L}$-formula φ. Notice that it is crucial that $\mathrm{c\_prv}_{\mathsf{T}}$ and $\mathrm{prv}_{\mathsf{T}}$ are $\mathscr{L}_{\mathsf{PA}}$-formulae, since otherwise we would have to specify how to interpret them in the standard model $\mathbb{N}$. Moreover, using COROLLARY 9.4, we obtain

$$\mathsf{P}_0: \qquad \mathbb{N} \vDash \mathrm{c\_prv}_{\mathsf{T}}(n, \#\varphi) \implies \mathsf{T} \vdash \mathrm{c\_prv}_{\mathsf{T}}(\underline{n}, \ulcorner\varphi\urcorner)\,,$$
$$\mathsf{P}_1: \qquad \mathbb{N} \vDash \neg\,\mathrm{c\_prv}_{\mathsf{T}}(n, \#\varphi) \implies \mathsf{T} \vdash \neg\,\mathrm{c\_prv}_{\mathsf{T}}(\underline{n}, \ulcorner\varphi\urcorner)\,.$$

In the following, we present two proofs of the FIRST INCOMPLETENESS THEOREM for gödelisable theories $\mathsf{T} \supseteq \mathsf{RA}$. The restriction to extensions of RA ensures that N_0–N_5, and hence also COROLLARY 9.4, hold.

Gödel's original proof uses the assumption of a slightly stronger property than just consistency: An $\mathscr{L}_{\mathsf{PA}}$-theory T is said to be **ω-consistent** if whenever $\mathsf{T} \vdash \exists x \varphi(x)$ for some $\mathscr{L}_{\mathsf{PA}}$-formula $\varphi(x)$, then there exists $n \in \mathbb{N}$ such that $\mathsf{T} \nvdash \neg\varphi(\underline{n})$.

FACT 10.10. *If T is an $\mathscr{L}_{\mathsf{PA}}$-theory with $\mathbb{N} \vDash \mathsf{T}$, then T is ω-consistent. In particular, PA and RA are ω-consistent.*

Proof. If $\mathsf{T} \vdash \exists x \varphi(x)$, then $\mathbb{N} \vDash \exists x \varphi(x)$. Hence, there is an $n \in \mathbb{N}$ with $\mathbb{N} \vDash \varphi(n)$, which shows that $\mathsf{T} + \varphi(\underline{n})$ is consistent and implies $\mathsf{T} \nvdash \neg\varphi(\underline{n})$. ⊣

THEOREM 10.11 (FIRST INCOMPLETENESS THEOREM, GÖDEL'S VERSION). *Let $\mathsf{T} \supseteq \mathsf{RA}$ be a gödelisable $\mathscr{L}_{\mathsf{PA}}$-theory. If T is ω-consistent, then T is incomplete.*

Proof. Observe that the proof of the DIAGONALISATION LEMMA still works if we replace PA by T. Take a sentence σ such that

$$\sigma \Leftrightarrow_{\mathsf{PA}} \neg\,\mathrm{prv}_{\mathsf{T}}(\ulcorner\sigma\urcorner).$$

Assume towards a contradiction that T is complete. Then we have that either $\mathsf{T} \vdash \sigma$ or $\mathsf{T} \vdash \neg\sigma$.

Case 1: $\mathsf{T} \vdash \sigma$. In this case, the argument is the same as in THEOREM 10.6.

Case 2: $\mathsf{T} \vdash \neg\sigma$. Since we can encode the proof of $\neg\sigma$ within T, we have

$$\mathsf{T} \vdash \mathrm{prv}_{\mathsf{T}}(\ulcorner \neg\sigma \urcorner)\,.$$

On the other hand, we have $\neg\sigma \Leftrightarrow_{\mathsf{T}} \neg\neg\,\mathrm{prv}_{\mathsf{T}}(\ulcorner \sigma \urcorner) \Leftrightarrow_{\mathsf{T}} \mathrm{prv}_{\mathsf{T}}(\ulcorner \sigma \urcorner)$, and therefore we also have

$$\mathsf{T} \vdash \mathrm{prv}_{\mathsf{T}}(\ulcorner \sigma \urcorner)\,.$$

So, by COROLLARY 10.2, we have $\mathsf{T} \vdash \mathrm{prv}_{\mathsf{T}}(\ulcorner \sigma \wedge \neg\sigma \urcorner)$, and by the ω-consistency of T, there is an $n \in \mathbb{N}$ such that

$$\mathsf{T} \nvdash \neg\,\mathrm{c\_prv}_{\mathsf{T}}(\underline{n}, \ulcorner \sigma \wedge \neg\sigma \urcorner)\,.$$

However, since T is consistent, we have $\mathsf{T} \nvdash \sigma \wedge \neg\sigma$, which implies

$$\mathbb{N} \vDash \neg\,\mathrm{c\_prv}_{\mathsf{T}}(n, \#(\sigma \wedge \neg\sigma))\,,$$

and by P_1 we obtain

$$\mathsf{T} \vdash \neg\,\mathrm{c\_prv}_{\mathsf{T}}(\underline{n}, \ulcorner \sigma \wedge \neg\sigma \urcorner)\,,$$

which is obviously a contradiction. ⊣

In [46], Rosser showed how to get rid of this dependency on ω-consistency by slightly modifying the provability predicate:

$$\begin{aligned} \mathrm{c\_prv}^{\mathrm{R}}_{\mathsf{T}}(c, x) &:\Longleftrightarrow \mathrm{c\_prv}_{\mathsf{T}}(c, x) \wedge \neg\exists c' < c(\mathrm{c\_prv}_{\mathsf{T}}(c', \mathrm{not}(x))) \\ \mathrm{prv}^{\mathrm{R}}_{\mathsf{T}}(x) &:\Longleftrightarrow \exists c(\mathrm{c\_prv}^{\mathrm{R}}_{\mathsf{T}}(c, x)) \end{aligned}$$

THEOREM 10.12 (FIRST INCOMPLETENESS THEOREM, USING ROSSER'S TRICK). *Let $\mathscr{L} \supseteq \mathscr{L}_{\mathsf{PA}}$ be a gödelisable language and let T be a gödelisable $\mathscr{L}$-theory. If T is consistent, then it is incomplete.*

Proof. As before, we apply the DIAGONALISATION LEMMA, this time to the formula $\neg\,\mathrm{prv}^{\mathrm{R}}(x)$. Thus we obtain an $\mathscr{L}$-sentence σ with

$$\sigma \Leftrightarrow_{\mathsf{PA}} \neg\,\mathrm{prv}^{\mathrm{R}}(\ulcorner \sigma \urcorner).$$

Again, we prove that neither σ nor $\neg\sigma$ is provable from T. Observe first that our assumption on σ implies

$$\sigma \Leftrightarrow_{\mathsf{PA}} \forall c(\mathrm{c\_prv}(c, \ulcorner \sigma \urcorner) \to \exists c' < c\,(\mathrm{c\_prv}(c', \ulcorner \neg\sigma \urcorner)))$$

since $\mathrm{not}(\ulcorner \sigma \urcorner) \equiv \ulcorner \neg\sigma \urcorner$. Assume towards a contradiction that T is complete. As before, we have two cases:

Case 1: $\mathsf{T} \vdash \sigma$. By P_0, there is an $n \in \mathbb{N}$ such that

$$\mathsf{T} \vdash \mathrm{c\_prv}_{\mathsf{T}}(\underline{n}, \ulcorner \sigma \urcorner)\,,$$

and by our above observation we have

$$\mathsf{T} \vdash \exists c' < \underline{n}(\text{c\_prv}_{\mathsf{T}}(c', \ulcorner \neg\sigma \urcorner)) .$$

Since T satisfies N_5, this means that there exists $k < n$ in $\mathbb{N}$ such that

$$\mathsf{T} \vdash \text{c\_prv}_{\mathsf{T}}(\underline{k}, \ulcorner \neg\sigma \urcorner) ,$$

and therefore, there is an $m \in \mathbb{N}$ with

$$\mathsf{T} \vdash \text{c\_prv}_{\mathsf{T}}(\underline{m}, \ulcorner \sigma \wedge \neg\sigma \urcorner) .$$

Hence, by $\mathbb{N}$-conformity of $\text{c\_prv}_{\mathsf{T}}$, we have

$$\mathbb{N} \vDash \text{c\_prv}_{\mathsf{T}}(m, \#(\sigma \wedge \neg\sigma)) ,$$

which implies

$$\mathsf{T} \vdash \sigma \wedge \neg\sigma.$$

This contradicts our assumption that T is consistent.

Case 2: $\mathsf{T} \vdash \neg\sigma$. In this case, there is a $c' \in \mathbb{N}$ such that

$$\mathsf{T} \vdash \text{c\_prv}_{\mathsf{T}}(\underline{c'}, \ulcorner \neg\sigma \urcorner) .$$

On the other hand, we have $\mathsf{T} \vdash \text{prv}^{\text{R}}_{\mathsf{T}}(\ulcorner \sigma \urcorner)$, and hence, there is a $c \in \mathbb{N}$ with

$$\mathsf{T} \vdash \text{c\_prv}^{\text{R}}_{\mathsf{T}}(\underline{c}, \ulcorner \sigma \urcorner) .$$

By definition of $\text{c\_prv}^{\text{R}}_{\mathsf{T}}$, we must have $c \leq c'$. Now, we can use N_5 to reach the same contradiction as in the first case. ⊣

Tarski's Theorem

The DIAGONALISATION LEMMA allows us to make self-referential statements such as the Gödel sentence which formalises to some extent the sentence "This sentence is not provable". Recall that we call an $\mathscr{L}_{\mathsf{PA}}$-sentence φ **true** in $\mathbb{N}$, if $\mathbb{N} \vDash \varphi$. Is it possible to express truth in the standard model $\mathbb{N}$ by a formula, i.e., is there a formula $\text{truth}(x)$ with one free variable x such that for every $\mathscr{L}_{\mathsf{PA}}$-sentence φ,

$$\mathbb{N} \vDash \text{truth}(\#\varphi) \iff \mathbb{N} \vDash \varphi \,?$$

Or equivalently, is there a formula $\text{truth}(x)$ such that for every $\mathscr{L}_{\mathsf{PA}}$-sentence φ,

$$\mathbb{N} \vDash \text{truth}(\#\varphi) \leftrightarrow \varphi \,?$$

Using the DIAGONALISATION LEMMA, we provide a negative answer to this question.

THEOREM 10.13 (TARSKI'S THEOREM). *There is no* $\mathscr{L}_{\mathsf{PA}}$*-formula* truth$(x)$ *with one free variable* x *such that* $\mathbb{N} \vDash \text{truth}(\#\varphi) \leftrightarrow \varphi$.

Proof. Assume towards a contradiction that such a formula truth exists. By the DIAGONALISATION LEMMA there exists an $\mathscr{L}_{\mathsf{PA}}$-sentence σ such that

$$\mathsf{PA} \vdash \sigma \leftrightarrow \neg\,\text{truth}(\ulcorner\sigma\urcorner)\,.$$

But then

$$\begin{aligned}\mathbb{N} \vDash \text{truth}(\#\sigma) \quad &\Longleftrightarrow \quad \mathbb{N} \vDash \sigma \\ &\Longleftrightarrow \quad \mathbb{N} \vDash \neg\,\text{truth}(\#\sigma),\end{aligned}$$

which is impossible. ⊣

Note that we have just solved the so-called **Liar Paradox** concerned with the sentence

"This sentence is false.",

which is obviously true if and only if it is false. Clearly, the above sentence corresponds to the sentence σ in the proof of TARSKI'S THEOREM. In order to express it in PA, one would need to be able to define truth in $\mathbb{N}$, which is impossible by TARSKI'S THEOREM.

NOTES

The FIRST INCOMPLETENESS THEOREM was first proven by Gödel [16] in 1931. Rather than using Peano Arithmetic in first-order logic, as we did, he based his proof on Type Theory in the system of Principia Mathematica [56] introduced by Russell and Whitehead. Gödel's original proof makes use of the stronger assumption of ω-consistency, which Rosser [46] showed to be negligible. The observation that all proof steps of the FIRST INCOMPLETENESS THEOREM can in fact be carried out in Robinson Arithmetic was made by Robinson [45] in 1950. Although TARSKI'S THEOREM is usually attributed to Tarski and was first published by him in [54], Gödel already mentioned this result in 1931 in a letter to Bernays; previously he had been trying to come up with a definition of a truth predicate (see [36]). Usually, gödelisable theories are called *recursive*, which means that there exists an algorithm terminating after finitely many steps that can decide whether $\varphi \in \mathsf{T}$ or $\varphi \notin \mathsf{T}$. More generally, a property $P(n)$ of natural numbers is said to be recursive, if there is an algorithm which decides in finitely many steps whether a given number n has the property P (i.e., whether or not $P(n)$ holds). With the so-called *Recursion Theory*, one can analyse the strength of various theories of Arithmetic very precisely.

EXERCISES

10.0 Prove $\mathsf{PA} \vdash \forall x(\text{term}(\text{gn}(x)))$.

10.1 Let φ and ψ be $\mathscr{L}_{\mathsf{PA}}$-formulae.

(a) Show that $\mathsf{PA} \vdash \text{prv}(\ulcorner\varphi\urcorner) \vee \text{prv}(\ulcorner\psi\urcorner) \to \text{prv}(\ulcorner\varphi \vee \psi\urcorner)$.

(b) Does the converse also hold?

10.2 A theory T with signature $\mathscr{L}_{\mathsf{PA}}$ is said to be **ω-incomplete** if it holds that $\mathsf{T} \vdash \varphi(\underline{n})$ for every $n \in \mathbb{N}$ but $\mathsf{T} \nvdash \forall x \varphi$.

(a) Show that PA is ω-incomplete.

(b) Show that every ω-complete $\mathscr{L}_{\mathsf{PA}}$-theory has an extension which is consistent but ω-inconsistent.

10.3 Let $\varphi(x, y)$ and $\psi(x, y)$ be $\mathscr{L}_{\mathsf{PA}}$-formulae with at most two free variables. Show that there are $\mathscr{L}_{\mathsf{PA}}$-sentences such that

$$\sigma \Leftrightarrow_{\mathsf{PA}} \varphi(\ulcorner \sigma \urcorner, \ulcorner \tau \urcorner) \quad \text{and} \quad \tau \Leftrightarrow_{\mathsf{PA}} \psi(\ulcorner \sigma \urcorner, \ulcorner \tau \urcorner).$$

Note that this is a generalisation of the DIAGONALISATION LEMMA.

10.4 A famous paradox, denoted as *Curry's Paradox*, states informally: "If this sentence is true, then $0 = 1$ holds". Explain why this is contradictory and formalise the paradox in PA. Use this to give an alternative proof of Gödel's version of the FIRST INCOMPLETENESS THEOREM.

Chapter 11
The Second Incompleteness Theorem

It follows from Gödel's COMPLETENESS THEOREM that a theory is consistent if and only if it has a model. In particular, the consistency of Peano Arithmetic follows from $\mathbb{N} \vDash \mathsf{PA}$. With the help of the provability relation prv, we are even able to express consistency of an arithmetical theory on the formal level, i.e., we can introduce a sentence $\mathrm{con}_{\mathsf{PA}}$ which expresses in $\mathbb{N}$ the consistency of PA. The SECOND INCOMPLETENESS THEOREM which we shall prove in this chapter states that $\mathsf{PA} \nvdash \mathrm{con}_{\mathsf{PA}}$, i.e., PA cannot prove its own consistency.

Outline of the Proof

Recall that a theory is consistent, if it cannot prove contradictions. In the case of PA, a simple contradiction is the sentence $\mathsf{0} = \mathsf{1}$. Thus, we have

$$\mathrm{Con}(\mathsf{PA}) \iff \mathsf{PA} \nvdash \mathsf{0} = \mathsf{1}\,.$$

As a formalised version of this statement, we define the $\mathscr{L}_{\mathsf{PA}}$-sentence $\mathrm{con}_{\mathsf{PA}}$ by stipulating

$$\mathrm{con}_{\mathsf{PA}} :\iff \neg\,\mathrm{prv}(\ulcorner \mathsf{0} = \mathsf{1} \urcorner).$$

Since $\mathbb{N} \vDash \mathrm{prv}(\ulcorner \varphi \urcorner)$ if and only if $\mathsf{PA} \vdash \varphi$, the consistency of PA implies that $\mathbb{N} \vDash \mathrm{con}_{\mathsf{PA}}$. In particular, this shows that $\mathsf{PA} \nvdash \neg\,\mathrm{con}_{\mathsf{PA}}$. The SECOND INCOMPLETENESS THEOREM states that $\mathrm{con}_{\mathsf{PA}}$ is independent of the axioms of PA.

L. Halbeisen, R. Krapf, *Gödel's Theorems and Zermelo's Axioms*,
https://doi.org/10.1007/978-3-030-52279-7_11

THEOREM 11.1 (SECOND INCOMPLETENESS THEOREM). $\mathsf{PA} \nvdash \mathrm{con}_{\mathsf{PA}}$.

As a matter of fact, we would like to mention that by the COMPLETENESS THEOREM, $\mathsf{PA} \nvdash \mathrm{con}_{\mathsf{PA}}$ implies that there exists a model $\mathbf{M} \vDash \mathsf{PA}$ in which $\mathrm{con}_{\mathsf{PA}}$ fails (i.e., $\mathbf{M} \vDash \neg \mathrm{con}_{\mathsf{PA}}$), which shows that with respect to $\mathbf{M}$, the sentence $\mathrm{con}_{\mathsf{PA}}$ is *not* equivalent to the statement Con(PA).

The proof of the SECOND INCOMPLETENESS THEOREM hinges on the following properties of the provability predicate, also called the *Hilbert-Bernays-Löb derivability conditions*, which state that for every $\mathscr{L}_{\mathsf{PA}}$-formula φ the following conditions hold:

D_0: If $\mathsf{PA} \vdash \varphi$ then $\mathsf{PA} \vdash \mathrm{prv}(\ulcorner \varphi \urcorner)$,

D_1: $\mathsf{PA} \vdash \mathrm{prv}(\ulcorner \varphi \to \psi \urcorner) \to (\mathrm{prv}(\ulcorner \varphi \urcorner) \to \mathrm{prv}(\ulcorner \psi \urcorner))$,

D_2: $\mathsf{PA} \vdash \mathrm{prv}(\ulcorner \varphi \urcorner) \to \mathrm{prv}(\ulcorner \mathrm{prv}(\ulcorner \varphi \urcorner) \urcorner)$.

Note that D_0 follows from COROLLARY 9.18 and D_1 is exactly the statement of Corollary 10.2.(a). Assuming D_0–D_2, the proof of the SECOND INCOMPLETENESS THEOREM becomes quite simple:

Proof of Theorem 11.1. Assume towards a contradiction that $\mathsf{PA} \vdash \mathrm{con}_{\mathsf{PA}}$, in other words, assume that $\mathsf{PA} \vdash \neg \mathrm{prv}(\ulcorner 0 = 1 \urcorner)$. Using the DIAGONALISATION LEMMA we can find an $\mathscr{L}_{\mathsf{PA}}$-sentence σ such that

$$\sigma \Leftrightarrow_{\mathsf{PA}} \neg \mathrm{prv}(\ulcorner \sigma \urcorner).$$

Now, observe that by Corollary 10.3 we have

$$\mathrm{prv}(\ulcorner 0 = 1 \urcorner) \;\Leftrightarrow_{\mathsf{PA}}\; \mathrm{prv}(\ulcorner \sigma \wedge \neg\sigma \urcorner) \;\Leftrightarrow_{\mathsf{PA}}\; \mathrm{prv}(\ulcorner \sigma \urcorner) \wedge \mathrm{prv}(\ulcorner \neg\sigma \urcorner).$$

Another application of COROLLARY 10.3 yields

$$\mathrm{prv}(\ulcorner \neg\sigma \urcorner) \Leftrightarrow_{\mathsf{PA}} \mathrm{prv}(\ulcorner \mathrm{prv}(\ulcorner \sigma \urcorner) \urcorner),$$

and therefore we have

$$\mathrm{prv}(\ulcorner 0 = 1 \urcorner) \Leftrightarrow_{\mathsf{PA}} \mathrm{prv}(\ulcorner \sigma \urcorner) \wedge \mathrm{prv}(\ulcorner \mathrm{prv}(\ulcorner \sigma \urcorner) \urcorner).$$

Furthermore, by D_2 we have $\mathsf{PA} \vdash \mathrm{prv}(\ulcorner \varphi \urcorner) \to \mathrm{prv}(\ulcorner \mathrm{prv}(\ulcorner \varphi \urcorner) \urcorner)$, and hence, by TAUTOLOGY (D.2) we obtain

$$\mathsf{PA} \vdash \mathrm{prv}(\ulcorner \sigma \urcorner) \to \big(\mathrm{prv}(\ulcorner \sigma \urcorner) \wedge \mathrm{prv}(\ulcorner \mathrm{prv}(\ulcorner \sigma \urcorner) \urcorner)\big),$$

and by L_3, this implies

$$\mathrm{prv}(\ulcorner \sigma \urcorner) \wedge \mathrm{prv}(\ulcorner \mathrm{prv}(\ulcorner \sigma \urcorner) \urcorner) \Leftrightarrow_{\mathsf{PA}} \mathrm{prv}(\ulcorner \sigma \urcorner).$$

Therefore, we obtain

$$\mathrm{prv}(\ulcorner 0 = 1 \urcorner) \Leftrightarrow_{\mathsf{PA}} \mathrm{prv}(\ulcorner \sigma \urcorner),$$

and consequently, we have

$$\mathrm{con}_{\mathsf{PA}} \Leftrightarrow_{\mathsf{PA}} \neg\,\mathrm{prv}(\ulcorner\sigma\urcorner) \Leftrightarrow_{\mathsf{PA}} \sigma,$$

which is a contradtiction to THEOREM 10.6 which states that $\mathsf{PA} \nvdash \sigma$. ⊣

Proving the Derivability Condition D_2

In order to complete our proof of Gödel's SECOND INCOMPLETENESS THEOREM, it remains to prove D_2. At a first glance, it looks very similar to the statement D_0. There is, however, a subtle difference between the two statements: While the implication in D_0 is just a meta-implication, i.e., an implication in the meta-logic, the implication in D_2 is a formal one. Note that it follows from D_0 that

$$\text{if } \mathsf{PA} \vdash \mathrm{prv}(\ulcorner\varphi\urcorner) \text{ then } \mathsf{PA} \vdash \mathrm{prv}(\ulcorner\mathrm{prv}(\ulcorner\varphi\urcorner)\urcorner),$$

which, however, is weaker than D_2. A first attempt would be to try to prove

$$\mathsf{PA} \vdash \alpha \rightarrow \mathrm{prv}(\ulcorner\alpha\urcorner)$$

for every $\mathscr{L}_{\mathsf{PA}}$-formula α. However, this is false in general, as the following example shows:

Example 11.2. Let σ denote the formula from the proof of the FIRST INCOMPLETENESS THEOREM, i.e., σ satisfies

$$\sigma \Leftrightarrow_{\mathsf{PA}} \neg\,\mathrm{prv}(\ulcorner\sigma\urcorner).$$

As a consequence of the proof of FIRST INCOMPLETENESS THEOREM, we have $\mathbb{N} \vDash \sigma$ but $\mathsf{PA} \nvdash \sigma$. Now, if $\mathsf{PA} \vdash \sigma \rightarrow \mathrm{prv}(\ulcorner\sigma\urcorner)$, then $\mathbb{N} \vDash \sigma \rightarrow \mathrm{prv}(\ulcorner\sigma\urcorner)$ and hence $\mathbb{N} \vDash \mathrm{prv}(\ulcorner\sigma\urcorner)$. By construction of the provability predicate, this would imply $\mathsf{PA} \vdash \sigma$, which is not the case, as we have seen.

This means that we have to slightly modify our approach. For this purpose, recall that we proved in PROPOSITION 9.3.(a) that every $\exists$-sentence which is true in the standard model $\mathbb{N}$ has a formal proof in PA. If we can transfer this result to PA, this would mean that we have

D_3: $\mathsf{PA} \vdash \alpha \rightarrow \mathrm{prv}(\ulcorner\alpha\urcorner)$ for every $\exists$-sentence α.

Clearly, once we have established D_3 we obtain D_2 by taking α to be the $\exists$-sentence $\mathrm{prv}(\ulcorner\varphi\urcorner)$ for some $\mathscr{L}_{\mathsf{PA}}$-formula φ. The most natural way to prove D_3 is by induction on the construction of the $\exists$-sentence α. This, however, turns out to be problematic, since in the formula construction of α there are also subformulae which are not sentences:

Example 11.3. Assume that we can prove $\mathsf{PA} \vdash \alpha \to \mathrm{prv}(\ulcorner\alpha\urcorner)$ for some $\mathscr{L}_{\mathsf{PA}}$-formula $\alpha \equiv (v_0 = v_1)$, i.e.,

$$\mathsf{PA} \vdash v_0 = v_1 \to \mathrm{prv}(\ulcorner v_0 = v_1 \urcorner).$$

Observe that $\mathrm{prv}(\ulcorner v_0 = v_1 \urcorner)$ does not contain any free variables. Now, since v_0 and v_1 are free variables in α, we obtain by substitution $\mathsf{PA} \vdash \mathsf{0} = \mathsf{0} \to \mathrm{prv}(\ulcorner v_0 = v_1 \urcorner)$. Therefore, using Modus Ponens, we get $\mathsf{PA} \vdash \mathrm{prv}(\ulcorner v_0 = v_1 \urcorner)$, which is clearly false in the standard model $\mathbb{N}$, since there is no formal proof of $v_0 = v_1$.

This problem can be solved by slightly modifying our provability predicate in such a way that free variables are permitted. Thus, we first want to adjust our Gödel coding such that (some) free variables can be preserved. The way to do this is by defining for some set V of variables

$$\lceil \nu \rceil_V :\equiv \begin{cases} \nu & \text{if } \nu \in V, \\ \ulcorner \nu \urcorner & \text{otherwise.} \end{cases}$$

Roughly speaking, variables $\nu \in V$ remain variables and all other variables become natural numbers, namely $\ulcorner \nu \urcorner$. Now, as in the case of gödelisation, we can inductively extend this definition to terms by stipulating:

$$\begin{aligned} \lceil \mathsf{0} \rceil_V &:\equiv \ulcorner \mathsf{0} \urcorner \\ \lceil \mathsf{s}\tau \rceil_V &:\equiv \mathrm{succ}(\ulcorner \mathsf{s} \urcorner, \lceil \tau \rceil_V) \\ \lceil \tau_1 + \tau_2 \rceil_V &:\equiv \mathrm{add}(\ulcorner + \urcorner, \lceil \tau_1 \rceil_V, \lceil \tau_2 \rceil_V) \\ \lceil \tau_1 \cdot \tau_2 \rceil_V &:\equiv \mathrm{add}(\ulcorner \cdot \urcorner, \lceil \tau_1 \rceil_V, \lceil \tau_2 \rceil_V) \end{aligned}$$

For formulae, one proceeds similarly. The only noteworthy cases are those of quantification:

$$\begin{aligned} \lceil \exists \nu \varphi \rceil_V &:\equiv \mathrm{ex}(\ulcorner \exists \urcorner \ulcorner \nu \urcorner, \lceil \varphi \rceil_{V \setminus \{\nu\}}), \\ \lceil \forall \nu \varphi \rceil_V &:\equiv \mathrm{ex}(\ulcorner \exists \urcorner \ulcorner \nu \urcorner, \lceil \varphi \rceil_{V \setminus \{\nu\}}). \end{aligned}$$

Thus, the set V contains all variables which remain free in $\lceil \varphi \rceil_V$. In particular, if $V \cap \mathrm{free}(\varphi)$ is the empty-set, then $\lceil \varphi \rceil_V$ is the same as $\ulcorner \varphi \urcorner$. The other special case is when V contains all indices of free variables in φ. In that case, we write $\lceil \varphi \rceil$ for $\lceil \varphi \rceil_V$ and say that $\lceil \varphi \rceil$ is the **pseudo-code** of φ.

Pseudo-coding is intended to mimic the usual process of Gödel coding. Hence, we will often substitute the free variables ν of $\lceil \varphi \rceil$ by the term $\mathrm{gn}(\nu)$ with free variable ν. For terms τ and formulae φ whose free variables are among $\{x_1, \ldots, x_n\}$, we will henceforth use the notation

$$\begin{aligned} \lceil \tau \rceil_V^{\mathrm{gn}} &:\equiv \lceil \tau \rceil_V \big(x_1/\,\mathrm{gn}(x_1), \ldots, x_n/\,\mathrm{gn}(x_n)\big), \\ \lceil \varphi \rceil_V^{\mathrm{gn}} &:\equiv \lceil \varphi \rceil_V \big(x_1/\,\mathrm{gn}(x_1), \ldots, x_n/\,\mathrm{gn}(x_n)\big). \end{aligned}$$

Note that $\lceil\varphi\rceil_V^{\text{gn}}$ has the same free variables as $\lceil\varphi\rceil_V$. Recall that for natural numbers $n \in \mathbb{N}$, LEMMA 10.4 implies $\mathsf{PA} \vdash \ulcorner\underline{n}\urcorner = \text{gn}(\underline{n})$. In particular, if we substitute each variable x_i by some natural number m_i, then $\ulcorner\varphi\urcorner$ and $\lceil\varphi\rceil^{\text{gn}}$ coincide, i.e.,

$$\ulcorner\varphi(x_1/\underline{m_1},\ldots,x_n/\underline{m_n})\urcorner \quad \text{is equal to} \quad \lceil\varphi\rceil^{\text{gn}}(x_1/\underline{m_1},\ldots,x_n/\underline{m_n})\,.$$

To see this, notice that on the left hand side, the variable x_i is first replaced by $\underline{m_i}$, and when computing $\ulcorner\varphi\urcorner$, $\underline{m_i}$ is replaced by $\ulcorner\underline{m_i}\urcorner$. On the right hand side, when computing $\lceil\varphi\rceil^{\text{gn}}$, the variable x_i is replaced by $\text{gn}(x_i)$, and then x_i — which is a free variable in $\text{gn}(x_i)$ — is replaced by $\underline{m_i}$. Thus, on the left hand side, x_i is replaced by $\ulcorner\underline{m_i}\urcorner$, and on the right hand side, x_i is replaced by $\text{gn}(\underline{m_i})$, and as mentioned above, $\ulcorner\underline{m_i}\urcorner$ is equal to $\text{gn}(\underline{m_i})$. A slightly stronger result is given by the following

FACT 11.4. *For terms τ and formulae φ whose free variables are among $\{x_1,\ldots,x_n\}$, we have*

$$\mathsf{PA} \vdash \ulcorner\tau(x_1/\underline{m_1},\ldots,x_n/\underline{m_n})\urcorner = \lceil\tau\rceil^{\text{gn}}(x_1/\underline{m_1},\ldots,x_n/\underline{m_n})\,,$$
$$\mathsf{PA} \vdash \ulcorner\varphi(x_1/\underline{m_1},\ldots,x_n/\underline{m_n})\urcorner = \lceil\varphi\rceil^{\text{gn}}(x_1/\underline{m_1},\ldots,x_n/\underline{m_n})\,.$$

The proof left as an exercise to the reader (see EXERCISE 11.1).

Our next goal is to prove the following

THEOREM 11.5. *If φ is an $\exists$-formula, then*

$$\mathsf{PA} \vdash \varphi \to \text{prv}(\lceil\varphi\rceil^{\text{gn}}). \tag{$*$}$$

Notice that for $\exists$-sentences φ, THEOREM 11.5 implies D_3. In order to prove THEOREM 11.5, we first need some auxiliary results whose proofs turn out to be quite technical. The following lemma essentially states that removing a variable x from V amounts to substituting in $\lceil\varphi\rceil_{V\setminus\{x\}}^{\text{gn}}$ each occurrence of $\ulcorner x\urcorner$ by the term $\text{gn}(x)$, thus obtaining $\lceil\varphi\rceil_V^{\text{gn}}$. While this seems to be completely obvious, its proof is highly non-trivial, since it requires us to unravel all the details of the formalised substitution function.

LEMMA 11.6. *Let V be a finite set of variables, let $x \in V$, and suppose that τ is an $\mathscr{L}_{\mathsf{PA}}$-term and that φ is an $\mathscr{L}_{\mathsf{PA}}$-formula with $x \in \text{free}(\varphi)$. Then we have:*

(a) $\mathsf{PA} \vdash \text{sb\_term}(\ulcorner x\urcorner, \text{gn}(x), \lceil\tau\rceil_{V\setminus\{x\}}^{\text{gn}}, \lceil\tau\rceil_V^{\text{gn}})$
(b) $\mathsf{PA} \vdash \text{sb\_fml}(\ulcorner x\urcorner, \text{gn}(x), \lceil\varphi\rceil_{V\setminus\{x\}}^{\text{gn}}, \lceil\varphi\rceil_V^{\text{gn}})$

Proof. We give a detailed proof of (a). Note that (b) is very similar, and since the proof is quite lengthy, we omit the proof of (b). A complete proof of both statements, in a slightly different context, is given in [53, Lem. 7.4–Lem. 7.6].

In order to prove (a), we proceed by induction on the construction of τ. The case when $\tau \equiv 0$ is trivial, since in that case, τ does not have any free variables. The other atomic case is when $\tau \equiv y$ is a variable. In that case, we need to distinguish between three possibilities: either $y \equiv x$ or, if $y \not\equiv x$, then either $y \in V$ or $y \notin V$.

Case 1. If $y \equiv x$, then $\lceil y \rceil_{V\setminus\{x\}} = \lceil y \rceil_{V\setminus\{y\}} = \ulcorner y \urcorner$, and, since $x \in V$, $\lceil y \rceil_V = y$, which implies $\lceil y \rceil_V^{\mathrm{gn}} = \mathrm{gn}(y)$. Then the claim follows, since by EXERCISE 10.0 we have $\mathsf{PA} \vdash \mathrm{sb\_term}(\ulcorner y \urcorner, \mathrm{gn}(y), \ulcorner y \urcorner, \mathrm{gn}(y))$.

Case 2. If $y \not\equiv x$ and $y \in V$, then $\lceil y \rceil_{V\setminus\{x\}} = \lceil y \rceil_V = y$ and therefore $\lceil y \rceil^{\mathrm{gn}}_{V\setminus\{x\}} = \lceil y \rceil^{\mathrm{gn}}_V = \mathrm{gn}(y)$. Since the variable x does not appear in $\tau \equiv y$, there is nothing to substitute. More precisely, we have

$$\mathsf{PA} \vdash \mathrm{sb\_term}(\ulcorner x \urcorner, \mathrm{gn}(x), \mathrm{gn}(y), \mathrm{gn}(y))$$

as desired.

Case 3. Suppose that $y \not\equiv x$ and $y \notin V$. Then $\lceil y \rceil_{V\setminus\{x\}} = \lceil y \rceil_V = \ulcorner y \urcorner$, and therefore $\lceil y \rceil^{\mathrm{gn}}_{V\setminus\{x\}} = \lceil y \rceil^{\mathrm{gn}}_V = \ulcorner y \urcorner$, and since $\ulcorner y \urcorner$ does not have any free variables, the claim trivially holds.

Let us now consider the cases when τ is not atomic. Suppose that $\tau \equiv \mathsf{s}\tau'$ for some term τ'. Clearly, all variables in τ' are also among V. By our inductive assumption, we have

$$\mathsf{PA} \vdash \mathrm{sb\_term}(\ulcorner x \urcorner, \mathrm{gn}(x), \lceil \tau' \rceil^{\mathrm{gn}}_{V\setminus\{x\}}, \lceil \tau' \rceil^{\mathrm{gn}}_V).$$

Note that by definition of sb_term we have in general

$$\mathsf{PA} \vdash \mathrm{sb\_term}(v, t_0, t, t') \to \mathrm{sb\_term}(v, t_0, \mathrm{succ}(t), \mathrm{succ}(t')).$$

In our case, this means that if we set

$$v :\equiv \ulcorner x \urcorner, \quad t_0 :\equiv \mathrm{gn}(x)\,, \quad t :\equiv \lceil \tau' \rceil^{\mathrm{gn}}_{V\setminus\{x\}}\,, \quad t' :\equiv \lceil \tau' \rceil^{\mathrm{gn}}_V$$

in the above formula, then we have $\lceil \tau \rceil^{\mathrm{gn}}_{V\setminus\{x\}} \equiv \mathrm{succ}(t)$ and $\lceil \tau \rceil^{\mathrm{gn}}_V \equiv \mathrm{succ}(t')$, and therefore, τ satisfies (a).

The cases when $\tau \equiv \tau_1 + \tau_2$ or $\tau \equiv \tau_1 \cdot \tau_2$ are shown similarly. ⊣

THEOREM 11.7. *For every $\mathscr{L}_{\mathsf{PA}}$-formula φ, we have:*

$$\mathsf{PA} \vdash \mathrm{prv}(\ulcorner \varphi \urcorner) \to \mathrm{prv}(\lceil \varphi \rceil^{\mathrm{gn}})$$

Note that for sentences φ, THEOREM 11.7 becomes trivial. On the other hand, if φ has free variables, then the statement still seems obvious, since it should not matter whether the free variables are gödelized at the same time as φ — as in the case of $\ulcorner \varphi \urcorner$ — or whether one gödelizes the formula such that the variables remain free, and afterwards substitutes the Gödel code of the variables — as in the case of $\lceil \varphi \rceil^{\mathrm{gn}}$. However, the proof is trickier than it might be expected, since one needs to use the properties of the formalised substitution function, which is, unfortunately, a very complicated function.

Proof of Theorem 11.7. First, observe that $\mathsf{PA} \vdash \forall x\, \mathrm{term}(\mathrm{gn}(x))$. This follows from an easy inductive argument: Firstly, since $\mathrm{gn}(\mathsf{0}) = \mathsf{0}$, it is clear that $\mathsf{PA} \vdash \mathrm{term}(\mathrm{gn}(\mathsf{0}))$. Now, if we inductively assume $\mathrm{c\_term}(c, \mathrm{gn}(x))$, then we also get $\mathrm{c\_term}\big(c^\frown\langle \mathrm{succ}(x)\rangle, \mathrm{gn}(\mathsf{s}x)\big)$, since $\mathrm{gn}(\mathsf{s}x) = \mathrm{succ}(x)$ by LEMMA 10.4. Moreover, by induction on x we can also show

$$\mathsf{PA} \vdash \forall v \forall x \neg\, \mathrm{var\_in\_term}(v, \mathrm{gn}(x)),$$

which proves that the formalized substitution $v/\mathrm{gn}(x)$ is always admissible.

Note that if we have $\mathsf{PA} \vdash \varphi(\nu)$ for some variable ν, then we have $\mathsf{PA} \vdash \varphi(\nu/\tau)$ whenever τ is a term such that the substitution ν/τ is admissible:

φ_0:	$\varphi(\nu)$	by assumption
φ_1:	$\forall\nu\varphi(\nu)$	from φ_0 using $(\forall)$
φ_2:	$\forall\nu\varphi(\nu) \to \varphi(\nu/\tau)$	instance of L_{10}
φ_3:	$\varphi(\tau)$	from φ_2 and φ_1 using (MP)

Now, if we transfer this proof to the formalised level, this implies that

$$\mathsf{PA} \vdash \mathrm{c\_prv}(c, f) \wedge \mathrm{sb\_fml}(v, t, f, f') \to \mathrm{c\_prv}(c', f'),$$

where

$$c' :\equiv c^\frown\Big\langle\, \mathrm{all}(v, f),\ \mathrm{imp}\big(\mathrm{all}(v, f), f'\big),\ \mathrm{mp}\Big(\mathrm{all}(v, f), \mathrm{imp}\big(\mathrm{all}(v, f), f'\big)\Big)\Big\rangle.$$

In particular, this yields

$$\mathsf{PA} \vdash \mathrm{prv}(f) \wedge \mathrm{sb\_fml}(v, t, f, f') \to \mathrm{prv}(f').$$

Now, let φ be an arbitrary $\mathscr{L}_{\mathsf{PA}}$-formula. We may assume that all free variables of φ are among $v_0, \ldots, v_n$ for some $n \in \mathbb{N}$. Using the above observation together with LEMMA 11.6, we obtain that $\mathsf{PA} \vdash \mathrm{prv}(\ulcorner\varphi\urcorner) \to \mathrm{prv}(\lceil\varphi\rceil^{\mathrm{gn}}_{\{v_0\}})$, and for each $k \in \{\mathbf{1}, \ldots, n\}$,

$$\mathsf{PA} \vdash \mathrm{prv}(\lceil\varphi\rceil^{\mathrm{gn}}_{\{v_0,\ldots,v_{k-1}\}}) \to \mathrm{prv}(\lceil\varphi\rceil^{\mathrm{gn}}_{\{v_0,\ldots,v_k\}}).$$

After F I N I T E L Y many applications of TAUTOLOGY (D.0), we obtain

$$\mathsf{PA} \vdash \mathrm{prv}(\ulcorner\varphi\urcorner) \to \mathrm{prv}\big(\lceil\varphi\rceil^{\mathrm{gn}}_{\{v_0,\ldots,v_n\}}\big),$$

and since $\lceil\varphi\rceil^{\mathrm{gn}}_{\{v_0,\ldots,v_n\}} \equiv \lceil\varphi\rceil^{\mathrm{gn}}$, this completes the proof. ⊣

The following results are easy consequences of THEOREM 11.7.

COROLLARY 11.8. *Let φ be an $\mathscr{L}_{\mathsf{PA}}$-formula. If $\mathsf{PA} \vdash \varphi$ then $\mathsf{PA} \vdash \mathrm{prv}(\lceil\varphi\rceil^{\mathrm{gn}})$.*

Proof. Note that from $\mathsf{PA} \vdash \varphi$ and D_0 we obtain $\mathsf{PA} \vdash \mathrm{prv}(\ulcorner\varphi\urcorner)$, and by THEOREM 11.7 we have $\mathsf{PA} \vdash \mathrm{prv}(\lceil\varphi\rceil^{\mathrm{gn}})$. ⊣

COROLLARY 11.9. *Let φ and ψ be arbitrary $\mathscr{L}_{\mathsf{PA}}$-formulae. Then $\mathsf{PA} \vdash \varphi \to \psi$ implies $\mathsf{PA} \vdash \mathrm{prv}(\lceil\varphi\rceil^{\mathrm{gn}}) \to \mathrm{prv}(\lceil\psi\rceil^{\mathrm{gn}})$. In particular, if $\varphi \Leftrightarrow_{\mathsf{PA}} \psi$ then $\mathrm{prv}(\lceil\varphi\rceil^{\mathrm{gn}}) \Leftrightarrow_{\mathsf{PA}} \mathrm{prv}(\lceil\psi\rceil^{\mathrm{gn}})$.*

Proof. Suppose that $\mathsf{PA} \vdash \varphi \to \psi$. An application of COROLLARY 11.8 yields

$$\mathsf{PA} \vdash \mathrm{prv}(\lceil\varphi \to \psi\rceil^{\mathrm{gn}}).$$

Recall that $\lceil\varphi \to \psi\rceil$ equals $\mathrm{imp}(\lceil\varphi\rceil, \lceil\psi\rceil)$, and therefore, $\lceil\varphi \to \psi\rceil^{\mathrm{gn}}$ equals $\mathrm{imp}(\lceil\varphi\rceil^{\mathrm{gn}}, \lceil\psi\rceil^{\mathrm{gn}})$. Moreover, by definition of formalised Modus Ponens we have

$$\mathsf{PA} \vdash \mathrm{mp}\big(\lceil\varphi\rceil^{\mathrm{gn}}, \lceil\varphi \to \psi\rceil^{\mathrm{gn}}, \lceil\psi\rceil^{\mathrm{gn}}\big)\,.$$

Hence, by LEMMA 10.1.(a) we obtain

$$\mathsf{PA} \vdash \big(\mathrm{prv}(\lceil\varphi\rceil^{\mathrm{gn}}) \wedge \mathrm{prv}(\lceil\varphi \to \psi\rceil^{\mathrm{gn}})\big) \to \mathrm{prv}(\lceil\psi\rceil^{\mathrm{gn}})\,,$$

which completes the proof. ⊣

Now we have assembled all ingredients for the proof of THEOREM 11.5:

Proof of Theorem 11.5. We have to show that for every $\exists$-formula φ_0,

$$\mathsf{PA} \vdash \varphi_0 \to \mathrm{prv}(\lceil\varphi_0\rceil^{\mathrm{gn}}). \qquad (*)$$

Notice that by COROLLARY 11.9 it suffices to check $(*)$ for strict $\exists$-formulae φ_0, i.e., for formulae built up from atomic formulae and negated atomic formulae using $\wedge$, $\vee$, existential quantification $\exists\nu$ and bounded universal quantification $\forall\nu < \tau$ (for some term τ). We proceed by induction on the construction of formulae φ_0.

We start with atomic formulae. Since $\mathscr{L}_{\mathsf{PA}}$ does not contain relation symbols, we only have to consider atomic formulae of the form $\tau_i = \tau_j$ for some $\mathscr{L}_{\mathsf{PA}}$-terms τ_i and τ_j. Moreover, by substitution it suffices to show that each atomic formula of the form $v_i = v_j$, where v_i and v_j are variables, satisfies $(*)$. For example, if $\varphi_0 \equiv \mathsf{s0} + v_0 = \mathsf{s0} \cdot \mathsf{s0}$, then, for $\varphi \equiv v_1 = v_2$, we have $\varphi_0 \equiv \varphi\big(v_1/\mathsf{s0} + v_0, v_2/\mathsf{s0} \cdot \mathsf{s0}\big)$. Therefore, let us consider the formula $v_i = v_j$. First, note that we obviously have $\mathsf{PA} \vdash v_i = v_i$, and hence, by COROLLARY 11.8 we have

$$\mathsf{PA} \vdash \mathrm{prv}(\lceil v_i = v_i\rceil^{\mathrm{gn}})\,.$$

Furthermore, since v_i and v_j are free variables in $\mathrm{prv}(\lceil v_i = v_j\rceil^{\mathrm{gn}})$, we can use EXERCISE 2.3 to obtain

$$\mathsf{PA} \vdash \big(v_i = v_i \wedge v_i = v_j\big) \to \big(\mathrm{prv}(\lceil v_i = v_i\rceil^{\mathrm{gn}}) \to \mathrm{prv}(\lceil v_i = v_j\rceil^{\mathrm{gn}})\big)\,.$$

Putting these facts together and using logical axioms, tautologies and twice Modus Ponens, we obtain the following formal proof:

$$
\begin{aligned}
\mathsf{PA} + v_i = v_j \; &\vdash v_i = v_i \\
&\vdash v_i = v_j \\
&\vdash v_i = v_i \wedge v_i = v_j \\
&\vdash \big(v_i = v_i \wedge v_i = v_j\big) \to \big(\mathrm{prv}(\ulcorner v_i = v_i \urcorner^{\mathrm{gn}}) \to \mathrm{prv}(\ulcorner v_i = v_j \urcorner^{\mathrm{gn}})\big) \\
&\vdash \mathrm{prv}(\ulcorner v_i = v_i \urcorner^{\mathrm{gn}}) \to \mathrm{prv}(\ulcorner v_i = v_j \urcorner^{\mathrm{gn}}) \\
&\vdash \mathrm{prv}(\ulcorner v_i = v_i \urcorner^{\mathrm{gn}}) \\
&\vdash \mathrm{prv}(\ulcorner v_i = v_j \urcorner^{\mathrm{gn}})
\end{aligned}
$$

Therefore, by the DEDUCTION THEOREM we obtain

$$\mathsf{PA} \vdash v_i = v_j \to \mathrm{prv}(\ulcorner v_i = v_j \urcorner^{\mathrm{gn}})$$

as desired.

For negated atomic formulae, we only have to show that each formula of the form $v_i \neq v_j$ satisfies $(*)$. Now, we obviously have

$$v_i \neq v_j \;\Leftrightarrow_{\mathsf{PA}}\; (v_i < v_j) \vee (v_j < v_i)\,,$$

where

$$(v_i < v_j) \vee (v_j < v_i) \;\Leftrightarrow_{\mathsf{PA}}\; \exists v_k\big((v_k < v_j \wedge v_k = v_i) \vee (v_k < v_i \wedge v_k = v_j)\big)\,.$$

Hence, the case of negated atomic formulae follows from the fact that formulae of the form $\exists v_i \varphi$ satisfy $(*)$, which will be shown below.

Suppose now that φ satisfies $(*)$. We have to verify that $\varphi(v_i/\tau)$ (where the substitution v_i/τ is admissible), and that $\exists v_i \varphi$ and $\forall v_i < v_j\, \varphi$ satisfy $(*)$.

- Suppose that $v_i \in \mathrm{free}(\varphi)$ and that τ is an $\mathscr{L}_{\mathsf{PA}}$-term such that the substitution v_i/τ is admissible. We have to show that $\varphi(v_i/\tau)$ satisfies $(*)$. For the sake of simplicity, we assume that v_i is the only free variable of φ. By assumption we have $\mathsf{PA} \vdash \varphi \to \mathrm{prv}(\ulcorner \varphi \urcorner^{\mathrm{gn}})$. Now, using Generalisation we obtain

  $$\mathsf{PA} \vdash \forall v_i\, \big(\varphi \to \mathrm{prv}(\ulcorner \varphi \urcorner^{\mathrm{gn}})\big)\,,$$

  and hence, by L_{10} and Modus Ponens we get

  $$\mathsf{PA} \vdash \varphi(\tau) \to \mathrm{prv}(\ulcorner \varphi \urcorner^{\mathrm{gn}})(v_i/\tau).$$

  Thus, it is enough to verify that $\mathsf{PA} \vdash \ulcorner \varphi \urcorner^{\mathrm{gn}}(v_i/\tau) = \ulcorner \varphi(\tau) \urcorner^{\mathrm{gn}}$. For this, we first prove that $\mathsf{PA} \vdash \mathrm{gn}(\tau) = \ulcorner \tau \urcorner^{\mathrm{gn}}$ by induction on the construction of the term τ: If $\tau \equiv \mathsf{0}$ then $\mathsf{PA} \vdash \ulcorner \mathsf{0} \urcorner^{\mathrm{gn}} = \lceil \mathsf{0} \rceil = \ulcorner \mathsf{0} \urcorner = \mathsf{0} = \mathrm{gn}(\mathsf{0})$. The case when τ is a variable is similar. We now verify our claim for $\tau \equiv \mathsf{s}\tau'$ and leave the other cases as an exercise to the reader. By induction, we may assume that $\mathsf{PA} \vdash \mathrm{gn}(\tau') = \ulcorner \tau' \urcorner^{\mathrm{gn}}$. Then

$$\begin{aligned}\mathsf{PA} &\vdash \ulcorner \mathbf{s}\tau' \urcorner^{\mathrm{gn}} = \mathrm{succ}(\ulcorner \tau' \urcorner^{\mathrm{gn}})\\ &\vdash \mathrm{succ}(\ulcorner \tau' \urcorner^{\mathrm{gn}}) = \mathrm{succ}(\mathrm{gn}(\tau'))\\ &\vdash \mathrm{succ}(\mathrm{gn}(\tau')) = \mathrm{gn}(\mathbf{s}\tau'),\end{aligned}$$

and by transitivity of the relation $=$ we have $\mathsf{PA} \vdash \ulcorner \mathbf{s}\tau' \urcorner^{\mathrm{gn}} = \mathrm{gn}(\mathbf{s}\tau')$ as desired.

As a consequence of $\mathsf{PA} \vdash \mathrm{gn}(\tau) = \ulcorner \tau \urcorner^{\mathrm{gn}}$, we obtain

$$\begin{aligned}\mathsf{PA} &\vdash \ulcorner \varphi \urcorner^{\mathrm{gn}}(v_i/\tau) = \ulcorner \varphi \urcorner(v_i/\,\mathrm{gn}(v_i))(v_i/\tau)\\ &\vdash \ulcorner \varphi \urcorner(v_i/\,\mathrm{gn}(v_i))(v_i/\tau) = \ulcorner \varphi \urcorner(v_i/\,\mathrm{gn}(\tau))\\ &\vdash \ulcorner \varphi \urcorner(v_i/\,\mathrm{gn}(\tau)) = \ulcorner \varphi \urcorner(v_i/\ulcorner \tau \urcorner^{\mathrm{gn}})\\ &\vdash \ulcorner \varphi \urcorner(v_i/\ulcorner \tau \urcorner^{\mathrm{gn}}) = \ulcorner \varphi(\tau) \urcorner^{\mathrm{gn}},\end{aligned}$$

and by transitivity of $=$ we obtain $\mathsf{PA} \vdash \ulcorner \varphi \urcorner^{\mathrm{gn}}(v_i/\tau) = \ulcorner \varphi(\tau) \urcorner^{\mathrm{gn}}$ as desired.

- Under the assumption

$$\mathsf{PA} \vdash \varphi \to \mathrm{prv}(\ulcorner \varphi \urcorner^{\mathrm{gn}})$$

we show that $\forall v_i < v_j\, \varphi$ (where $i \not\equiv j$) satisfies $(*)$. Let v_j' be a variable which does not occur in φ. Since

$$\forall v_i < v_j\, \varphi \;\Leftrightarrow_{\mathsf{PA}}\; \exists v_j' \left(v_j' = v_j \wedge \forall v_i < v_j'\, \varphi\right),$$

we may assume without loss of generality that v_j does not occur in φ. Furthermore, let $\psi(v_j)$ denote the formula

$$\forall v_i < v_j\, \varphi \to \mathrm{prv}(\ulcorner \forall v_i < v_j\, \varphi \urcorner^{\mathrm{gn}}).$$

It suffices to show that $\mathsf{PA} \vdash \forall v_j\, \psi(v_j)$. So, by PA_6 it is enough to show that $\mathsf{PA} \vdash \psi(\mathsf{0})$, and for all v_j, $\mathsf{PA} \vdash \psi(v_j) \to \psi(\mathbf{s}v_j)$.

Notice that, since $\forall v_i < \mathsf{0}\, \varphi$ is a tautology, by COROLLARY 11.8 we have $\mathsf{PA} \vdash \psi(\mathsf{0})$. For the induction step, assume that $\mathsf{PA} \vdash \psi(v_j)$. Recall that by LEMMA 8.6 we have

$$v_i < \mathbf{s}v_j \;\Leftrightarrow_{\mathsf{PA}}\; v_i < v_j \vee v_i = v_j\,,$$

and hence,

$$\forall v_i < \mathbf{s}v_j\, \varphi \;\Leftrightarrow_{\mathsf{PA}}\; \forall v_i < v_j\, \varphi \,\wedge\, \varphi(v_i/v_j)\,,$$

where the substitution v_i/v_j is admissible because v_j does not occur in φ. Since $\mathsf{PA} \vdash \psi(v_j)$, by LEMMA 10.1.(b) it suffices to show that

$$\mathsf{PA} \vdash \varphi(v_i/v_j) \to \mathrm{prv}(\ulcorner \varphi(v_i/v_j) \urcorner^{\mathrm{gn}},$$

which follows from the previous case, using our assumption that φ satisfies $(*)$.

- Now, we show that $\exists v_i\varphi$ satisfies $(*)$. Since by L_{11} we have $\mathsf{PA} \vdash \varphi \to \exists v_i\varphi$, we can apply COROLLARY 11.9 and obtain

$$\mathsf{PA} \vdash \text{prv}(\ulcorner\varphi\urcorner^{\text{gn}}) \to \text{prv}(\ulcorner\exists v_i\varphi\urcorner^{\text{gn}}).$$

  Therefore, by TAUTOLOGY (D.0) we have

$$\mathsf{PA} \vdash \varphi \to \text{prv}(\ulcorner\exists v_i\varphi\urcorner^{\text{gn}}),$$

  and by Generalisation we obtain

$$\mathsf{PA} \vdash \forall v_i\,\big(\varphi \to \text{prv}(\ulcorner\exists v_i\varphi\urcorner^{\text{gn}})\big).$$

  Now, since v_i does not occur as a free variable in $\text{prv}(\ulcorner\exists v_i\varphi\urcorner^{\text{gn}})$, by L_{13} and Modus Ponens we finally obtain

$$\mathsf{PA} \vdash \exists v_i\varphi \to \text{prv}(\ulcorner\exists v_i\varphi\urcorner^{\text{gn}})$$

  as desired.

Finally, suppose that φ and ψ both satisfy $(*)$. We have to show that $\varphi \wedge \psi$ and $\varphi \vee \psi$ also satisfy $(*)$.

- In order to see that $\varphi \wedge \psi$ satisfies $(*)$, notice first that

$$\text{and}\big(\ulcorner\varphi\urcorner^{\text{gn}}, \ulcorner\psi\urcorner^{\text{gn}}\big) = \ulcorner\varphi \wedge \psi\urcorner^{\text{gn}}.$$

  Therefore, by LEMMA 10.1 we have

$$\mathsf{PA} \vdash \text{prv}\big(\ulcorner\varphi\urcorner^{\text{gn}}\big) \wedge \text{prv}\big(\ulcorner\psi\urcorner^{\text{gn}}\big) \to \text{prv}\big(\ulcorner\varphi \wedge \psi\urcorner^{\text{gn}}\big).$$

  Using our assumption that φ and ψ both satisfy $(*)$, it follows that $\varphi \wedge \psi$ also satisfies $(*)$.
- The case $\varphi \vee \psi$ is similar and thus left as an exercise to the reader.

$\dashv$

Concluding Remarks

To summarise, we have shown that $\mathsf{PA} \nvdash \text{con}_{\mathsf{PA}}$, where

$$\text{con}_{\mathsf{PA}} \equiv \neg\,\text{prv}(\ulcorner 0 = 1\urcorner).$$

In other words, we have shown that

$$\mathsf{PA} \nvdash \neg\,\text{prv}(\ulcorner 0 = 1\urcorner).$$

Now, by the COMPLETENESS THEOREM we know that there exists a model $\mathbf{M} \vDash \mathsf{PA}$ such that

$$\mathbf{M} \vDash \text{prv}(\ulcorner 0 = 1\urcorner).$$

Since $\mathsf{PA} \vdash \neg(0 = 1)$, we have $\mathbf{M} \vDash \neg(0 = 1)$, which shows that

$$\mathbf{M} \vDash \neg(0 = 1) \wedge \operatorname{prv}(\ulcorner 0 = 1 \urcorner).$$

A slightly more general result can be obtained from LÖB'S THEOREM.

Löb's Theorem

Recall that by LEMMA 9.17 we have, for every $\mathscr{L}_{\mathsf{PA}}$-formula φ, $\mathbb{N} \vDash \operatorname{prv}(\#\varphi)$ if and only if $\mathsf{PA} \vdash \varphi$. In particular, this implies that $\mathbb{N} \vDash \operatorname{prv}(\#\varphi) \to \varphi$. In other words, in the standard model $\mathbb{N}$, each "provable" formula is true, where "provable" in $\mathbb{N}$ is meant with respect to the provability predicate prv. This applies because in $\mathbb{N}$ one can retrieve the proof of φ from its code. Another consequence of $\mathbb{N} \vDash \operatorname{prv}(\#\varphi) \to \varphi$ is that for every $\mathscr{L}_{\mathsf{PA}}$-formula φ,

$$\mathbb{N} \nvDash \neg\varphi \wedge \operatorname{prv}(\#\varphi).$$

This leads to the natural question whether this is true in any model of PA. LÖB'S THEOREM gives a negative answer to this question.

THEOREM 11.10 (LÖB'S THEOREM). *Suppose that φ is an $\mathscr{L}_{\mathsf{PA}}$-sentence. Then $\mathsf{PA} \vdash \operatorname{prv}(\ulcorner\varphi\urcorner) \to \varphi$ implies $\mathsf{PA} \vdash \varphi$.*

Proof. Assume that $\mathsf{PA} \vdash \operatorname{prv}(\ulcorner\varphi\urcorner) \to \varphi$ and let σ be an $\mathscr{L}_{\mathsf{PA}}$-sentence such that

$$\sigma \Leftrightarrow_{\mathsf{PA}} \operatorname{prv}(\ulcorner\sigma\urcorner) \to \varphi.$$

In order to see that such a sentence σ exists, let $\psi(v_0) :\equiv \operatorname{prv}(v_0) \to \varphi$. Then by the DIAGONALISATION LEMMA, there exists an $\mathscr{L}_{\mathsf{PA}}$-sentence σ such that $\sigma \Leftrightarrow_{\mathsf{PA}} \psi(\ulcorner\sigma\urcorner)$.

CLAIM. $\mathsf{PA} \vdash \sigma$, *or equivalently,* $\mathsf{PA} \vdash \operatorname{prv}(\ulcorner\sigma\urcorner) \to \varphi$.

Proof of Claim. By our assumption we have $\mathsf{PA} \vdash \operatorname{prv}(\ulcorner\varphi\urcorner) \to \varphi$. Therefore, it suffices to check that $\mathsf{PA} \vdash \operatorname{prv}(\ulcorner\sigma\urcorner) \to \operatorname{prv}(\ulcorner\varphi\urcorner)$. Note that by COROLLARY 10.3.(a) we have $\operatorname{prv}(\ulcorner\sigma\urcorner) \Leftrightarrow_{\mathsf{PA}} \operatorname{prv}(\ulcorner\operatorname{prv}(\ulcorner\sigma\urcorner) \to \varphi\urcorner)$. Moreover, D_1 implies

$$\mathsf{PA} \vdash \operatorname{prv}\big(\ulcorner\operatorname{prv}(\ulcorner\sigma\urcorner) \to \varphi\urcorner\big) \to \Big(\operatorname{prv}\big(\ulcorner\operatorname{prv}(\ulcorner\sigma\urcorner)\urcorner\big) \to \operatorname{prv}(\ulcorner\varphi\urcorner)\Big).$$

Now, if we assume $\operatorname{prv}(\ulcorner\sigma\urcorner)$, then by D_2 we obtain $\operatorname{prv}(\ulcorner\operatorname{prv}(\ulcorner\sigma\urcorner)\urcorner)$, and by the preceding observations, Modus Ponens, and the DEDUCTION THEOREM, we finally obtain $\operatorname{prv}(\ulcorner\varphi\urcorner)$. $\dashv_{\text{Claim}}$

Using the above claim, we have $\mathsf{PA} \vdash \sigma$ and therefore, we get $\mathsf{PA} \vdash \operatorname{prv}(\ulcorner\sigma\urcorner)$ by D_0. So, by $\mathsf{PA} \vdash \operatorname{prv}(\ulcorner\sigma\urcorner) \to \varphi$ and Modus Ponens we get $\mathsf{PA} \vdash \varphi$ as desired. $\dashv$

LÖB'S THEOREM has some remarkable consequences. For example, if we use the DIAGONALISATION LEMMA to obtain a sentence σ such that $\sigma \Leftrightarrow_{\mathsf{PA}} \mathrm{prv}(\ulcorner\sigma\urcorner)$, then it follows that $\mathsf{PA} \vdash \sigma$. Hence, if we replace the sentence stating "I am unprovable" by "I am provable" — the so-called *truth-teller sentence* — then this does not yield an undecidable statement.

LÖB'S THEOREM also implies that if $\mathsf{PA} \nvdash \varphi$ for some $\mathscr{L}_{\mathsf{PA}}$-formula φ, then $\mathsf{PA} \nvdash \mathrm{prv}(\ulcorner\varphi\urcorner) \to \varphi$, i.e., $\mathsf{PA} \nvdash \varphi \vee \neg\,\mathrm{prv}(\ulcorner\varphi\urcorner)$. This illustrates the difference between truth and provability in non-standard models of PA: For every formula φ with $\mathsf{PA} \nvdash \varphi$, there are models $\mathbf{M}$ such that $\mathbf{M} \vDash \neg\varphi \wedge \mathrm{prv}(\ulcorner\varphi\urcorner)$. In $\mathbf{M}$, the code of the "proof" of φ is of non-standard length and does henceforth not code an actual proof of φ. In this sense, Gödel's FIRST INCOMPLETENESS THEOREM can be viewed as a special case of LÖB'S THEOREM, by taking φ to be the formula σ with $\sigma \Leftrightarrow_{\mathsf{PA}} \neg\,\mathrm{prv}(\ulcorner\sigma\urcorner)$.

NOTES

The question of whether arithmetic can be shown to be consistent was the second of Hilbert's famous list [25] of 23 open problems of mathematics. While Gödel's SECOND INCOMPLETENESS THEOREM published in [16] in 1931 gives a negative answer to Hilbert's second problem, Gentzen [13] provided in 1936 a consistency proof of PA in primitive recursive arithmetic with the additional principle of quantifier-free transfinite induction up to the ordinal number ε_0. The proof of the SECOND INCOMPLETENESS THEOREM using pseudo-coding which we presented here follows Świerczkowski [53]. However, Świerczkowski worked in the theory of hereditarily finite sets, which is equivalent to PA. His proof was actually formalised and proof-checked using the interactive theorem prover Isabelle by Paulson [40] in 2013. The derivability conditions D_0–D_2, although already used by Gödel, were first introduced by Hilbert and Bernays [27] and re-formulated in its current form by Löb [31]. In the same paper, Löb also proved his theorem as an answer to a question posed by Henkin [23] in 1952.

EXERCISES

11.0 Give an alternative proof of the SECOND INCOMPLETENESS THEOREM using LÖB'S THEOREM.

11.1 Prove FACT 11.4.

Hint: Use induction on term and formula construction, respectively.

11.2 Prove that all formulae of the form $v_i + v_j = v_k$ and $v_i \cdot v_j = v_k$ satisfy $(*)$ in THEOREM 11.5.

11.3 Prove that if φ and ψ satisfy $(*)$ in THEOREM 11.5, then so does the disjunction $\varphi \vee \psi$.

11.4 Let φ be an $\mathscr{L}_{\mathsf{PA}}$-formula.

(a) Show that LÖB'S THEOREM is provable within PA, i.e.,

$$\mathsf{PA} \vdash \mathrm{prv}(\ulcorner \mathrm{prv}(\ulcorner\varphi\urcorner) \to \varphi \urcorner) \to \mathrm{prv}(\ulcorner\varphi\urcorner).$$

Hint: Set $\psi :\equiv \mathrm{prv}(\ulcorner \mathrm{prv}(\ulcorner\varphi\urcorner) \to \varphi \urcorner) \to \mathrm{prv}(\ulcorner\varphi\urcorner)$ and prove $\mathsf{PA} \vdash \mathrm{prv}(\ulcorner\psi\urcorner) \to \psi$.

(b) Prove $\mathsf{PA} \vdash \neg\,\mathrm{prv}(\ulcorner\varphi\urcorner) \to \neg\,\mathrm{prv}(\ulcorner\neg\,\mathrm{prv}(\ulcorner\varphi\urcorner)\urcorner)$.

(c) Conclude that the SECOND INCOMPLETENESS THEOREM is provable within PA.

11.5 Use EXERCISE 11.4 to prove the following generalisation of LÖB'S THEOREM: For for all $\mathscr{L}_{\mathsf{PA}}$-formulae φ and ψ,

$$\mathsf{PA} \vdash \mathrm{prv}(\ulcorner\varphi\urcorner) \wedge \big(\mathrm{prv}(\ulcorner\psi\urcorner) \to \psi\big) \implies \mathsf{PA} \vdash \mathrm{prv}(\ulcorner\varphi\urcorner) \to \psi.$$

11.6 Prove that for every $\mathscr{L}_{\mathsf{PA}}$-formula φ,

$$\mathsf{PA} \vdash \mathrm{prv}(\ulcorner\varphi \leftrightarrow \mathrm{con}_{\mathsf{PA}}\urcorner) \to (\mathrm{prv}(\ulcorner\varphi\urcorner) \leftrightarrow \neg\,\mathrm{con}_{\mathsf{PA}}),$$

and give an interpretion of this result in the standard model.

11.7 Let $\mathrm{con}^{R}_{\mathsf{PA}}$ denote the formula $\neg\,\mathrm{prv}^{\mathrm{R}}(\ulcorner 0 = 1\urcorner)$, i.e., the formula obtained from $\mathrm{con}_{\mathsf{PA}}$ by replacing the provability predicate prv by $\mathrm{prv}^{\mathrm{R}}$. Show that $\mathsf{PA} \vdash \mathrm{con}^{R}_{\mathsf{PA}}$. Note that this implies that $\mathrm{prv}^{\mathrm{R}}$ does not satisfy either D_1 or D_2.

Chapter 12
Completeness of Presburger Arithmetic

In Chapter 10, we have seen that Peano Arithmetic PA is incomplete. Moreover, if we omit the Induction Schema and replace it by the axiom $\forall x(x = 0 \vee \exists y(x = \mathsf{s}y))$, stating that every number is either 0 or has a predecessor, then the resulting theory, called Robinson Arithmetic, is also incomplete. There are, however, other natural ways to weaken the axioms of PA: One could, for example, drop one of the function symbols + or $\cdot$ as well as the corresponding axioms. In the former case, this leads to **Skolem Arithmetic**, and in the latter case to **Presburger Arithmetic**. In this chapter, we will only consider Presburger Arithmetic, denoted by PrA.

In the proof of the First Incompleteness Theorem, we introduced the β-function which allows to express exponentiation in terms of addition and multiplication. A natural question that arises in this context is whether multiplication might already be expressible in terms of the successor function and addition. If this is the case, then we can carry out the proof of the First Incompleteness Theorem in PrA, and obtain that PrA is incomplete. However, we will prove below that PrA is complete, which implies that we cannot express multiplication in terms of addition and successors.

Basic Arithmetic in Presburger Arithmetic

As already mentioned above, the language of Presburger Arithmetic PrA is given by $\mathscr{L}_{\mathsf{PrA}} = \{0, \mathsf{s}, +, \}$, where, as in $\mathscr{L}_{\mathsf{PA}}$, 0 is a constant symbol, s is a unary function symbol, and + is a binary function symbol. The axioms of PrA are simply given by the axioms of PA except PA_4 and PA_5. More precisely, the axioms of PrA are

PA_0: $\neg\exists x(\mathsf{s}x = 0)$
PA_1: $\forall x\forall y(\mathsf{s}x = \mathsf{s}y \to x = y)$
PA_2: $\forall x(x + 0 = x)$
PA_3: $\forall x\forall y(x + \mathsf{s}y = \mathsf{s}(x + y))$

L. Halbeisen, R. Krapf, *Gödel's Theorems and Zermelo's Axioms*,
https://doi.org/10.1007/978-3-030-52279-7_12

together with the Induction Schema, i.e., if φ is an $\mathscr{L}_{\mathsf{PrA}}$-formula such that $x \in \text{free}(\varphi)$, then

PA_6: $(\varphi(0) \wedge \forall x(\varphi(x) \rightarrow \varphi(\mathsf{s}(x)))) \rightarrow \forall x \varphi(x)$.

Presburger, who first investigated PrA and proved its completeness, originally axiomatised the theory in a distinct manner: For example, he did not use the Induction Schema, but also postulated the existence of negative numbers and hence subtraction. In particular, he included the axiom $\forall x \forall y \exists z (x + z = y)$.

Clearly, in PrA one can prove all standard results of arithmetic which do not involve multiplication. In particular, we can define the relations $<$ and $\leq$ in the same way as in PA. Furthermore, as in Peano Arithmetic, we are able to define the terms $\underline{n}$ for all $n \in \mathbb{N}$, where the terms $\underline{n}$ are called *natural numbers*. Moreover, we can prove the properties N_0-N_5 stated in PROPOSITION 9.1.

In the subsequent sections, it will become clear that it is impossible to define multiplication using addition and the successor function. However, it is possible to define the multiplication with a natural number of the form $\underline{n}$ for $n \in \mathbb{N}$. Using the Recursion Principle, we define

$$\begin{aligned} \underline{\mathbf{0}} \cdot x &:\equiv 0, \text{ and} \\ \underline{n + \mathbf{1}} \cdot x &:\equiv \underline{n} \cdot x + x. \end{aligned}$$

Note that for the sake of simplicity, we usually write $\underline{n}x$ rather than $\underline{n} \cdot x$.

LEMMA 12.1. *Let $n, m \in \mathbb{N}$. Multiplication with natural numbers satisfies the associativity and distributivity laws:*

$$\begin{aligned} &\mathsf{PrA} \vdash \forall x \forall y (\underline{n}(x + y) = \underline{n}x + \underline{n}y) \\ &\mathsf{PrA} \vdash \forall x (\underline{m + n}x = \underline{m}x + \underline{n}x) \\ &\mathsf{PrA} \vdash \forall x (\underline{mn}x = \underline{m} \cdot (\underline{n}x)) \end{aligned}$$

Furthermore, multiplication with natural numbers respects the two binary relations $=$ and $<$:

$$\begin{aligned} &\mathsf{PrA} \vdash \forall x \forall y (\underline{n}x = \underline{n}y \leftrightarrow x = y) \\ &\mathsf{PrA} \vdash \forall x \forall y (\underline{n}x < \underline{n}y \leftrightarrow x < y) \end{aligned}$$

Proof. The proof is similar to the proof of PROPOSITION 9.1 and uses metainduction in $\mathbb{N}$. We only prove the first statement, since all proofs are similar. For $n = \mathbf{0}$, we obviously have $0(x + y) = 0 = 0 \cdot x + 0 \cdot y$. Suppose now that the claim holds for some $n \in \mathbb{N}$. Then

$$\begin{aligned} \underline{n + \mathbf{1}}(x + y) &= \underline{n}(x + y) + (x + y) = (\underline{n}x + \underline{n}y) + (x + y) \\ &= (\underline{n}x + x) + (\underline{n}y + y) = \underline{n + \mathbf{1}} \cdot x + \underline{n + \mathbf{1}} \cdot y, \end{aligned}$$

where the second equality follows from our induction hypothesis. ⊣

For $n \in \mathbb{N}$ with $n \geq \mathbf{2}$, we define

$$x \equiv_n y :\Longleftrightarrow \exists z \big(\underline{n}z + x = y \ \vee \ \underline{n}z + y = x \big).$$

Furthermore, we abbreviate $\neg(x \equiv_n y)$ by $x \not\equiv_n y$. Formulae of the form $x \equiv_n y$ are called **congruences**. It is straightforward to check that congruences are — on the formal level — equivalence relations, i.e.,

$$\begin{aligned} &\mathsf{PrA} \vdash \forall x (x \equiv_n x)\,, \\ &\mathsf{PrA} \vdash \forall x \forall y (x \equiv_n y \leftrightarrow y \equiv_n x)\,, \\ &\mathsf{PrA} \vdash \forall x \forall y \forall x \big(x \equiv_n y \wedge y \equiv_n z \ \rightarrow \ x \equiv_n z \big)\,. \end{aligned}$$

In fact, as the name already suggests, they define congruence relations with respect to $+$:

$$\mathsf{PrA} \vdash \forall x \forall y \forall z \big(x \equiv_n y \ \leftrightarrow \ x + z \equiv_n y + z \big)\,.$$

The following result is a version of division with remainder, where the divisor is in $\mathbb{N}$.

LEMMA 12.2. *For every natural number $n \geq \mathbf{2}$, we have*

$$\mathsf{PrA} \vdash \forall x \exists y \left(\bigvee_{k=\mathbf{0}}^{n-\mathbf{1}} \underline{n}y + \underline{k} = x \right).$$

In particular,

$$\mathsf{PrA} \vdash \forall x \left(\bigvee_{k=\mathbf{0}}^{n-\mathbf{1}} x \equiv_n \underline{k} \right).$$

Proof. We proceed by induction on x. For $x = \mathsf{0}$, the statement is trivial. Suppose that $x = \underline{n}y + \underline{k}$ for some $k < n$ and some y. Then $\mathsf{s}x = \mathsf{s}(\underline{n}y + \underline{k}) = \underline{n}y + \underline{k + \mathbf{1}}$. Now, if $k + \mathbf{1} < n$, we are done. Otherwise, $k + \mathbf{1} = n$ and hence $x = \underline{n}y + \underline{n} = \underline{n}(y + 1)$. ⊣

Quantifier Elimination

The idea for proving that PrA is complete, consists of proving, in a language extension of $\mathscr{L}_{\mathsf{PrA}}$, that every sentence is logically equivalent to a quantifier-free one. Such sentences can easily be shown to be $\mathbb{N}$-conform, and hence, they can be either proven or disproven, depending on whether they are satisfied in $\mathbb{N}$. In this section, we will prove a more general result, which we will then apply to PrA. We say that a theory $\mathbf{\Phi}$ in some language $\mathscr{L}$ **admits quantifier elimination**, if for every $\mathscr{L}$-formula φ there is a quantifier-free formula ψ such that

$$\mathbf{\Phi} \vdash \varphi \leftrightarrow \psi\,.$$

The key point is to note that in order to prove that a theory admits quantifier elimination, it suffices to check that a single existential quantifier can be eliminated:

THEOREM 12.3 (QUANTIFIER ELIMINATION THEOREM). *Let* $\mathbf{\Phi}$ *be a theory in some language* $\mathscr{L}$ *such that the following holds:*

(a) *If* φ *is an atomic* $\mathscr{L}$*-formula, then* $\neg\varphi$ *is logically equivalent to a disjunction of conjunctions of atomic* $\mathscr{L}$*-formulae.*

(b) *For every* $\mathscr{L}$*-formula* φ *of the form* $\varphi \equiv \exists\nu(\varphi_1 \wedge \ldots \wedge \varphi_n)$, *where each* φ_i *is an atomic* $\mathscr{L}$*-formula, there is a quantifier-free* $\mathscr{L}$*-formula* ψ *with* $\text{free}(\psi) = \text{free}(\varphi) \setminus \{\nu\}$ *such that* $\mathbf{\Phi} \vdash \varphi \leftrightarrow \psi$.

Then $\mathbf{\Phi}$ *admits quantifier elimination.*

Proof. The following steps show how to transform any $\mathscr{L}$-formula into an equivalent quantifier-free one using the above statements (a) and (b).

Step 1. Using THEOREM 2.13, we can transform any $\mathscr{L}$-sentence into an $\mathscr{L}$-sentence in PNF.

Step 2. Using TAUTOLOGY (R), we can eliminate all universal quantifiers, i.e., every $\mathscr{L}$-formula in PNF is equivalent to an $\mathscr{L}$-formula of the form

$$(\neg)\exists\nu_1 \ldots (\neg)\exists\nu_n\varphi,$$

where $\nu_1, \ldots, \nu_n$ are variables and φ is quantifier-free.

Step 3. Given an $\mathscr{L}$-sentence of the form $(\neg)\exists\nu_1 \ldots (\neg)\exists\nu_n\varphi$ as above, one can transform the quantifier-free part φ into DNF by the DISJUNCTIVE NORMAL FORM THEOREM, i.e., all of the conjuncts are atomic or negated atomic formulae. Moreover, using (a) we can replace each negated atomic formula by a disjunction of conjunctions of atomic formulae and can thus transform the quantifier-free part into DNF in such a way that all conjuncts are atomic.

Step 4. From Step 3 we obtain an $\mathscr{L}$-formula of the form

$$(\neg)\exists\nu_1 \ldots (\neg)\exists\nu_n\Big(\big(\varphi_{1,1} \wedge \ldots \wedge \varphi_{1,k_1}\big) \vee \cdots \vee \big(\varphi_{m,1} \wedge \cdots \wedge \varphi_{m,k_m}\big)\Big),$$

where each $\varphi_{i,j}$ is an atomic or negated atomic $\mathscr{L}$-formula. Using TAUTOLOGY (U.2) this is equivalent to

$$(\neg)\exists\nu_1 \ldots (\neg)\exists\nu_{n-1}(\neg)\big(\exists\nu_n(\varphi_{1,1} \wedge \ldots \wedge \varphi_{1,k_1}) \vee \cdots \vee \exists\nu_n(\varphi_{m,1} \wedge \cdots \wedge \varphi_{m,k_m})\big).$$

Now using (a) and (b), each of the formulae $\exists\nu_n(\varphi_{i,1} \wedge \cdots \wedge \varphi_{i,k_i})$ is equivalent to a corresponding quantifier-free $\mathscr{L}$-formula ψ_i.

Step 5. In the case that there is a negation symbol $\neg$ in front of the existential quantifier $\exists\nu_n$, we use (a) to eliminate the negation, and then return to Step 3 in order to restore the DNF.

Steps 3–5 have to be repeated F I N I T E L Y many times until no more quantifiers are left. Thus, the above described algorithm yields a quantifier-free disjunction of conjunctions of almost atomic formulae, as desired. ⊣

Note that one could simplify THEOREM 12.3 by omitting (a) and requiring instead in (b) that each φ_i is either atomic or the negation of an atomic formula.

Completeness of Presburger Arithmetic

We will now show that PrA is complete. Using the previous result, one might be tempted to first show that PrA admits quantifier elimination and then show that quantifier-free $\mathscr{L}_{\mathsf{PrA}}$-sentences can be either proven or disproven. While the second step will be verified in LEMMA 12.8, the first one is not possible: An example is the $\mathscr{L}_{\mathsf{PrA}}$-formula

$$\exists y(x = 2y),$$

stating that $x \equiv_2 0$ is not equivalent to a quantifier-free formula; another example for an $\mathscr{L}_{\mathsf{PA}}$-formula which is not equivalent to a quantifier-free formula is the formula $\exists z(x + z = y)$, stating that $x \leq y$ (see EXERCISE 12.0).

However, this problem can be overcome by extending the language $\mathscr{L}_{\mathsf{PrA}}$ to admit the binary relations $<$ and $\equiv_m$ for $m \in \mathbb{N}$: Let $\mathscr{L}_{\mathsf{PrA}^*}$ denote the language $\mathscr{L}_{\mathsf{PrA}} \cup \{<\} \cup \{\equiv_m \mid m \in \mathbb{N}\}$. By THEOREM 6.1 we get that the completeness of PrA with respect to $\mathscr{L}_{\mathsf{PrA}}$ is equivalent to the completeness of PrA with respect to the extended language $\mathscr{L}_{\mathsf{PrA}^*}$.

In the following paragraphs, we will show that PrA admits quantifier elimination with respect to $\mathscr{L}_{\mathsf{PrA}^*}$. The first step is to introduce a normal form for equations and congruences with respect to a fixed variable:

LEMMA 12.4. *Let ν be a variable. Then, for $n, m \in \mathbb{N}$, every atomic $\mathscr{L}_{\mathsf{PrA}^*}$-formula is logically equivalent to a formula of the form*

$$\underline{n}\nu + \tau = \tau', \quad \underline{n}\nu + \tau \equiv_m \tau', \quad \underline{n}\nu + \tau < \tau', \quad \tau' < \underline{n}\nu + \tau,$$

where ν does not occur in τ, τ'.

Proof. Since all cases are similar, we may assume that φ is an equation. We prove by induction on the term construction that for every $\mathscr{L}_{\mathsf{PrA}^*}$-term τ there exist $n \in \mathbb{N}$ and an $\mathscr{L}_{\mathsf{PrA}^*}$-term τ' such that $\mathsf{PrA} \vdash \tau = \underline{n}\nu + \tau'$.

- Suppose first that τ is atomic. If $\tau \equiv 0$, then obviously $\mathsf{PrA} \vdash \tau = \underline{0}\nu + 0$. If $\tau \equiv \nu$, then $\mathsf{PrA} \vdash \tau = \underline{\mathbf{1}}\nu + 0$, and if $\tau \equiv w$ for some variable $w \not\equiv \nu$, then we can set $n \equiv \mathbf{0}$ and $\tau' \equiv w$.
- Assume now that $\tau \equiv \mathsf{s}\tau'$. By induction, we may assume that $\mathsf{PrA} \vdash \tau' = \underline{n}\nu + \tau''$ for some $n \in \mathbb{N}$ and some $\mathscr{L}_{\mathsf{PrA}^*}$-term τ'' such that ν does not occur in τ''. Then $\mathsf{PrA} \vdash \tau = \mathsf{s}\tau' = \mathsf{s}(\underline{n}\nu + \tau'') = \underline{n}\nu + \mathsf{s}\tau''$ by PA_3.

- Finally, let $\tau \equiv \tau_1 + \tau_2$, where $\mathsf{PrA} \vdash \tau_1 = \underline{n}\nu + \tau_1'$ and $\mathsf{PrA} \vdash \tau_2 = \underline{m}\nu + \tau_2'$, where $n, m \in \mathbb{N}$ and τ_1', τ_2' are terms such that ν does not occur in τ_1' and τ_2'. Then $\mathsf{PrA} \vdash \tau = \tau_1 + \tau_2 = (\underline{n}\nu + \tau_1') + (\underline{m}\nu + \tau_2') = \underline{n+m}\nu + \tau'$, where $\tau' \equiv \tau_1' + \tau_2'$.

It follows that every equation is equivalent to an equation of the form $\underline{n}\nu + \tau = \underline{m}\nu + \tau'$. Without loss of generality, we may assume that $n \geq m$, and hence $\mathsf{PrA} \vdash \underline{n} \geq \underline{m}$. Therefore, by LEMMA 12.1, we have

$$\mathsf{PrA} \vdash \underline{n}\nu + \tau = \underline{m}\nu + \tau' \leftrightarrow \underline{n-m}\nu + \tau = \tau',$$

which completes the proof. ⊣

We say that an atomic $\mathscr{L}_{\mathsf{PrA}^*}$-formula φ is in **ν-normal form**, if it is of the form

$$\underline{n}\nu + \tau = \tau', \quad \underline{n}\nu + \tau \equiv_m \tau', \quad \underline{n}\nu + \tau < \tau', \quad \tau' < \underline{n}\nu + \tau\,,$$

where $n, m \in \mathbb{N}$ and ν does not occur in τ, τ'. In that case, we call the number $n \in \mathbb{N}$ the **ν-coefficient** of φ.

LEMMA 12.5. *Let $\varphi_1, \ldots, \varphi_n$ be atomic $\mathscr{L}_{\mathsf{PrA}^*}$-formulae in ν-normal form. Then $\varphi_1 \wedge \ldots \wedge \varphi_n$ is logically equivalent to a conjunction of atomic $\mathscr{L}_{\mathsf{PrA}^*}$-formulae in ν-normal form, each of whose ν-coefficient is either* $\mathbf{0}$ *or* $\mathbf{1}$.

Proof. Without loss of generality, we may assume that each φ_i has a ν-coefficient $k_i \in \mathbb{N}$ with $k_i > \mathbf{0}$. Note that it is easy to check that $x\, R\, y \Leftrightarrow_{\mathsf{PrA}} \underline{k}x\, R\, \underline{k}y$, where R is either $=$, $<$, or $>$, and $k \in \mathbb{N}$ with $k \not\equiv \mathbf{0}$; a similar property also holds for congruences (see EXERCISE 12.1). In particular, by replacing k_i by $k \equiv \mathrm{lcm}^{\mathbb{N}}(k_1, \ldots, k_n)$ (see EXERCISE 9.3), we may assume that each formula φ_i has the same ν-coefficient k. Now if $\varphi_i \equiv \underline{k}\nu + \tau_i\, R_i\, \tau_i'$, then we can replace $\underline{k}\nu$ by w and obtain

$$\varphi_1 \wedge \ldots \wedge \varphi_m \;\Leftrightarrow_{\mathsf{PrA}}\; \psi_1 \wedge \ldots \wedge \psi_n \wedge w \equiv_k \mathsf{0},$$

where φ_i is the formula $w + \tau_i \sim_i \tau_i'$. ⊣

LEMMA 12.6. *Let ν be a variable. If $\varphi_1, \ldots, \varphi_n$ are atomic $\mathscr{L}_{\mathsf{PrA}^*}$-formulae such that either φ_1 is an equation or each φ_i is a congruence, then there are atomic $\mathscr{L}_{\mathsf{PrA}^*}$-formulae $\psi_1, \ldots, \psi_n$ such that ψ_i is of the same type as φ_i, ν does not occur in $\psi_2, \ldots, \psi_n$, and $\varphi_1 \wedge \ldots \wedge \varphi_n \Leftrightarrow_{\mathsf{PrA}} \psi_1 \wedge \ldots \wedge \psi_n$.*

Proof. By induction, we may assume that $n = \mathbf{2}$. By LEMMA 12.5, we may further suppose that each φ_i is in ν-normal form with ν-coefficient $\mathbf{1}$. There are two cases:

Case 1. $\varphi_1 \equiv \nu + \tau_1 = \tau_1'$ and $\varphi_2 \equiv \nu + \tau_2\, R\, \tau_2'$ for some terms $\tau_1, \tau_1', \tau_2, \tau_2'$ in which ν does not occur and R is either $=$, $<$, $>$ or $\equiv_m$ for some $m \in \mathbb{N}$. In this case, one can show that

$$\varphi_1 \wedge \varphi_2 \;\Leftrightarrow_{\mathsf{PrA}}\; \psi_1 \wedge \psi_2,$$

where $\psi_1 \equiv \varphi_1$ and $\psi_2 \equiv (\tau_1' + \tau_2)\, R\, (\tau_1 + \tau_2')$. Indeed, suppose that $\varphi_1 \wedge \varphi_2$ holds. Then we have

$$\tau_1' + \tau_2 = (\nu + \tau_1) + \tau_2 = \tau_1 + (\nu + \tau_2) \quad \text{and} \quad \big(\tau_1 + (\nu + \tau_2)\big)\, R\, \big(\tau_1 + \tau_2'\big).$$

Hence, we have $\psi_1 \wedge \psi_2$ as desired. The converse is similar.

Case 2. φ_1 is the formula $\nu + \tau_1 \equiv_{m_1} \tau_1'$ and φ_2 is $\nu + \tau_2 \equiv_{m_2} \tau_2'$. Then by EXERCISE 12.1 and by applying LEMMA 12.5 to scale the ν-coefficients, we may suppose that $m_1 \equiv m_2$. The rest of the proof is the same as for the first case. ⊣

THEOREM 12.7. *The theory* PrA *admits quantifier elimination with respect to the language* $\mathscr{L}_{\mathsf{PrA}^*}$.

Proof. We will check that PrA satisfies the assumptions of THEOREM 12.3 with respect to the extended language $\mathscr{L}_{\mathsf{PrA}^*}$.

For the first condition, note that

$$\neg(\tau = \tau') \;\Leftrightarrow_{\mathsf{PrA}}\; \tau < \tau' \,\vee\, \tau' < \tau\,,$$
$$\neg(\tau < \tau') \;\Leftrightarrow_{\mathsf{PrA}}\; \tau = \tau' \,\vee\, \tau' < \tau\,,$$
$$\neg(\tau \equiv_m \tau') \;\Leftrightarrow_{\mathsf{PrA}}\; \bigvee_{k=\mathbf{1}}^{m-\mathbf{1}} (\tau + \underline{k} \equiv_m \tau') \quad \text{for every } m \in \mathbb{N},$$

where the last equivalence follows from LEMMA 12.2.

We now turn to the second assumption. Let $\varphi_1, \ldots, \varphi_n$ be atomic $\mathscr{L}_{\mathsf{PrA}^*}$-formulae. We have to show that $\varphi \equiv \exists\nu(\varphi_1 \wedge \ldots \wedge \varphi_n)$ is logically equivalent to a quantifier-free $\mathscr{L}_{\mathsf{PrA}^*}$-formula ψ such that $\text{free}(\psi) \equiv \text{free}(\varphi) \setminus \{\nu\}$. Due to TAUTOLOGY (T.2) and LEMMA 12.6, we may suppose that each φ_i is in ν-normal form with ν-coefficient $\mathbf{1}$. We distinguish between the following four cases:

Case 1. There is an equation φ_i among $\varphi_1, \ldots, \varphi_n$. By F I N I T E L Y many applications of LEMMA 12.6 in combination with TAUTOLOGY (T.2), it suffices to check that $\exists\nu\varphi_i$ is logically equivalent to a quantifier-free $\mathscr{L}_{\mathsf{PrA}^*}$-formula, which is the case since $\exists\nu(\nu + \tau = \tau') \Leftrightarrow_{\mathsf{PrA}} \tau = \tau' \vee \tau < \tau'$.

Case 2. Each of the formulae $\varphi_1, \ldots, \varphi_n$ is a congruence. Then by LEMMA 12.6 and TAUTOLOGY (T.2), it suffices to check that the quantifier in $\exists\nu(\nu + \tau \equiv_n \tau')$ can be eliminated. This is obviously possible, since by LEMMA 12.2, there are $k, l \in \mathbb{N}$ such that $\tau \equiv_n \underline{k}$ and $\tau' \equiv \underline{l}$, and therefore, we can choose $\nu = \underline{l + n - k}$ in order to obtain a true formula.

Case 3. All formulae among $\varphi_1, \ldots, \varphi_n$ are inequalities. Since $<$ is a linear relation, we may order the lower and upper bounds in the following sense: For example, if $\nu + \tau_i < \tau_i'$ and $\nu + \tau_j < \tau_j'$ are two inequalities, then one can view them as upper bounds for ν, since if there is such a ν, then $\nu < \tau_i' - \tau_i$ and $\nu < \tau_j' - \tau_j$, where subtraction is defined as in LEMMA 8.9. Now, by linearity of $<$, we have either $\tau_i' - \tau_i \leq \tau_j' - \tau_j$ or $\tau_j' - \tau_j < \tau_i' - \tau_i$. In other words, $\varphi_i \wedge \varphi_j$ is equivalent to

$$(\tau_i' + \tau_j \leq \tau_i + \tau_j' \wedge \nu + \tau_i < \tau_i') \vee (\tau_i + \tau_j' < \tau_i' + \tau_j \wedge \nu + \tau_j < \tau_j'),$$

where φ_i is the stronger bound in the first disjunct, and φ_j is the stronger bound in the second one. In a similar way, we can order the upper bounds. Using the distributive laws as well as TAUTOLOGY (U.2), we may thus suppose that there is at most one lower and one upper bound. Moreover, note that

$$\exists \nu(\nu + \tau < \tau') \Leftrightarrow_{\mathsf{PrA}} \tau < \tau',$$
$$\exists \nu(\tau' < \nu + \tau) \Leftrightarrow_{\mathsf{PrA}} \mathsf{0} = \mathsf{0}\,.$$

On the other hand,

$$\exists \nu(\nu + \tau_1 < \tau_1' \wedge \tau_2' < \nu + \tau_2) \Leftrightarrow_{\mathsf{PrA}} \tau_1 + \tau_2' + \mathsf{1} < \tau_1' + \tau_2.$$

Therefore, the existential quantifier can be eliminated in both cases.

Case 4. There is at least one congruence and no equation among $\varphi_1, \ldots, \varphi_n$. As in the second case, without loss of generality we may assume that there is exactly one congruence and at most one inequality of each type. Without loss of generality, we only handle the case that there is exactly one lower and one upper bound, i.e., $\varphi \equiv \varphi_1 \wedge \varphi_2 \wedge \varphi_3$, where $\varphi_2 \equiv \tau_2' < \nu + \tau_2$ and $\varphi_3 \equiv \nu + \tau_3 < \tau_3'$. Then $\exists \nu \varphi$ is logically equivalent to the formula ψ with

$$\psi \equiv \bigvee_{k=1}^{m} \Big(\tau_3 + \tau_2' + \underline{k} < \tau_3' + \tau_2 \ \wedge \ \tau_1 + \tau_2' + \underline{k} \equiv_m \tau_1' + \tau_2\Big).$$

Note that if there is a ν such that φ holds, then it is of the form $\nu = \tau_2' - \tau_2 + x$ for some $x > \mathsf{0}$ such that $\nu < \tau_3' - \tau_3$, with the additional requirement that the congruence φ_1 be satisfied; one can then take x to be the smallest such solution, i.e., x is among $\underline{1}, \ldots, \underline{m}$. Clearly, $\nu \notin \mathrm{free}(\psi)$, and hence, ψ is as desired. ⊣

LEMMA 12.8. *For every quantifier-free $\mathscr{L}_{\mathsf{PrA}^*}$-sentence φ, we have*

$$\textit{either} \quad \mathsf{PrA} \vdash \varphi \quad \textit{or} \quad \mathsf{PrA} \vdash \neg\varphi\,.$$

Proof. We will first check the claim for atomic $\mathscr{L}_{\mathsf{PrA}^*}$-sentences by proving that for every $\mathscr{L}_{\mathsf{PrA}^*}$-term τ which does not contain any variables, there is an $n \in \mathbb{N}$ such that $\mathsf{PrA} \vdash \tau = \underline{n}$. We proceed by induction on term construction:

- If $\tau \equiv \mathsf{0}$, then there is nothing to check.
- If $\tau \equiv \mathsf{s}\tau'$ and $\mathsf{PrA} \vdash \tau' = \underline{n}$, then $\mathsf{PrA} \vdash \tau = \mathsf{s}\tau' = \mathsf{s} = \underline{\mathsf{s}n}$ by N_0.
- If we have $\tau \equiv \tau_1 + \tau_2$, $\mathsf{PrA} \vdash \tau_1 = \underline{n_1}$, and $\mathsf{PrA} \vdash \tau_2 = \underline{n_2}$, then N_1 implies $\mathsf{PrA} \vdash \tau = \tau_1 + \tau_2 = \underline{n_1} + \underline{n_2} = \underline{n_1 + n_2}$.

Therefore, every atomic $\mathscr{L}_{\mathsf{PrA}^*}$-sentence is equivalent to a formula of the form $\underline{m} = \underline{n}, \underline{m} < \underline{n}$ or $\underline{m} \equiv_k \underline{n}$ for some $k \geq \mathbf{2}$, and is therefore $\mathbb{N}$-conform. Since $\mathbb{N}$-conformity is preserved under negation, conjunctions and disjunctions, the claim follows. ⊣

COROLLARY 12.9. *The theory* PrA *is complete.*

Proof. This follows immediately from THEOREM 12.7 and LEMMA 12.8. ⊣

Note that the proof of the completeness of PrA actually provides a *decision procedure* for $\mathscr{L}_{\mathsf{PrA}}$-sentences. In fact, there is an algorithm which computes, with respect to a given $\mathscr{L}_{\mathsf{PrA}}$-sentence φ, a quantifier-free $\mathscr{L}_{\mathsf{PrA}}$-sentence ψ such that $\varphi \Leftrightarrow_{\mathsf{PrA}} \psi$, which, by LEMMA 12.8, can easily be decided in the sense that *either* $\mathsf{PrA} \vdash \psi$ *or* $\mathsf{PrA} \vdash \neg\psi$. In order to illustrate the algorithm, we provide an explicit example:

Example 12.10. We illustrate the quantifier elimination process using the example

$$\forall x \exists y \exists z (x < \underline{\mathbf{2}} z \land \underline{\mathbf{3}} z < x + y \land z \equiv_{\mathbf{2}} 0).$$

In a first step, we have to eliminate the quantifier $\exists z$. In order to achieve this, we first have to uniformise the z-coefficient using LEMMA 12.5, obtaining the equivalent formula

$$\forall x \exists y \exists z \Big(\underline{\mathbf{3}} x < \underline{\mathbf{6}} z \land \underline{\mathbf{6}} z < \underline{\mathbf{2}} x + \underline{\mathbf{2}} y \land \underline{\mathbf{6}} z \equiv_{\mathbf{12}} 0 \Big).$$

Now, we can replace $\underline{\mathbf{6}} z$ by w, which yields

$$\forall x \exists y \exists w \Big(\underline{\mathbf{3}} x < w \land w < \underline{\mathbf{2}} x + \underline{\mathbf{2}} y \land w \equiv_{\mathbf{12}} 0 \land w \equiv_{\mathbf{12}} 0 \Big).$$

Clearly, the first congruence is a consequence of the second one and can thus be removed. Using the second case in the proof of THEOREM 12.7, we can eliminate the variable w, and by further transformations we obtain

$$\begin{aligned} &\forall x \exists y \Big(\bigvee_{k=\mathbf{1}}^{\mathbf{12}} (\underline{\mathbf{3}} x + \underline{k} < \underline{\mathbf{2}} x + \underline{\mathbf{2}} y \land \underline{\mathbf{3}} x + \underline{k} \equiv_{\mathbf{6}} 0) \Big) \\ \Leftrightarrow_{\mathsf{PrA}}\ &\forall x \Big(\bigvee_{k=\mathbf{1}}^{\mathbf{6}} (\underline{\mathbf{3}} x + \underline{k} \equiv_{\mathbf{6}} 0 \land \exists y (x + \underline{k} < \underline{\mathbf{2}} y)) \Big) \\ \Leftrightarrow_{\mathsf{PrA}}\ &\forall x \Big(\bigvee_{k=\mathbf{1}}^{\mathbf{6}} \underline{\mathbf{3}} x + \underline{k} \equiv_{\mathbf{6}} 0 \Big), \end{aligned}$$

which is provable by PrA due to LEMMA 12.2. For the second equivalence, note that $\exists y (x + \underline{k} < \underline{\mathbf{2}} y)$ holds, which can be seen by taking $y = x + \underline{k}$. Following the algorithm, one would now have to replace the universal quantifier by a negated existential quantifier and negate the disjunction, and then restore the disjunctive normal form. Since this is very laborious, we will not pursue this further.

Let $\mathbb{N}^*$ be the $\mathscr{L}_{\mathsf{PrA}}$-structure $\big(\mathbb{N}, \mathsf{s}, +, \mathbf{0}\big)$, i.e., $\mathbb{N}^*$ is the same as $\mathbb{N}$, except that we do not have the binary function $\cdot^{\mathbb{N}}$ in $\mathbb{N}^*$. Since $\mathbb{N}$ is a model of PA, we find that $\mathbb{N}^*$ is a model of PrA. Now, since PrA is complete, we obtain that $\mathbf{Th}(\mathbb{N}^*)$ coincides with $\mathbf{Th}(\mathsf{PrA})$, i.e., for every $\mathscr{L}_{\mathsf{PrA}}$-sentence σ we have

$$\mathbb{N}^* \vDash \sigma \quad \Longleftrightarrow \quad \mathsf{PrA} \vdash \sigma \,.$$

The reader might now wonder why one does not take Presburger Arithmetic rather than Peano Arithmetic as the standard axiomatisation of arithmetic. Even though PrA is weaker than PA, it has the advantage that it is complete. However, the disadvantage of PrA is, that it is much too weak to serve as a proper axiomatisation of arithmetic. In fact, not even multiplication can be defined within PrA. This is a consequence of the following result, which states that every set of natural numbers defined by some $\mathscr{L}_{\mathsf{PrA}}$-formula is *eventually periodic*.

LEMMA 12.11. *Let $\varphi(x)$ be an $\mathscr{L}_{\mathsf{PrA}}$-formula with one free variable x. Then*

$$\mathbb{N}^* \vDash \exists p > \mathsf{0}\, \exists n_0\, \forall n \geq n_0 \left(\varphi(n) \leftrightarrow \varphi(n+p)\right).$$

Proof. By THEOREM 12.7, it suffices to check the claim for quantifier-free $\mathscr{L}_{\mathsf{PrA}^*}$-formulae φ. We proceed by induction on the construction of the formula φ.

- If φ is an atomic formula, then, for some $m, n, p, k \in \mathbb{N}$, φ is logically equivalent to one of the following formulae:

$$\begin{aligned} \underline{m}v + \underline{n} &= \underline{l}\,, \\ \underline{m}v + \underline{n} &< \underline{l}\,, \\ \underline{m}v + \underline{n} &> \underline{l}\,, \\ \underline{m}v + \underline{n} &\equiv_k \underline{l}\,. \end{aligned}$$

  This follows immediately from LEMMATA 12.4 and 12.8. The first three cases are trivial, since one can choose $n_0 \equiv l$ and $p \equiv \mathbf{1}$ (in the first two cases we have $\neg\varphi(n_1)$ for all $n_1 \geq n_0$), and in the third case $\varphi(n)$ holds. Finally, suppose that φ is $\underline{m}v + \underline{n} \equiv_k \underline{l}$. Without loss of generality, we may assume that $n \equiv \mathbf{0}$. If l does not divide $\gcd(k, m)$, then φ is never satisfied and hence the claim is trivial. Otherwise, by EXERCISE 12.2, we may assume that $m \equiv \mathbf{1}$ and choose $p \equiv k$ and $n_0 \equiv \mathbf{0}$.
- If the claim holds for φ, then by TAUTOLOGY (H.0) it also holds for $\neg\varphi$ with the same witnesses $p, n_0 \in \mathbb{N}$ as for φ.
- Suppose that the claim holds for φ and ψ with witnesses n_0, p_0 and n_1, p_1, respectively, i.e.,

$$\mathbb{N}^* \vDash \forall n \geq n_0 \big(\varphi(n) \leftrightarrow \varphi(n+p_0)\big) \text{ and } \mathbb{N}^* \vDash \forall n \geq n_1 \big(\psi(n) \leftrightarrow \psi(n+p_1)\big).$$

Now, let $n_2 :\equiv \max(n_0, n_1)$ and $p_2 :\equiv \operatorname{lcm}(p_0, p_1)$. Then, in $\mathbb{N}^*$ we have $\varphi(n) \leftrightarrow \varphi(n + p_2)$ and $\psi(n) \leftrightarrow \psi(n + p_2)$ for all $n \geq n_2$, and hence, the claim follows for $\varphi \vee \psi$ and $\varphi \wedge \psi$ by TAUTOLOGY (H.2) and (H.3), respectively.

⊣

THEOREM 12.12. *Multiplication is not definable in* PrA, *i.e., in* PrA *we cannot define a binary function* mult(· , ·) *which satisfies the axioms* PA_4 *and* PA_5.

Proof. Assume toward a contradiction that there exists an $\mathscr{L}_{\mathsf{PrA}}$-formula $\varphi(x, y, z)$ such that $\mathsf{PrA} \vdash \forall x \, \forall y \, \exists! z \, \varphi(x, y, z)$ and

$$\operatorname{mult}(x, y) = z \;:\Longleftrightarrow\; \varphi(x, y, z)\,.$$

If there was such a formula φ, then $\psi(z) :\equiv \exists x \varphi(x, x, z)$ would be a formula defining z to be the square of some x. By LEMMA 12.11, there are $p, n_0 \in \mathbb{N}$ with $p > 0$ such that

$$\mathbb{N}^* \vDash \forall n \geq n_0 \left(\psi(n) \leftrightarrow \psi(n + p)\right).$$

This, however, is impossible, since ψ is not eventually periodic: To see this, take, for example, $m := \max(n_0, p)$. Then $\mathbb{N}^* \vDash \psi(m^2)$, but $\mathbb{N}^* \nvDash \psi(m^2 + p)$, since $m \geq p$ implies

$$m^2 < (m^2 + p) < (m + \mathbf{1})^2,$$

and there are no squares between m^2 and $(m + \mathbf{1})^2$. ⊣

The previous theorem is the reason why Presburger Arithmetic is not considered as the standard axiomatisation of arithmetic. While multiplication cannot be expressed using the successor function and addition, exponentiation can be introduced using the successor function, addition and multiplication, as we have seen in Chapter 9. The main difference between PrA and PA lies in the fact that in PA allows *recursive definitions* using Gödel's β-function.

Non-standard models of PrA

In what follows, we recapitulate which axioms of PrA are actually necessary in order to prove its completeness. In fact, the proof only uses some particular instances of the Induction Schema; namely those which are concerned with equations and congruences. In order to axiomatise PrA, we surely need the axioms PA_0–PA_3. Moreover, in order to prove all required statements about equations, inequalities and congruences, we add the following axioms:

PrA_4: $\forall x \forall y (x + y = y + x)$

PrA_5: $\forall x \forall y \forall z (x + (y + z) = (x + y) + z)$

PrA_6: $\forall x \big(x = 0 \vee \exists y (x = \mathsf{s} y)\big)$

PrA_7: $\forall x \forall y (x < y \vee x = y \vee y < x)$

PrA_8: $\forall x \big(\bigvee_{k=\mathbf{0}}^{n-\mathbf{1}} x \equiv_n \underline{k} \big)$ for every $n \in \mathbb{N}$

Note hat PrA_8 is actually an axiom scheme, just like the Induction Schema. However, it is considerably simpler than the Induction Schema in the sense that it is indexed by standard natural numbers instead of formulae. Each of these axioms is used in the proof of the completeness of PrA. For example, the commutative and associative laws of addition, i.e., PrA_4 and PrA_5, are used in the proof of LEMMA 12.1. Moreover, by analysing all steps in the proof, it becomes clear that the axioms PA_0–PA_3 and PrA_4–PrA_8 suffice. By completeness, we thus obtain that the theory given by these axioms coincides with PrA.

Now that we are in possession of a simplified system of axioms, we can describe non-standard models of PrA. Note that since PA extends PrA, every non-standard model of PA is also a non-standard model of PrA. However, there are also non-standard models of PrA which have a simpler structure than non-standard models of PA. The simplest non-standard model of PrA is surely the $\mathscr{L}_{\mathsf{PrA}}$-structure $\mathbf{M}$ with domain M consisting of all rational polynomials in the indeterminate X of degree at most $\mathbf{1}$, such that *either* the leading coefficient is positive and the constant coefficient is an integer, *or* the leading coefficient equals $\mathbf{0}$ and the constant coefficient is a natural number. More formally, M consists of all polynomials of the form

$$qX + a \in \mathbb{Q}[X],$$

where $q \in \mathbb{Q}$, $q \geq \mathbf{0}$, and $a \in \mathbb{Z}$ or $a \in \mathbb{N}$, depending on whether $q > \mathbf{0}$ or $q = \mathbf{0}$. The interpretations of $\mathbf{0}$, s, and $+$ are the obvious ones. Note that for the ordering $<$ we then obtain

$$\mathbf{M} \vDash pX + a < qX + b \iff p <^{\mathbb{Q}} q \vee \big(p = q \wedge a <^{\mathbb{Z}} b\big),$$

which is essentially the lexicographic ordering. Since this is a linear order, it follows that $\mathbf{M} \vDash \mathsf{PrA}_7$. The other axioms, except for PrA_8, are obviously satisfied. In order to see that PrA_8 also holds in $\mathbf{M}$, note that

$$\mathbf{M} \vDash qX + a = \underline{n} \cdot \Big(\frac{q}{n} X\Big) + a \equiv_n a,$$

and hence, by taking $k \in \{\mathbf{0}, \dots, n - \mathbf{1}\}$ to be the modulus of a, we obtain that $\mathbf{M} \vDash qX + a \equiv_n \underline{k}$.

Another model of PrA is obtained by admitting all rational polynomials

$$\sum_{k=\mathbf{0}}^{n} a_k = a_n X^n + \ldots + a_1 X + a_0 \in \mathbb{Q}[X]$$

with positive leading coefficient $a_n > \mathbf{0}$ and integer constant coefficient $a_0 \in \mathbb{Z}$, where $a_0 \in \mathbb{N}$ in the case when $n = \mathbf{0}$. Notice that both models are *not* models of PA: In PA, one can define exponentiation, and hence, there exists, e.g., the number $(\mathbf{1}X)^{\mathbf{1}X}$, which obviously does not have a polynomial representation.

Notes

The method of quantifier elimination was already used by Skolem in 1919 to prove the completeness of the first-order theory of a class of boolean algebrae. Presburger Arithmetic is named after the Polish mathematician Mojżesz Presburger, who proved its consistency, completeness and decidability in 1929 (see [42]). However, his original proof is given for the integers rather than the natural numbers. The proof which is presented here follows the one presented in [8]. Skolem independently discovered Presburger's result in 1930 (see [50]) and further showed the same to be true for Skolem Arithmetic, i.e., the theory of natural numbers with multiplication but without addition.

Exercises

12.0 Show that there is no quantifier-free $\mathscr{L}_{\mathsf{PrA}}$-formula which is logically equivalent to the formula $\exists y(x = \mathbf{2}y)$, and conclude that Presburger Arithmetic does not admit quantifier elimination.

12.1 Prove that $\tau \equiv_n \tau' \Leftrightarrow_{\mathsf{PrA}} \underline{m}\tau \equiv_{mn} \underline{m}\tau'$ for all $\mathscr{L}_{\mathsf{PrA}}$-terms τ and τ' and for every $m \in \mathbb{N}$.

12.2 Use Bézout's Lemma (see Exercise 8.2) in the standard model $\mathbb{N}$ to prove that every formula of the form $\underline{m}v + \underline{n} \equiv_k \underline{l}$ for $m, n, k \in \mathbb{N}$ such that $\gcd(m, k) = \mathbf{1}$ is logically equivalent to a formula of the form $v \equiv_l \underline{l'}$ for some $l' \in \mathbb{N}$. Note that this essentially consists of proving that m has an inverse element modulo k.

12.3 Let $\mathscr{L} \equiv \{c_n \mid n \in \mathbb{N}\}$ and let $\mathsf{T} \equiv \{c_n \neq c_m \mid n, m \in \mathbb{N}, n \not\equiv m\}$. Show that T admits quantifier elimination.

12.4 Show that the theory DLO (see Exercise 3.5) admits quantifier elimination and conclude that DLO is complete.

12.5 Show that the theory $\mathbf{Th}(\mathbb{N}, <, \mathsf{s}, \mathbf{0})$ admits quantifier elimination and conclude that addition $+$ is not definable in $\mathbf{Th}(\mathbb{N}, <, \mathsf{s}, \mathbf{0})$.

12.6 Let $\mathscr{L} = \{\sim\} \cup \{c_n \mid n \in \mathbb{N}\}$ and let M_n be an infinite subset of $\mathbb{N}$ for every $n \in \mathbb{N}$ such that $\mathbb{N}$ is the disjoint union of the M_n's. Let T be the theory of one equivalence relation with infinitely many infinite equivalence classes, i.e., consisting of the axioms

- $\forall x(x \sim x)$
- $\forall x, y(x \sim y \to y \sim x)$
- $\forall x, y, z(x \sim y \wedge y \sim z \to x \sim z)$
- $\forall x \exists x_1 \ldots \exists x_n(x \sim x_1 \wedge \ldots \wedge x \sim x_n)$ for each $n \in \mathbb{N}$
- $\forall x \exists x_1 \ldots \exists x_n(\bigwedge_{i=1}^n \neg(x \sim x_i) \wedge \bigwedge_{i \neq j} \neg(x_i \sim x_j))$ for each $n \in \mathbb{N}$

Show that T admits quantifier elimination.

12.7 Prove that the divisibility relation is not definable in PrA.

Part IV
The Axiom System ZFC

In this part, we first present the axioms of Set Theory, including the Axiom of Choice. Then we discuss the consistency of this axiomatic system and provide standard as well as non-standard models of Set Theory. In the last three chapters, we use Set Theory to prove the LÖWENHEIM-SKOLEM THEOREMS, to construct models of PA, and to construct different models of the real numbers.

Chapter 13
The Axioms of Set Theory (ZFC)

In this chapter, we shall present and discuss the axioms of Zermelo-Fraenkel Set Theory including the Axiom of Choice, denoted ZFC. It will turn out that within this axiom system, we can develop all of first-order mathematics, and therefore, the axiom system ZFC serves as a foundation of mathematics. We will start with Zermelo's first axiomatisation of Set Theory and will show how basic mathematics can be developed within this system. Then we will introduce Zermelo's Axiom of Choice, Fraenkel's Axiom Schema of Replacement, and the Axiom of Foundation. Finally, we will discuss the notions of ordinal and cardinal numbers.

Before we begin presenting the axioms of Set Theory, let us say a few words about Set Theory in general: The signature of Set Theory $\mathscr{L}_{\mathsf{ST}}$ contains only one non-logical symbol, namely the binary **membership relation** denoted by $\in$, i.e., $\mathscr{L}_{\mathsf{ST}} = \{\in\}$. Furthermore, there exists just one type of objects, namely *sets*. However, to make life easier, instead of $\in(a, b)$ we write $a \in b$ (or also $b \ni a$ on rare occasions) and say that "a is an element of b", or that "a belongs to b". Furthermore, we write $a \notin b$ as an abbreviation of $\neg(a \in b)$. Later we will extend the signature of Set Theory $\mathscr{L}_{\mathsf{ST}}$ by defining some constants (like $\emptyset$ and ω), relations (like $\subseteq$), and operations (like the power set operation $\mathscr{P}$), but as we know from Chapter 6, all that can be expressed in Set Theory using defined constants, functions, and relations, can also be expressed by formulae containing the non-logical binary relation symbol $\in$ only.

L. Halbeisen, R. Krapf, *Gödel's Theorems and Zermelo's Axioms*,
https://doi.org/10.1007/978-3-030-52279-7_13

Zermelo's Axiom System (Z)

In 1905, Zermelo began to axiomatise Set Theory. In 1908, he published his first axiomatic system consisting of the following seven axioms:

1. Axiom der Bestimmtheit
   which corresponds to the Axiom of Extensionality
2. Axiom der Elementarmengen
   which includes the Axiom of Empty Set as well as the Axiom of Pairing
3. Axiom der Aussonderung
   which corresponds to the Axiom Schema of Separation
4. Axiom der Potenzmenge
   which corresponds to the Axiom of Power Set
5. Axiom der Vereinigung
   which corresponds to the Axiom of Union
6. Axiom der Auswahl
   which corresponds to the Axiom of Choice
7. Axiom des Unendlichen
   which corresponds to the Axiom of Infinity

The axioms 1–5 and axiom 7 (i.e., all axioms except the Axiom of Choice) form the so-called *Zermelo's axiom system*, denoted by Z, which will be discussed below.

Let us start with the axiom which states the existence of a set, namely the so-called *empty set*.

0. The Axiom of Empty Set

$$\exists x \forall z (z \notin x).$$

This axiom postulates the existence of a set without any elements, i.e., an empty set.

1. The Axiom of Extensionality

$$\forall x \forall y \big(\forall z (z \in x \leftrightarrow z \in y) \rightarrow x = y\big).$$

This axiom says that any sets x and y having the same elements are equal. Notice that the converse—which is: $x = y$ implies that x and y have the same elements—is just a consequence of the logical axiom L_{15}.

The Axiom of Extensionality also shows that the empty set, postulated by the Axiom of Empty Set, is unique: If x_0 and x_1 are empty sets, then we have $\forall z(z \notin x_0 \wedge z \notin x_1)$, which implies $\forall z(z \in x_0 \leftrightarrow z \in x_1)$, and therefore, $x_0 = x_1$. Thus, with the Axiom of Empty Set and the Axiom of Extensionality we can prove $\exists! x \, \forall z(z \notin x)$, and therefore, we can denote the unique empty set by the constant symbol $\emptyset$.

Similarly, we define the binary relation symbol $\subseteq$, called **subset**, by stipulating

$$x \subseteq y :\Longleftrightarrow \forall z(z \in y \rightarrow z \in x).$$

Notice that for every x we have $\emptyset \subseteq x$. Furthermore, we define the binary relation symbol $\subsetneq$, called **proper subset**, by stipulating

$$x \subsetneq y :\Longleftrightarrow x \subseteq y \wedge x \neq y.$$

So far, we have at least one set, namely the empty set $\emptyset$, for which we have $\emptyset \subseteq \emptyset$.

2. The Axiom of Pairing

$$\forall x \forall y \exists u \forall z \big(z \in u \leftrightarrow (z = x \vee z = y)\big)$$

Notice that by the Axiom of Extensionality, the set u is uniquely defined by the sets x and y. Therefore, we can define the binary function symbol $\{\cdot, \cdot\}$ by stipulating

$$\{x, y\} = u :\Longleftrightarrow \forall z \big(z \in u \leftrightarrow (z = x \vee z = y)\big).$$

Notice that by the Axiom of Extensionality we have $\{x, x\} = \{x\}$. Thus, by the Axiom of Pairing, if x is a set, then also $\{x\}$ is a set. Now, starting with $\emptyset$, an iterated application of the Axiom of Pairing yields for example the sets $\emptyset, \{\emptyset\}, \{\{\emptyset\}\}, \{\{\{\emptyset\}\}\}, \ldots$, as well as $\{\emptyset, \{\emptyset\}\}, \{\{\emptyset\}, \{\emptyset, \{\emptyset\}\}\}, \ldots$.

Notice also that by the Axiom of Extensionality we have $\{x, y\} = \{y, x\}$. Therefore, it does not matter in which order the elements of a 2-element set are written down. However, with the Axiom of Pairing we can easily define **ordered pairs**, denoted by $\langle x, y\rangle$, as follows:

$$\langle x, y\rangle := \big\{\{x\}, \{x, y\}\big\}$$

It is not hard to show that $\langle x, y\rangle = \langle x', y'\rangle$ if and only if $x = x'$ *and* $y = y'$. Thus, we can define the binary function symbol $\langle \cdot, \cdot \rangle$ by stipulating

$$\langle x, y\rangle = u :\Longleftrightarrow \forall z \Big(z \in u \leftrightarrow \big(z = \{x\} \vee z = \{x, y\}\big)\Big).$$

Similarly, one could also define ordered triples, ordered quadruples, et cetera, but the notation becomes quite hard to read. However, once we have more axioms at hand, we can easily define arbitrarily large tuples.

3. The Axiom of Union

$$\forall x \exists u \forall z \big(z \in u \leftrightarrow \exists w \in x (z \in w)\big)$$

With this axiom we can define the unary function symbol $\bigcup$, called **union**, by stipulating

$$\bigcup x = u :\Longleftrightarrow \forall z \big(z \in u \leftrightarrow \exists w \in x (z \in w)\big).$$

Informally, for all sets x there exists the union of x which consists of all sets which belong to at least one element of x. For example, $x = \bigcup\{x\}$.

Similarly, we define the binary function symbol $\cup$ by stipulating

$$x \cup y = u :\Longleftrightarrow u = \bigcup\{x, y\}.$$

The set $x \cup y$ is called the **union** of x and y.

Now, with the Axiom of Union and the Axiom of Pairing, and by stipulating $x + 1 := x \cup \{x\}$, we can build, e.g., the following sets:

$$\begin{aligned}
0 &:= \emptyset \\
1 &:= 0 + 1 = 0 \cup \{0\} = \{0\} \\
2 &:= 1 + 1 = 1 \cup \{1\} = \{0, 1\} \\
3 &:= 2 + 1 = 2 \cup \{2\} = \{0, 1, 2\},
\end{aligned}$$

and so on. This construction leads to the following definition: A set x such that $\forall y (y \in x \rightarrow (y \cup \{y\}) \in x)$ is called **inductive**. More formally, we define the unary relation symbol ind by stipulating

$$\operatorname{ind}(x) :\Longleftrightarrow \forall y \Big(y \in x \rightarrow \big(y \cup \{y\}\big) \in x\Big).$$

Obviously, the empty set $\emptyset$ is inductive, i.e., $\operatorname{ind}(\emptyset)$; but of course, this definition only makes sense if also some other inductive sets exist. However, in order to make sure that non-empty inductive sets exist as well, we need the following axiom.

4. The Axiom of Infinity

$$\exists I \big(\emptyset \in I \wedge \operatorname{ind}(I)\big)$$

Informally, the Axiom of Infinity postulates the existence of a non-empty inductive set containing $\emptyset$. All the sets $0, 1, 2, \ldots$ constructed above—which we recognise as natural numbers—must belong to every inductive set. Thus, if there was a set which contains just the natural numbers, it would be the "smallest" inductive set containing the empty set. In order to construct this set, we need some more axioms.

5. The Axiom Schema of Separation

For each formula $\varphi(z, \mathbf{p})$ with free$(\varphi) \subseteq \{z, \mathbf{p}\}$, the following formula is an axiom:

$$\forall x \forall \mathbf{p} \exists y \forall z \big(z \in y \leftrightarrow \big(z \in x \wedge \varphi(z, \mathbf{p})\big)\big),$$

where $\mathbf{p}$ is an abbreviation for $p_1, \ldots, p_n$, and correspondingly $\forall \mathbf{p}$ stands for $\forall p_1 \ldots \forall p_n$. One can think of the sets $p_1, \ldots, p_n$ as parameters of φ, which are usually some fixed sets. Informally, for each set x and every $\mathscr{L}_{\mathsf{ST}}$-formula $\varphi(z)$,

$$\{z \in x : \varphi(z)\}$$

is a set. Notice that the Axiom Schema of Separation just allows us to separate sets with a given property from a given set, but not to build collections of sets with a given property. For example, for a set x and $\varphi(z) \equiv z \notin z$, $\{z \in x : \varphi(z)\}$ is a set, but the collection $\{z : \varphi(z)\}$ is not a set.

As a first application of the Axiom Schema of Separation, we define the *intersection* of two sets x_0 and x_1: We use x_0 as a parameter and define $\varphi(z, x_0) \equiv z \in x_0$. Then, by the Axiom Schema of Separation, there exists a set $y = \{z \in x_1 : \varphi(z, x_0)\}$, i.e.,

$$z \in y \leftrightarrow (z \in x_1 \wedge z \in x_0).$$

In other words, for any sets x_0 and x_1, the collection of all sets which belong to both x_0 and x_1 is a set. This set is called the **intersection** of x_0 and x_1 and is denoted by $x_0 \cap x_1$. More formally, we define the binary function symbol $\cap$ by stipulating

$$x_0 \cap x_1 = y :\Longleftrightarrow \forall z \big(z \in y \leftrightarrow z \in x_1 \wedge z \in x_0\big).$$

In general, for non-empty sets x we define the unary function symbol $\bigcap$ by stipulating

$$\bigcap x = y :\Longleftrightarrow y = \big\{u \in \bigcup x : \forall z \in x\,(u \in z)\big\},$$

which is the intersection of all sets which belong to x. In order to see that $\bigcap x$ is a set which is uniquely determined by x, let $\varphi(z, x) \equiv \forall y \in x\,(z \in y)$ and apply the Axiom Schema of Separation to $\bigcup x$. Notice that $x \cap y = \bigcap\{x, y\}$.

Another example is $\varphi(z, y) \equiv z \notin y$, where y is a parameter. In this case, $\{z \in x : z \notin y\}$ is a set, denoted by $x \setminus y$, which is called the **set-theoretic difference** of x and y. More formally, we define the binary function symbol $\setminus$ by stipulating

$$x \setminus y = u :\Longleftrightarrow \forall z \big(z \in u \leftrightarrow z \in x \wedge z \notin y\big).$$

The next axiom gives us for any set x the set of all subsets of x.

6. The Axiom of Power Set

$$\forall x \exists y \forall z (z \in y \leftrightarrow z \subseteq x)$$

Informally, the Axiom of Power Set states that for each set x there is a set $\mathscr{P}(x)$, called the **power set** of x, which consists of all subsets of x. More formally, we define the unary function symbol $\mathscr{P}$ by stipulating

$$\mathscr{P}(x) = y :\Longleftrightarrow \forall z (z \in y \leftrightarrow z \subseteq x).$$

The Set ω

As an application of the axioms which we have so far, we define the smallest non-empty inductive set containing $\emptyset$, denoted by ω, which will be the smallest set containing the natural numbers (see Chapter 16): By the Axiom of Infinity, there exists an non-empty inductive set I_0. Now, with the Axiom of Power Set and the Axiom Schema of Separation, we can define the set

$$\omega := \bigcap \big\{ X \in \mathscr{P}(I_0) : \emptyset \in X \wedge \operatorname{ind}(X) \big\}.$$

We have to show that the set ω is the smallest set which is inductive and contains $\emptyset$: By definition, ω is inductive and contains $\emptyset$. Now, let I be an inductive set with $\emptyset \in I$, and let $X_0 := \omega \cap I$. On the one hand, X_0 is inductive and $\emptyset \in X_0$. On the other hand, since $X_0 \subseteq \omega$, we have $X_0 \in \mathscr{P}(I_0)$, which implies that $\omega \subseteq X_0$. Therefore, ω is the unique inductive set containing $\emptyset$, which is contained in every inductive set containing $\emptyset$.

Later in Chapter 16, we shall see that ω is the domain of the standard model of Peano Arithmetic PA.

Functions, Relations, and Models

With the axioms which we have so far (i.e., with Zermelo's axiom system Z), we can define notions like functions and relations.

Cartesian Products and Functions

Let us first define Cartesian products: For arbitrary sets A and B we define the binary function symbol $\times$ by stipulating

$$A \times B := \big\{ \langle x, y \rangle : x \in A \wedge y \in B \big\},$$

where $\langle x, y \rangle = \{\{x\}, \{x, y\}\}$. The set $A \times B$ is called **Cartesian product** of A and B. Thus, the Cartesian product of two sets A and B is a subset of $\mathscr{P}(\mathscr{P}(A \cup B))$.

Now, we define **functions** $f : A \to B$ which map the elements of a set A to elements of a set B as certain subsets of $A \times B$. The set of all such functions is denoted by ${}^{A}B$, where we define

$${}^{A}B := \{f \subseteq A \times B : \forall x \in A\, \exists! y \in B(\langle x, y\rangle \in f)\}.$$

For $f \in {}^{A}B$ (i.e., $f : A \to B$), we usually write $f(x) = y$ instead of $\langle x, y\rangle \in f$ and say that y is the **image** of x under f. Moreover, the set A is called the **domain** of f, denoted by $\operatorname{dom} f$. If $S \subseteq A$, then the **image** of S under f is denoted by $f[S] = \{f(x) : x \in S\}$, and $f|_S = \{\langle x, y\rangle \in f : x \in S\}$ is the restriction of f to S. Furthermore, for a function $f : A \to B$, $f[A]$ is called the **range** of f, denoted by $\operatorname{ran}(f)$.

Here are some special functions:

- A function $f : A \to B$ is **surjective**, or **onto**, if

$$\forall y \in B\, \exists x \in A\big(f(x) = y\big)\,.$$

  In order to emphasise the fact that the function f is surjective, one can write $f : A \twoheadrightarrow B$.

- A function $f : A \to B$ is **injective**, or **one-to-one**, if we have

$$\forall x_1 \in A\, \forall x_2 \in A\big(f(x_1) = f(x_2) \to x_1 = x_2\big).$$

  In order to emphasise the fact that f is injective, one can write $f : A \hookrightarrow B$.

- A function $f : A \to B$ is **bijective** if it is injective and surjective. If $f : A \to B$ is bijective, then

$$\forall y \in B\, \exists! x \in A\big(\langle x, y\rangle \in f\big),$$

  which implies that

$$f^{-1} := \{\langle y, x\rangle : \langle x, y\rangle \in f\} \in {}^{B}A$$

  is a function which is even bijective. Therefore, if a bijective function exists from A to B, then there is also one from B to A and we sometimes just say that there is a **bijection** between A and B. Notice that if $f : A \hookrightarrow B$ is injective, then f is a bijection between A and $f[A]$.

- If f is a function from A to B and g is a function from B to C, then the composition $g\circ f$ is a function from A to C, where

$$g\circ f := \Big\{\langle x, z\rangle \in A \times C : \exists y \in B\big(\langle x, y\rangle \in f \wedge \langle y, z\rangle \in g\big)\Big\}.$$

- If f is a function with domain $\alpha \in \Omega$, then we call f a **sequence** of length α. If $f(\beta) =: x_\beta$ for $\beta < \alpha$, then we may write $f = \langle x_\beta : \beta < \alpha\rangle$.

Cartesian Products and Relations

Let us turn back to Cartesian products: Assume that we have assigned a non-empty set A_ι to each $\iota \in I$ (for some set I). Then, for $A := \bigcup\{A_\iota : \iota \in I\}$, the set

$$\bigtimes_{\iota \in I} A_\iota := \left\{ f \in {}^I A : \forall \iota \in I\big(f(\iota) \in A_\iota\big) \right\}$$

is called the Cartesian product of the sets A_ι ($\iota \in I$). Notice that if all sets A_ι are equal to a given set A, then $\bigtimes_{\iota \in I} A_\iota = {}^I A$. If $I = n$ for some $n \in \omega$, in abuse of notation we write A^n instead of ${}^n A$ by identifying ${}^n A$ with the set

$$A^n = \underbrace{A \times \ldots \times A}_{n\text{-times}}.$$

Let us now consider subsets of finite Cartesian products: For any set A and any $n \in \omega$, a set $R \subseteq A^n$ is called an **n-ary relation** on A. If $n = 2$, then $R \subseteq A \times A$ is also called a **binary relation**. For binary relations R we usually write xRy instead of $\langle x, y\rangle \in R$.

Here are some order relations:

- A binary relation R on A is a **linear ordering** on A, if for any elements $x, y \in A$ we have xRy or $x = y$ or yRx, where these three cases are mutually exclusive.

- A linear ordering R on A is a **well-ordering** on A, if every non-empty subset $S \subseteq A$ has an R-minimal element, i.e., there exists a $x_0 \in S$ such that for each $y \in S$ we have $x_0 Ry$. Notice that, since R is a linear ordering, the R-minimal element x_0 is unique. If there is a well-ordering R on A, then we say that A is *well-orderable*. The question whether each set is well-orderable has to be postponed until we have the Axiom of Choice.

Other important binary relations are the so-called equivalence relations: Let S be an arbitrary non-empty set. A binary relation $\sim$ on S is an **equivalence relation** if it is

- *reflexive* (i.e., for all $x \in S$: $x \sim x$),
- *symmetric* (i.e., for all $x, y \in S$: $x \sim y \leftrightarrow y \sim x$), and
- *transitive* (i.e., for all $x, y, z \in S$: $x \sim y \wedge y \sim z \;\rightarrow\; x \sim z$).

The **equivalence class** of an element $x \in S$, denoted by $[x]^{\tilde{}}$, is the set $\{y \in S : x \sim y\}$. If it is clear from the context which relation $\sim$ is meant, we simply write $[x]$ for $[x]^{\tilde{}}$. We would like to recall the fact that for any $x, y \in S$ we have *either* $[x]^{\tilde{}} = [y]^{\tilde{}}$ *or* $[x]^{\tilde{}} \cap [y]^{\tilde{}} = \emptyset$. A set $A \subseteq S$ is a set of **representatives** if A has exactly one element in common with each equivalence class $[x]^{\tilde{}}$. We would like to mention that the existence of a set of representatives generally relies on the Axiom of Choice.

Zermelo-Fraenkel Set Theory with Choice (ZFC)

In 1922, Fraenkel and Skolem independently improved and extended Zermelo's original axiomatic system, and the final version was again presented by Zermelo in 1930. The two axioms which we have to add to Zermelo's system from 1908 are the Axiom Schema of Replacement and the Axiom of Foundation. In this section, we will present the remaining axioms of the so-called *Zermelo–Fraenkel Set Theory* with the Axiom of Choice, denoted by ZFC, which consists of Zermelo's axiom system Z together with the Axiom Schema of Replacement, the Axiom of Foundation, and the Axiom of Choice.

7. *The Axiom Schema of Replacement*

For every first-order formula $\varphi(x, y, \mathbf{p})$ with $\text{free}(\varphi) = \{x, y, \mathbf{p}\}$, where $\mathbf{p}$ denotes a sequence of parameters, the following formula is an axiom:

$$\forall A\, \forall \mathbf{p}\big(\forall x \in A\, \exists! y\, \varphi(x, y, \mathbf{p}) \rightarrow \exists B\, \forall x \in A\, \exists y \in B\, \varphi(x, y, \mathbf{p})\big)$$

In order to reformulate the Axiom Schema of Replacement, we introduce the notion of a *class function*: Let $\varphi(x, y)$ be a formula with $\text{free}(\varphi) = \{x, y\}$ such that

$$\forall x\, \exists! y\, \varphi(x, y)\,.$$

Then the unary function symbol F, defined by stipulating

$$F(x) = y \;:\Longleftrightarrow\; \varphi(x, y),$$

is called a **class function**. Now, the Axiom Schema of Replacement states that for every set A and for each class function F,

$$F[A] = \big\{F(x) : x \in A\big\}$$

is a set. More informally, images of sets under functions are sets.

With the Axiom Schema of Replacement, we can now define arbitrary Cartesian products: Let F be a class function and let I be an arbitrary set. Furthermore, for every $\iota \in I$ let $A_\iota := F(\iota)$ and let $A := \bigcup F[I]$. Then the set

$$\bigtimes_{\iota \in I} A_\iota := \Big\{f \in {}^I\!A : \forall \iota \in I\big(f(\iota) \in A_\iota\big)\Big\}$$

is called the Cartesian product of the sets A_ι ($\iota \in I$). As a matter of fact, we would like to mention that with the axioms we have so far, we cannot prove that Cartesian products $\bigtimes_{\iota \in I} A_\iota$ of non-empty sets A_ι are non-empty.

We would also like to mention that with the Axiom Schema of Replacement, the Axiom of Empty Set and the Axiom Schema of Separation are redundant (see EXERCISE 13.0).

8. The Axiom of Foundation

$$\forall x\big(\exists z(z \in x) \to \exists y \in x(y \cap x = \emptyset)\big)$$

As a consequence of the Axiom of Foundation, we see that there is no infinite descending sequence $x_0 \ni x_1 \ni x_2 \ni \cdots$, since otherwise, the set $\{x_0, x_1, x_2, \ldots\}$ would contradict the Axiom of Foundation. In particular, there is no set x such that $x \in x$ and there are also no cycles like $x_0 \in x_1 \in \cdots \in x_n \in x_0$. As a matter of fact, we would like to mention that if one assumes the Axiom of Choice, then the non-existence of such infinite descending sequences can be proved to be equivalent to the Axiom of Foundation.

The axiom system containing the axioms 0–8 is called **Zermelo–Fraenkel Set Theory** and is denoted by ZF.

9. The Axiom of Choice (AC)

$$\forall \mathscr{F}\, \exists f\, \big(f \text{ is a function from } \mathscr{F} \text{ to } \bigcup \mathscr{F} \wedge \big(\emptyset \notin \mathscr{F} \to \forall x \in \mathscr{F}\,(f(x) \in x)\big)\big),$$

or equivalently,

$$\forall \mathscr{F}\Big(\emptyset \notin \mathscr{F} \to \exists f\Big(f \in {}^{\mathscr{F}}\bigcup \mathscr{F} \wedge \forall x \in \mathscr{F}\big(f(x) \in x\big)\Big)\Big).$$

Informally, every family of non-empty sets has a choice function.

One can show that AC is equivalent to the statement that Cartesian products of non-empty sets are non-empty. More formally, let $\mathscr{F} = \{A_\iota : \iota \in I\}$ be a family of non-empty sets (i.e., for each $\iota \in I$, $A_\iota \neq \emptyset$). Then the Cartesian product

$$\bigtimes_{\iota \in I} A_\iota$$

is non-empty. In order to see this, let f be a choice function of $\mathscr{F}$. Then

$$\Big\{\big\langle \iota, f(A_\iota)\big\rangle : \iota \in I\Big\} \in \bigtimes_{\iota \in I} A_\iota\,,$$

and hence, $\bigtimes_{\iota \in I} A_\iota$ is non-empty.

ZF together with the Axiom of Choice AC is denoted by ZFC. Later on, we shall see that the axiom system ZFC is a foundation of first-order mathematics.

Well-Ordered Sets and Ordinal Numbers

In 1904, Zermelo [57] published his first proof of the so-called Well-Ordering Principle, which states that every set can be well-ordered, and in 1908 he published a second proof (see [58]). In the proof presented below, we essentially follow Zermelo's first proof, but first we have to introduce the notion of ordinal numbers.

Ordinal Numbers

One of the most important concepts in Set Theory is the notion of *ordinal number*, which can be seen as a transfinite extension of the natural numbers. In order to define the concept of ordinal numbers, we must first give some definitions: Let $z \in x$. Then z is called an **$\in$-minimal element** of x, denoted by $\min_{\in}(z,x)$, if $\forall y(y \notin z \vee y \notin x)$, or equivalently, for any y in z we have $y \notin x$, or more formally,

$$\min_{\in}(z,x) \;:\Longleftrightarrow\; z \in x \;\wedge\; \forall y(y \in z \to y \notin x).$$

A set x is **ordered by** $\in$ if for any sets $y_1, y_2 \in x$ we have $y_1 \in y_2$ or $y_1 = y_2$ or $y_1 \ni y_2$, where the three cases do not have to be mutually exclusive. More formally,

$$\operatorname{ord}_{\in}(x) \;:\Longleftrightarrow\; \forall y_1, y_2 \in x \left(y_1 \in y_2 \vee y_1 = y_2 \vee y_1 \ni y_2\right).$$

Now, a set x is called **well-ordered by** $\in$ if it is ordered by $\in$ and if every non-empty subset of x has an $\in$-minimal element. More formally,

$$\operatorname{wo}_{\in}(x) \;:\Longleftrightarrow\; \operatorname{ord}_{\in}(x) \;\wedge\; \forall y \in \mathscr{P}(x)\left(y \neq \emptyset \to \exists z \in y \min_{\in}(z,y)\right).$$

Furthermore, a set x is called **transitive** if each element of x is a subset of x, i.e.,

$$\operatorname{trans}(x) \;:\Longleftrightarrow\; \forall y(y \in x \to y \subseteq x).$$

Notice that if x is transitive and $z \in y \in x$, then this implies $z \in x$. A set is called an **ordinal number**, or just an **ordinal**, if it is transitive and well-ordered by $\in$, i.e.,

$$\operatorname{ordinal}(x) :\Longleftrightarrow\; \operatorname{trans}(x) \;\wedge\; \operatorname{wo}_{\in}(x).$$

Ordinal numbers are usually denoted by Greek letters like α, β, γ, λ, et cetera, and the collection of all ordinal numbers is denoted by Ω. We will see later that Ω is not a set. However, we can consider "$\alpha \in \Omega$" as an abbreviation of $\operatorname{ordinal}(x)$, which is just a property of α, and thus, there is no harm in using the symbol Ω in this way, even though Ω is *not* an object of the set-theoretic universe.

Now, we would be ready to prove the following result (see, e.g., Halbeisen [21, Thm. 3.3]).

FACT 13.1.

(a) *If $\alpha \in \Omega$, then either $\alpha = \emptyset$ or $\emptyset \in \alpha$.*

(b) *If* $\alpha \in \Omega$, *then* $\alpha \notin \alpha$.

(c) *If* $\alpha, \beta \in \Omega$, *then* $\alpha \in \beta$ *or* $\alpha = \beta$ *or* $\alpha \ni \beta$, *where these three cases are mutually exclusive.*

(d) *If* $\alpha \in \beta \in \Omega$, *then* $\alpha \in \Omega$.

(e) *If* $\alpha \in \Omega$, *then also* $\alpha + 1 \in \Omega$, *where* $\alpha + 1 := \alpha \cup \{\alpha\}$.

(f) Ω *is transitive and is well-ordered by* $\in$. *More precisely,* Ω *is transitive and ordered by* $\in$, *and every non-empty collection* $C \subseteq \Omega$ *has an* $\in$*-minimal element.*

(g) *If* $\alpha, \beta \in \Omega$ *and* $\alpha \in \beta$, *then* $\alpha + 1 \subseteq \beta$. *In other words,* $\alpha + 1$ *is the least ordinal which contains* α.

(h) *For every ordinal* $\alpha \in \Omega$, *we have either* $\alpha = \bigcup \alpha$ *or there exists a* $\beta \in \Omega$ *such that* $\alpha = \beta + 1$.

Notice that if Ω is a set, then by (f), Ω is an ordinal number, and therefore $\Omega \in \Omega$, which contradicts (b). Thus, the collection of all ordinals Ω is not a set, but a so-called *class*. Since $\in$ constitutes a linear ordering by (c), we use the following notation:

$$\alpha < \beta \;:\Longleftrightarrow\; \alpha \in \beta$$
$$\alpha \leq \beta \;:\Longleftrightarrow\; \alpha < \beta \vee \alpha = \beta$$

The last two facts, i.e., (g) and (h), lead to the following definitions: An ordinal α is called a **successor ordinal** if there exists an ordinal β such that $\alpha = \beta + 1$; otherwise, it is called a **limit ordinal**. In particular, $\emptyset$ is a limit ordinal. Notice that $\alpha \in \Omega$ is a limit ordinal if and only if $\bigcup \alpha = \alpha$.

With these definitions one can show that ω, defined above as the least non-empty inductive set, is in fact the least non-empty limit ordinal. In particular, we have $\bigcup \omega = \omega$.

Now we are ready to prove the following

THEOREM 13.2. *The Well-Ordering Principle is equivalent to the Axiom of Choice.*

Proof. ($\Rightarrow$) Let $\mathscr{F}$ be any family of non-empty sets and let $<$ be any well-ordering on $\bigcup \mathscr{F}$. Define $f : \mathscr{F} \to \bigcup \mathscr{F}$ by stipulating $f(x)$ to be the $<$-minimal element of x.

($\Leftarrow$) Let M be a set. If $M = \emptyset$, then M is well-ordered and we are done. Therefore, assume that $M \neq \emptyset$ and let $\mathscr{P}^*(M) := \mathscr{P}(M) \setminus \{\emptyset\}$. Furthermore, let

$$f : \mathscr{P}^*(M) \to M$$

be an arbitrary but fixed choice function for the family $\mathscr{P}^*(M)$, which exists by the Axiom of Choice. Now, an injective function

$$w_\alpha : \alpha \hookrightarrow M$$

from some ordinal $\alpha \in \Omega$ into M is called an ***f*-set** if for all $\gamma \in \alpha$ we have

$$w_\alpha(\gamma) = f\big(M \setminus \{w_\alpha(\delta) : \delta \in \gamma\}\big).$$

For example, $w_1 = \{\langle 0, f(M)\rangle\}$ is an f-set – in fact, w_1 is the unique f-set with domain $\{0\}$. In general, for every $\alpha \in \Omega$ there is at most one f-set w_α with domain α. In order to see this, assume that w_α and w'_α are two distinct f-sets with domain α. Because w_α and w'_α are distinct and $\alpha \in \Omega$, there exists an $\in$-minimal $\gamma \in \alpha$ such that $w_\alpha(\gamma) \neq w'_\alpha(\gamma)$, but since for all $\delta \in \gamma$ we have $w_\alpha(\delta) = w'_\alpha(\delta)$, which contradicts the fact that

$$w_\alpha(\gamma) = f\big(M \setminus \{w_\alpha(\delta) : \delta \in \gamma\}\big) = f\big(M \setminus \{w'_\alpha(\delta) : \delta \in \gamma\}\big) = w'_\alpha(\gamma).$$

Thus, if there exists an f-set w_α for some $\alpha \in \Omega$, then this f-set w_α is the unique f-set with $\operatorname{dom}(w_\alpha) = \alpha$. Moreover, if w_β and w_α are f-sets and $\beta \in \alpha$, then $w_\alpha|_\beta = w_\beta$ (i.e., the restriction of w_α to β is equal to w_β).

Because every f-set w_α induces a well-ordering on $\operatorname{ran}(w_\alpha) \subseteq M$, by the Axiom Schema of Separation, the collection of all f-sets is a set, say S. Now, on S we define the ordering $\prec$ as follows: For two distinct f-sets w_α and w_β, let

$$w_\alpha \prec w_\beta \iff \alpha \in \beta.$$

Since the class Ω is well-ordered by $\in$, S is well-ordered by $\prec$. Let $w := \bigcup S$ and let

$$M' := \big\{x \in M : \exists \gamma \in \operatorname{dom}(w)\big(w(\gamma) = x\big)\big\}.$$

Then $M' = M$ and $w \in S$, since otherwise, w can be extended to the f-set

$$w \cup \big\{\big\langle \operatorname{dom}(w), f(M \setminus M')\big\rangle\big\},$$

which is a contradiction to the definition of S. Therefore, the injective function $w : \operatorname{dom}(w) \hookrightarrow M$ is surjective. In other words, there exists an ordinal $\alpha \in \Omega$ such that w is a bijection between α and M. Finally, define the binary relation $<$ on M by stipulating

$$x < y :\iff w^{-1}(x) \in w^{-1}(y).$$

Then, since α is well-ordered by $\in$, M is well-ordered by $<$. ⊣

Ordinal Arithmetic

The next result is the TRANSFINITE RECURSION THEOREM, a very powerful tool which is used, e.g., to define ordinal arithmetic (see below) or to build the cumulative hierarchy of sets (see Chapter 14).

THEOREM 13.3 (TRANSFINITE RECURSION THEOREM). *Let F be a class function which is defined for all sets. Then there is a unique class function G defined on Ω such that for each $\alpha \in \Omega$ we have*

$$G(\alpha) = F(G|_\alpha), \quad \text{where } G|_\alpha = \{\langle \beta, G(\beta) \rangle : \beta \in \alpha\}.$$

By transfinite recursion we are able to define addition, multiplication, and exponentiation of arbitrary ordinal numbers (see EXERCISE 13.1):

***Ordinal Addition*:** For arbitrary ordinals $\alpha \in \Omega$, we define

(a) $\alpha + 0 := \alpha$,
(b) $\alpha + (\beta + 1) := (\alpha + \beta) + 1$ for all $\beta \in \Omega$,
(c) and if $\beta \in \Omega$ is non-empty and a limit ordinal, then $\alpha + \beta := \bigcup_{\delta \in \beta}(\alpha + \delta)$.

Notice that, e.g., $1 + \omega = \omega \neq \omega + 1$, which shows that addition of ordinals is in general not commutative.

***Ordinal Multiplication*:** For arbitrary ordinals $\alpha \in \Omega$, we define

(a) $\alpha \cdot 0 := 0$,
(b) $\alpha \cdot (\beta + 1) := (\alpha \cdot \beta) + \alpha$ for all $\beta \in \Omega$,
(c) and if $\beta \in \Omega$ is a limit ordinal, then $\alpha \cdot \beta := \bigcup_{\delta \in \beta}(\alpha \cdot \delta)$.

Notice that, e.g., $2 \cdot \omega = \omega \neq \omega + \omega = \omega \cdot 2$, which shows that multiplication of ordinals is in general not commutative.

***Ordinal Exponentiation*:** For arbitrary ordinals $\alpha \in \Omega$, we define

(a) $\alpha^0 := 1$,
(b) $\alpha^{\beta+1} := \alpha^\beta \cdot \alpha$ for all $\beta \in \Omega$,
(c) and if $\beta \in \Omega$ is non-empty and a limit ordinal, then $\alpha^\beta := \bigcup_{\delta \in \beta}(\alpha^{\delta+1})$.

By definition, we obtain that addition, multiplication, and exponentiation of ordinals are binary operations on Ω. Moreover, one can prove that addition and multiplication of ordinals are also associative and that the left distributive law holds (but not the right distributive law). In order to prove a property for all ordinals, the following generalisation of the induction principle for natural numbers is a very powerful tool:

THEOREM 13.4 (TRANSFINITE INDUCTION PRINCIPLE). *Suppose that $\varphi(x)$ is an $\mathscr{L}_{\mathsf{ST}}$-formula and suppose that the following conditions hold:*

(a) $\varphi(0)$
(b) $\forall \alpha \in \Omega(\varphi(\alpha) \to \varphi(\alpha + 1))$
(c) $(\forall \beta < \alpha\, \varphi(\beta)) \to \varphi(\alpha)$ *if $\alpha \in \Omega$ is a limit ordinal.*

Then $\varphi(\alpha)$ holds for all ordinals $\alpha \in \Omega$.

Example 13.5. We prove the left distributive law of ordinal arithmetic, where we assume that the associativity of addition has already been shown. Let $\alpha, \beta \in \Omega$ be fixed ordinals.

(a) By definition, we have $\alpha \cdot (\beta + 0) = \alpha \cdot \beta = (\alpha \cdot \beta) + (\alpha \cdot 0)$.

(b) Assume that $\alpha \cdot (\beta + \gamma)$ holds. Then we obtain

$$\begin{aligned} \alpha \cdot \big(\beta + (\gamma + 1)\big) &= \alpha \cdot \big((\beta + \gamma) + 1\big) \\ &= \alpha \cdot (\beta + \gamma) + \alpha \\ &= \big((\alpha \cdot \beta) + (\alpha \cdot \gamma)\big) + \alpha \\ &= (\alpha \cdot \beta) + \big((\alpha \cdot \gamma) + \alpha\big) \\ &= (\alpha \cdot \beta) + \big(\alpha \cdot (\gamma + 1)\big). \end{aligned}$$

(c) Suppose that γ is a limit ordinal and that $\alpha \cdot (\beta + \delta) = (\alpha \cdot \beta) + (\alpha \cdot \delta)$ for all $\delta < \gamma$. Then we have

$$\begin{aligned} \alpha \cdot (\beta + \gamma) = \alpha \cdot \bigcup_{\delta<\gamma} (\beta + \delta) = \bigcup_{\delta<\gamma} \alpha \cdot (\beta + \delta) = \bigcup_{\delta<\gamma} \big((\alpha \cdot \beta) + (\alpha \cdot \delta)\big) \\ = (\alpha \cdot \beta) + \bigcup_{\delta<\gamma} \alpha \cdot \delta = (\alpha \cdot \beta) + (\alpha \cdot \gamma). \end{aligned}$$

Hence, by the TRANSFINITE INDUCTION PRINCIPLE, the left distributive law holds for all ordinals.

Let us consider the set ω again. The ordinals belonging to ω are called **natural numbers**. Since ω is the smallest non-empty limit ordinal, all natural numbers, except 0, are successor ordinals. Thus, for each $n \in \omega$ we have either $n = 0$ or there is an $m \in \omega$ such that $n = m + 1$. Since by definition,

$$k < n \iff k \in n$$

for each $n \in \omega$ we have $n = \{k \in \omega : k < n\}$, i.e., $n = \{0, 1, \dots, n-1\}$. In particular, for every $n \in \omega$, n is a set containing exactly n elements.

With ordinal addition, multiplication, and exponentiation we can define sums, products, and powers of natural numbers within ZF. In fact, we can define these operations in Z already (see EXERCISE 13.5).

Cardinal Numbers and Cardinal Arithmetic

One can show (see, e.g., Halbeisen [21, Prp. 3.20]) that for each well-ordering $<$ of a set A there exists a unique ordinal α and a unique bijective function $f : A \to \alpha$ such that for all $x, y \in A$,

$$x < y \iff f(x) \in f(y).$$

The unique ordinal α which corresponds to a well-ordering $<$ of A is called the **order type** of the well-ordering $<$.

In the presence of AC, we are now able to define cardinal numbers as ordinals: For any set A we define the cardinality of A, denoted by $|A|$, by stipulating

$$|A| := \min\{\alpha \in \Omega : \alpha \text{ is the order type of a well-ordering of } A\}.$$

By definition we have

$$|A| = \min\{\alpha \in \Omega : \text{there is a bijection between } \alpha \text{ and } A\}.$$

In order to see that this definition makes sense, notice that by AC, every set A is well-orderable, and that by the above remark, every well-ordering on A corresponds to exactly one ordinal. Therefore, for each set A, the set of all order types of well-orderings of A is a non-empty set of ordinals. Let $C \subseteq \Omega$ be this set of ordinals. Then, by FACT 13.1.(f), C has an $\in$-minimal element $\min C$, which shows that $|A|$ is indeed an ordinal.

For example, we have $|n| = n$ for every $n \in \omega$, and $|\omega| = \omega$; but in general, for $\alpha \in \Omega$, we do not have $|\alpha| = \alpha$. For example, $|\omega + 1| \neq \omega + 1$, since $|\omega + 1| = \omega$ and $\omega \neq \omega + 1$. However, there are also other ordinals α besides $n \in \omega$ and ω itself for which we have $|\alpha| = \alpha$. This leads to the following definition:

An ordinal number $\kappa \in \Omega$ such that $|\kappa| = \kappa$ is called a **cardinal number**, or just a **cardinal**. Cardinal numbers are usually denoted by Greek letters like κ, λ, μ, et cetera, or by $\aleph$'s. For example, the cardinal number ω is denoted by $\aleph_0$, which is the cardinality of countably infinite sets.

A cardinal κ is **infinite** if $\kappa \notin \omega$, otherwise, it is **finite**. In other words, a cardinal is finite if and only if it is a natural number.

Since cardinal numbers are just a special kind of ordinal, they are well-ordered by $\in$. However, for cardinal numbers κ and λ we usually write $\kappa < \lambda$ instead of $\kappa \in \lambda$, i.e.,

$$\kappa < \lambda \iff \kappa \in \lambda.$$

The next result implies that there are arbitrarily large cardinal numbers.

THEOREM 13.6 (CANTOR'S THEOREM). *For every set A, $|A| < |\mathscr{P}(A)|$.*

Proof. Let A be an arbitrary set. Obviously, we have $|A| \leq |\mathscr{P}(A)|$. If we had $|A| = |\mathscr{P}(A)|$, then there would be a bijection between A and $\mathscr{P}(A)$. In particular, there would be a surjection $A \twoheadrightarrow \mathscr{P}(A)$. Therefore, in order to prove $|A| < |\mathscr{P}(A)|$, it is enough to show that there is no surjection $f : A \twoheadrightarrow \mathscr{P}(A)$.

If $A = \emptyset$, then $\mathscr{P}(A) = \{\emptyset\}$ and $f = \emptyset$; hence, f is not a surjection.

If $A \neq \emptyset$, consider the set

$$\Gamma := \{x \in A : x \notin f(x)\}.$$

On the one hand, since $\Gamma \subseteq A$, $\Gamma \in \mathscr{P}(A)$. On the other hand, for each $x \in A$ we have

$$x \in \Gamma \iff x \notin f(x),$$

and therefore, there is no $x \in A$ such that $f(x) = \Gamma$, which shows that f is not surjective. $\dashv$

Let κ be a cardinal. The smallest cardinal number which is greater than κ is denoted by κ^+, i.e.,

$$\kappa^+ = \min\{\alpha \in \Omega : \kappa < |\alpha|\}.$$

Notice that by THEOREM 13.6, $\kappa < 2^\kappa$ for every cardinal κ. In particular, for every cardinal κ, $\{\alpha \in \Omega : \kappa < |\alpha|\}$ is non-empty and therefore κ^+ exists.

A cardinal μ is called a **successor cardinal** if there exists a cardinal κ such that $\mu = \kappa^+$; otherwise, it is called a **limit cardinal**. In particular, every positive integer $n \in \omega$ is a successor cardinal and ω is the smallest non-zero limit cardinal. By induction on $\alpha \in \Omega$ we define $\aleph_{\alpha+1} := \aleph_\alpha^+$, where $\aleph_0 := \omega$, and $\aleph_\alpha := \bigcup_{\delta\in\alpha} \aleph_\delta$ for limit ordinals α; notice that $\bigcup_{\delta\in\alpha} \aleph_\delta$ is a cardinal (see EXERCISE 13.2). In particular, $\aleph_\omega$ is the smallest uncountable limit cardinal and $\aleph_1 = \aleph_0^+$ is the smallest uncountable cardinal. The collection $\{\aleph_\alpha : \alpha \in \Omega\}$ is the class of all infinite cardinals, i.e., for every infinite cardinal κ there is an $\alpha \in \Omega$ such that $\kappa = \aleph_\alpha$. Notice that the collection of cardinals is—like the collection of ordinals—a proper *class* and not a *set*.

Cardinal addition, multiplication, and exponentiation are defined as follows:

***Cardinal Addition*:** For cardinals κ and μ, let

$$\kappa + \mu := |(\kappa \times \{0\}) \cup (\mu \times \{1\})|.$$

***Cardinal Multiplication*:** For cardinals κ and μ, let

$$\kappa \cdot \mu := |\kappa \times \mu|.$$

***Cardinal Exponentiation*:** For cardinals κ and μ, let

$$\kappa^\mu := |{}^\mu\kappa|.$$

As a consequence of these definitions we get the following

FACT 13.7. *Addition and multiplication of cardinals are associative and commutative, and we have the distributive law for multiplication over addition, and for all cardinals κ, λ, μ, we have*

$$\kappa^{\lambda+\mu} = \kappa^\lambda \cdot \kappa^\mu, \quad \kappa^{\mu\cdot\lambda} = (\kappa^\lambda)^\mu, \quad (\kappa \cdot \lambda)^\mu = \kappa^\mu \cdot \lambda^\mu.$$

Note that there is a bijection

$$f : \mathscr{P}(\kappa) \to {}^\kappa 2$$

given by

$$f(X)(\lambda) := \begin{cases} 1, & \lambda \in X \\ 0, & \lambda \notin X \end{cases}$$

for $\lambda < \kappa$. Hence $2^\kappa = |\mathscr{P}(\kappa)|$, and therefore, THEOREM 13.6 states that for every cardinal κ we have $\kappa < 2^\kappa$.

The Continuum Hypothesis (CH) states that $2^{\aleph_0} = \aleph_1$, and the Generalised Continuum Hypothesis (GCH) states that $2^{\aleph_\alpha} = \aleph_{\alpha+1}$ for all $\alpha \in \Omega$.

PROPOSITION 13.8. *For any ordinal numbers $\alpha, \beta \in \Omega$, we have*

$$\aleph_\alpha + \aleph_\beta = \aleph_\alpha \cdot \aleph_\beta = \aleph_{\alpha \cup \beta} = \max\{\aleph_\alpha, \aleph_\beta\}.$$

In particular, for every infinite cardinal κ and for every $n \in \omega$, we have $\kappa^n = \kappa$.

For a proof see, e.g., Halbeisen [21, Thm. 3.25].

For a cardinal κ, let $\operatorname{fin}(\kappa)$ denote the set of all finite subsets of κ, and let $\operatorname{seq}(\kappa)$ denote the set of all finite sequences which we can build with elements of κ. As a consequence of PROPOSITION 13.8, we have (see EXERCISE 13.4)

FACT 13.9. *For every infinite cardinal κ, we have $\kappa = |\operatorname{fin}(\kappa)| = |\operatorname{seq}(\kappa)|$.*

NOTES

In 1905, Zermelo began to axiomatise Set Theory, and in 1908 he published his first axiomatic system consisting of the seven above-mentioned axioms. In 1930, he presented in [60] his second axiomatic system, which he called the ZF-system, where he incorporated ideas of Fraenkel [10], Skolem [49], and von Neumann [37, 38, 39]. In fact, he added the Axiom Schema of Replacement (which was already used implicitly by Cantor in 1899) and the Axiom of Foundation to his former system, cancelled the Axiom of Infinity and did not explicitly mention the Axiom of Choice. More details can be found, e.g., in the notes of Halbeisen [21, Ch. 3].

EXERCISES

13.0 (a) Show that the Axiom of Empty Set follows from the Axiom Schema of Replacement.

(b) Show that the Axiom Schema of Separation follows from the Axiom Schema of Replacement.

Hint: Let A be a set and let $\varphi(x)$ be a formula with $\operatorname{free}(\varphi) = \{x\}$. Furthermore, let $\psi(x, y)$ be the formula

$$\big(\varphi(x) \wedge y = x\big) \vee \big(\neg\varphi(x) \wedge y = \{A\}\big).$$

Then $\psi(x, y)$ is a class function F and

$$F[A] \setminus \big\{\{A\}\big\} = \big\{x \in A : \varphi(x)\big\}.$$

13.1 (a) Define the addition of ordinals by transfinite recursion .

Hint: For each $\alpha \in \Omega$, define a class function F_α by stipulating $F_\alpha(x) := \emptyset$ if x is *not* a function; if x *is* a function, then let

$$F_\alpha(x) = \begin{cases} \alpha & \text{if } x = \emptyset, \\ x(\beta) \cup \{x(\beta)\} & \text{if } \operatorname{dom}(x) = \beta + 1 \text{ and } \beta \in \Omega, \\ \bigcup_{\delta \in \beta} x(\delta) & \text{if } \operatorname{dom}(x) = \beta \text{ and } \beta \in \Omega \setminus \{\emptyset\} \text{ is a limit ordinal}, \\ \emptyset & \text{otherwise.} \end{cases}$$

(b) Define multiplication of ordinals by transfinite recursion.

(c) Define exponentiation of ordinals by transfinite recursion.

13.2 Show that for limit ordinals $\alpha \in \Omega$, $\bigcup_{\delta \in \alpha} \aleph_\delta$ is a cardinal.

Hint: Let $\lambda := \bigcup_{\delta \in \alpha} \aleph_\delta$. Then λ is an ordinal, and if $|\lambda| < \lambda$, then there is a $\delta \in \alpha$ such that $|\lambda| = \aleph_\delta$.

13.3 Prove FACT 13.7.

13.4 (a) Show that if κ is an infinite cardinal, then $\kappa = |\operatorname{seq}(\kappa)|$.

Hint: Notice that

$$|\operatorname{seq}(\kappa)| = \Big|\bigcup_{n \in \omega} \kappa^n\Big| = \aleph_0 \cdot \kappa.$$

(b) Show that if κ is an infinite cardinal, then $\kappa = |\operatorname{fin}(\kappa)|$.

13.5 Show that addition, multiplication, and exponentiation of natural numbers (i.e., of elements of ω) can be defined within the axiom system Z. In particular, show that addition, multiplication, and exponentiation of ordinals in ω can be defined without the Axiom Schema of Replacement (i.e., without the help of the TRANSFINITE RECURSION THEOREM).

13.6 Prove that the following statements for a set x are equivalent:

(a) $\operatorname{ordinal}(x)$

(b) $\operatorname{trans}(x) \wedge \forall y \in x\big(\operatorname{trans}(y)\big)$

(c) $\operatorname{ord}_\in(x) \wedge \operatorname{trans}(x)$

Chapter 14
Models of Set Theory

Zermelo writes in [59, p. 262] that he was not able to show that the seven axioms for Set Theory given in that article are consistent. Even though it is essential whether a theory is consistent or not, we know that whenever a theory is strong enough to prove the axioms of PA, then there is no way to prove its consistency within this theory (see Chapter 11). Therefore, since we can prove within ZF that PA is consistent, we cannot prove the consistency of ZF within ZF. On the the other hand, we know that every consistent theory has a model. In particular, if ZF is consistent, then it has a model.

In what follows, we first show what models of ZF look like, then we briefly discuss briefly non-standard models of ZF, and finally we give a construction of Gödel's model of ZFC, which shows that AC is relatively consistent with ZF.

The Cumulative Hierarchy of Sets

Let us assume that ZF is consistent. Then, by GÖDEL'S COMPLETENESS THEOREM 5.5, we know that there is a model $\mathbf{M}$ of ZF. Surprisingly, all models of ZF have the same simple structure. Therefore, assume that $\mathbf{M} \vDash \mathsf{ZF}$. Within $\mathbf{M}$, by induction on $\alpha \in \Omega$ we define the sets

$$\begin{aligned} \mathrm{V}_0 &= \emptyset, \\ \mathrm{V}_\alpha &= \bigcup_{\beta \in \alpha} \mathrm{V}_\beta \quad \text{if } \alpha \text{ is a limit ordinal}, \\ \mathrm{V}_{\alpha+1} &= \mathscr{P}(\mathrm{V}_\alpha), \end{aligned}$$

and let

$$\mathbf{V} = \bigcup_{\alpha \in \Omega} \mathrm{V}_\alpha .$$

L. Halbeisen, R. Krapf, *Gödel's Theorems and Zermelo's Axioms*,
https://doi.org/10.1007/978-3-030-52279-7_14

Notice that by construction, for each $\alpha \in \Omega$, V_α is a *set* in the model $\mathbf{M}$. Again by induction on $\alpha \in \Omega$ one can easily show that the sets V_α have the following properties:

- Each V_α is transitive.
- If $\alpha \in \beta$, then $\mathrm{V}_\alpha \subsetneq \mathrm{V}_\beta$.
- $\mathrm{V}_\alpha \cap \Omega = \alpha$.

These facts are visualised by the following figure:

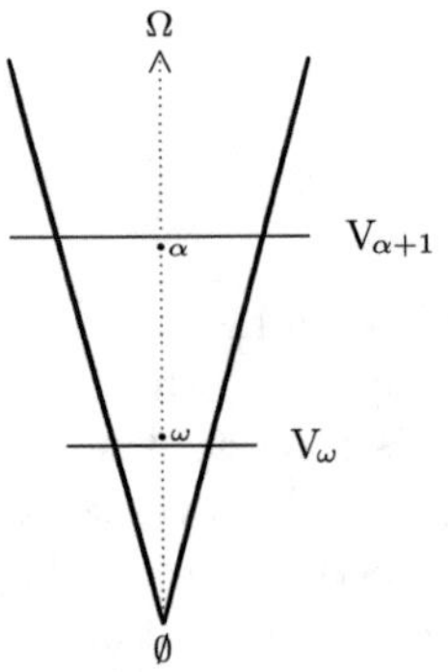

By definition, $\mathbf{V}$ is a collection of sets of $\mathbf{M}$, and since the class of ordinals Ω is contained in $\mathbf{V}$, $\mathbf{V}$ is a class in $\mathbf{M}$.

Before we can prove that the class $\mathbf{V}$, called the **cumulative hierarchy**, is equal to $\mathbf{M}$ (i.e., $\mathbf{V}$ contains *all* sets of $\mathbf{M}$), we have to introduce the notion of *transitive closure*: Let S be an arbitrary set. By induction on $n \in \omega$, we define

$$S_0 = S, \qquad S_{n+1} = \bigcup S_n,$$

and finally

$$\mathrm{TC}(S) = \bigcup_{n\in\omega} S_n,$$

where $\bigcup_{n\in\omega} S_n := \bigcup\{S_n : n \in \omega\}$. For example $x_1 \in S_1$ if and only if $\exists x_0 \in S_0(x_0 \ni x_1)$, and in general, $x_{n+1} \in S_{n+1}$ if and only if $\exists x_0 \in S_0 \cdots \exists x_n \in S_n(x_0 \ni x_1 \ni \cdots \ni x_{n+1})$. Notice that by the Axiom of Foundation, every descending sequence of the form $x_0 \ni x_1 \ni \cdots$ must be finite. More precisely, every descending sequence $x_0 \ni x_1 \ni \cdots$ is of the form $x_0 \ni x_1 \ni \cdots \ni x_n$ for some $n \in \omega$.

By construction, $\mathrm{TC}(S)$ is transitive, i.e., $x \in \mathrm{TC}(S)$ implies $x \subseteq \mathrm{TC}(S)$, and we further have $S \subseteq \mathrm{TC}(S)$. Moreover, since every transitive set T must satisfy $\bigcup T \subseteq T$, it follows that the set $\mathrm{TC}(S)$ is the smallest transitive set which contains S. Thus,

$$\mathrm{TC}(S) = \bigcap\{T : T \supseteq S \text{ and } T \text{ is transitive}\}.$$

Consequently, the set $\mathrm{TC}(S)$ is called the **transitive closure** of S.

THEOREM 14.1. *For every set x in the model $\mathbf{M}$, there is an ordinal α such that $x \in \mathrm{V}_\alpha$. In particular, every set in $\mathbf{M}$ belongs to $\mathbf{V}$, and vice versa.*

Proof. Assume towards a contradiction that there exists a set x in the model $\mathbf{M}$ which does not belong to $\mathbf{V}$. Let $\bar{x} := \mathrm{TC}(\{x\})$ and let $w := \{z \in \bar{x} : z \notin \mathbf{V}\}$, i.e., $w = \bar{x} \setminus \{z' \in \bar{x} : \exists \alpha \in \Omega\, (z' \in \mathrm{V}_\alpha)\}$. Since $x \in w$ we have $w \neq \emptyset$, and by the Axiom of Foundation there is a $z_0 \in w$ such that $(z_0 \cap w) = \emptyset$. Since $z_0 \in w$ we have $z_0 \notin \mathbf{V}$, which implies that $z_0 \neq \emptyset$; but for all $u \in z_0$ there is a least ordinal α_u such that $u \in \mathrm{V}_{\alpha_u}$. By the Axiom Schema of Replacement, $\{\alpha_u : u \in z_0\}$ is a set, and moreover, $\alpha = \bigcup\{\alpha_u : u \in z_0\} \in \Omega$. This implies that $z_0 \subseteq \mathrm{V}_\alpha$ and consequently we get $z_0 \in \mathrm{V}_{\alpha+1}$, which contradicts the fact that $z_0 \notin \mathbf{V}$ and therefore completes the proof. Hence, every set in $\mathbf{M}$ belongs to $\mathbf{V}$, and since, by definition, every set in $\mathbf{V}$ belongs to $\mathbf{M}$, we have that $\mathbf{M} = \mathbf{V}$. ⊣

As an immediate consequence of THEOREM 14.1, we get that every model of ZF has the structure of the cumulative hierarchy.

Non-Standard Models of ZF

On the one hand, in Chapter 7 we have constructed the standard model of PA with domain $\mathbb{N}$, on which we later defined the linear ordering $<$. On the other hand, in Chapter 13 we have constructed from ZF the set ω with the linear ordering given by the membership relation $\in$. Now, if, in a model $\mathbf{V}$ of ZF, the structure $(\omega, \in)$ is isomorphic to the structure $(\mathbb{N}, <)$, then we call $\mathbf{V}$ a **standard model** of ZF; otherwise, $\mathbf{V}$ is called a **non-standard model** of ZF. Recall that in ZF, a set x is called *finite* if and only if there exists a bijection between x and some element of ω. This leads to an alternative definition of standard models of ZF: A model $\mathbf{V}$ of ZF is a standard model if and only if each set x which is finite in $\mathbf{V}$ is also finite with respect to the metamathematical notion of F I N I T E N E S S.

Before we show the existence of non-standard models of ZF, we would like to mention that we do not have a criteria to decide whether a model $\mathbf{V}$ of ZF is standard. In particular, when we assumed earlier that $\mathbf{M} \vDash$ ZF, it might have been that $\mathbf{M}$ was a non-standard model, and therefore we find that the non-standard models of ZF also have the structure of the cumulative hierarchy.

Now, we will show that if ZF is consistent, then there exists a non-standard model of ZF. For this purpose, we first extend the signature $\mathscr{L}_{\mathsf{ST}} = \{\in\}$ by adding countably many new constant symbols $\mathsf{c}_0, \mathsf{c}_1, \mathsf{c}_2, \ldots$, i.e., the new signature is $\{\in, \mathsf{c}_0, \mathsf{c}_1, \mathsf{c}_2, \ldots\}$. Then, we extend the axioms ZF by adding the formulae

$$\underbrace{\mathsf{c}_1 \in \mathsf{c}_0}_{\varphi_0}, \quad \underbrace{\mathsf{c}_2 \in \mathsf{c}_1}_{\varphi_1}, \quad \underbrace{\mathsf{c}_3 \in \mathsf{c}_2}_{\varphi_2}, \quad \ldots,$$

and let Ψ be the collection of these formulae. Now, if ZF has a model $\mathbf{V}$ and Φ is any finite subset of Ψ, then, by interpreting the finitely many c_i's $\mathsf{c}_0, \ldots, \mathsf{c}_n$ appearing in Ψ in a suitable way, e.g., by stipulating

$$\mathsf{c}_i := n - i,$$

$\mathbf{V}$ is also a model of $\mathsf{ZF} \cup \Phi$, which implies that $\mathsf{ZF} \cup \Phi$ is consistent. Thus, by the COMPACTNESS THEOREM, $\mathsf{ZF} \cup \Psi$ is also consistent and therefore has a model, say $\mathbf{V}^*$. Since $\mathbf{V}^* \vDash \mathsf{ZF} \cup \Psi$, we get that the Axiom of Foundation holds in $\mathbf{V}^*$. In particular, there must be a set $z \in \mathrm{TC}\left(\mathsf{c}_0^{\mathbf{V}^*}\right)$ such that

$$\mathbf{V}^* \vDash z \cap \mathrm{TC}\left(\mathsf{c}_0^{\mathbf{V}^*}\right) = \emptyset,$$

which implies that z must be different from all the sets $\mathsf{c}_n^{\mathbf{V}^*}$. On the other hand, by the Axiom of Foundation, the length of a decreasing sequence of the form

$$\mathsf{c}_0^{\mathbf{V}^*} \ni \mathsf{c}_1^{\mathbf{V}^*} \ni \mathsf{c}_2^{\mathbf{V}^*} \ni \cdots \ni z$$

must be finite in the sense of $\mathbf{V}^*$. In other words, the length of such a decreasing sequence must be an element of ω in the model $\mathbf{V}^*$, denoted by $\omega^{\mathbf{V}^*}$, which shows that $\omega^{\mathbf{V}^*}$ contains sets which are not finite with respect to the metamathematical notion of F I N I T E N E S S. In particular, the structures $(\omega^{\mathbf{V}^*}, \in)$ and $(\mathbb{N}, <)$ are not isomorphic, and hence, $\mathbf{V}^*$ is a non-standard model of ZF.

As a matter of fact, we would like to mention that from the consistency of ZF we obtain the existence of non-standard models of ZF, but without using the metamathematical notion of F I N I T E N E S S, we do not obtain the existence of standard models of ZF.

Gödel's Incompleteness Theorems for Set Theory

In this section, we indicate how Gödel's Incompleteness Theorems can be transferred to ZF and ZFC, respectively. In Chapter 16, we shall see that within ZF we can construct a model of PA with domain ω. In particular, we obtain

$$\mathrm{Con}(\mathsf{ZF}) \Longrightarrow \mathrm{Con}(\mathsf{PA}).$$

Therefore, with respect to ω, we can define within ZF the non-logical symbols of $\mathscr{L}_{\mathsf{PA}}$ and can extend the language $\mathscr{L}_{\mathsf{ST}}$ to the language $\mathscr{L} := \mathscr{L}_{\mathsf{ST}} \cup \mathscr{L}_{\mathsf{PA}} \cup \{\omega\}$. Furthermore, we can extend the theory ZF to the theory $\mathsf{T} := \mathsf{ZF} \cup \mathsf{PA}$. Now, since $\mathscr{L} \supseteq \mathscr{L}_{\mathsf{PA}}$ is a gödelisable language and T is a gödelisable $\mathscr{L}$-theory, we can apply THEOREM 10.12 and obtain that if ZF is consistent, then it is incomplete — the same applies to ZFC.

Moreover, within ZF we can define the $\mathscr{L}_{\mathsf{ST}}$-sentence

$$\mathrm{con}_{\mathsf{ZF}} \;:\Longleftrightarrow\; \neg\,\mathrm{prv}_{\mathsf{ZF}}(\ulcorner\emptyset = \{\emptyset\}\urcorner)$$

and show that if ZF is consistent, then $\mathsf{ZF} \nvdash \mathrm{con}_{\mathsf{ZF}}$ — where the same applies to ZFC.

Summing up, we obtain the following

THEOREM 14.2 (GÖDEL'S INCOMPLETENESS THEOREMS FOR SET THEORY).

(a) *If* ZF *is consistent, then it is incomplete.*

(b) *If* ZF *is consistent, then* $\mathsf{ZF} \nvdash \mathrm{con}_{\mathsf{ZF}}$.

Correspondingly, the same holds for ZFC.

As a matter of fact, we would like to mention that for T as above, on the one hand we have

$$\mathrm{Con}(\mathsf{T}) \implies \mathrm{Con}(\mathsf{PA}),$$

but on the other hand, for $\mathrm{con}_{\mathsf{PA}} :\Longleftrightarrow \neg\,\mathrm{prv}_{\mathsf{PA}}(\ulcorner 0 = 1\urcorner)$ we have

$$\mathrm{Con}(\mathsf{T}) \;\not\Longrightarrow\; \mathsf{T} \nvdash \mathrm{con}_{\mathsf{PA}}.$$

Absoluteness

Earlier in this chapter, we constructed within a model $\mathbf{M} \vDash \mathsf{ZF}$ the cumulative hierarchy of sets $\mathbf{V}$ and showed that $\mathbf{M} = \mathbf{V}$. Therefore, the only model of ZF within a given model $\mathbf{M} \vDash \mathsf{ZF}$ which contains the cumulative hierarchy is $\mathbf{V}$. It is natural to ask whether we also find some proper sub-models within $\mathbf{V}$ which are models of ZFC, or at least of some fragment of ZFC. It is clear that such sub-models cannot contain the entire cumulative hierarchy of sets. In order to investigate sub-models $\mathbf{M} \subseteq \mathbf{V}$ of fragments of ZFC, the notion of absoluteness will be crucial.

If M is the domain of a sub-model $\mathbf{M} \subseteq \mathbf{V}$, then M is a collection of sets of $\mathbf{V}$. However, M cannot be a set in $\mathbf{V}$. In order to define M within $\mathbf{V}$, we introduce the notion of a *class*. By a **class** we define any collection of sets of the form $\{x : \varphi(x)\}$ for some $\mathscr{L}_{\mathsf{ST}}$-formula φ, possibly with parameters. Note that every set (i.e., every element of $\mathbf{V}$) is a class (e.g., for $X \in \mathbf{V}$ let $\varphi(x) \equiv x \in X$). If M is a class in $\mathbf{V}$, we can **relativise** a formula $\varphi(\nu_1, \dots, \nu_n)$ to the model $\mathbf{M} = (M, \in)$ such that for all $x_1, \dots, x_n \in M$, the relativised formula $\varphi^{\mathbf{M}}(\nu_1, \dots, \nu_n)$ has the following property:

$$\varphi^{\mathbf{M}}(x_1, \dots, x_n) \iff \mathbf{M} \vDash \varphi(x_1, \dots, x_n)$$

To do this, by THEOREM 1.7, we may assume that φ only contains $\neg$ and $\wedge$ as logical operators and $\exists$ as quantifier. We define $\varphi^{\mathbf{M}}$ recursively as follows:

$$\begin{aligned}
(x_i \in x_j)^{\mathbf{M}} &:\iff \mathbf{V} \vDash x_i \in x_j \\
(x_i = x_j)^{\mathbf{M}} &:\iff \mathbf{V} \vDash x_i = x_j \\
(\neg\varphi)^{\mathbf{M}} &:\iff \mathbf{V} \vDash \neg\varphi^{\mathbf{M}} \\
(\varphi \wedge \psi)^{\mathbf{M}} &:\iff \mathbf{V} \vDash \varphi^{\mathbf{M}} \wedge \psi^{\mathbf{M}} \\
(\exists \nu\, \varphi)^{\mathbf{M}} &:\iff \exists x \in M : \mathbf{V} \vDash \varphi^{\mathbf{M}}(x)
\end{aligned}$$

If M and N are two classes in $\mathbf{V}$ such that $M \subseteq N$, then an $\mathscr{L}_{\mathsf{ST}}$-formula $\varphi(\nu_1, \ldots, \nu_n)$ is called **absolute** between the models $\mathbf{M} = (M, \in)$ and $\mathbf{N} = (N, \in)$, if for all $x_1, \ldots, x_n \in M$,

$$\varphi^{\mathbf{M}}(x_1, \ldots, x_n) \iff \varphi^{\mathbf{N}}(x_1, \ldots, x_n).$$

If $\mathbf{N} = \mathbf{V}$, then we say that φ is **absolute** for $\mathbf{M}$.

As for $\mathscr{L}_{\mathsf{PA}}$-formulae, we say that an $\mathscr{L}_{\mathsf{ST}}$-formula φ is a **Δ-formula** if it is built up from atomic formulae using $\neg$, $\wedge$, $\vee$, and bounded quantification, i.e., $\forall \nu \in \tau$ and $\exists \nu \in \tau$ for some term τ.

Example 14.3. The formula $x \subseteq y$ can be expressed by a Δ-formula, since

$$x \subseteq y \iff \forall z \in x (z \in y).$$

The formula stating that x is an ordinal is equivalent to a Δ-formula, since $\mathrm{trans}(x)$ is a Δ-formula and by EXERCISE 13.6.(b) we have

$$\mathrm{ordinal}(x) \iff \mathrm{trans}(x) \wedge \forall y \in x \big(\mathrm{trans}(y)\big).$$

FACT 14.4. *Let M be a class and let $\mathbf{M} = (M, \in)$. Then every $\mathscr{L}_{\mathsf{ST}}$-formula which is logically equivalent to a Δ-formula is absolute for $\mathbf{M}$.*

The following result is useful, since it provides easy criteria for the axioms of ZF being valid in transitive classes (a proof can be found in Kunen [30]).

LEMMA 14.5. *Let M be a transitive class (i.e., M is a class and for all x, y we have $x \in y \in M \to x \in M$), and let $\mathbf{M} = (M, \in)$. Then the following statements hold:*

(a) $\mathbf{M} \vDash$ *Axiom of Extensionality*
(b) $\mathbf{M} \vDash$ *Axiom of Foundation*
(c) $\mathbf{M} \vDash$ *Axiom of Pairing* $\iff \forall x, y \in M\big(\{x, y\} \in M\big)$
(d) $\mathbf{M} \vDash$ *Axiom of Union* $\iff \forall x \in M\big(\bigcup x \in M\big)$
(e) $\mathbf{M} \vDash$ *Axiom of Infinity* $\iff \omega \in M$
(f) $\mathbf{M} \vDash$ *Axiom of Power Set* $\iff \forall x \in M\big(\mathscr{P}(x) \cap M \in M\big)$

Gödel's Constructible Model L

In this section, we present Gödel's constructible universe **L**, which essentially consists of all sets which can be "described" and is therefore the smallest model of Set Theory. More precisely, in each step of the construction of **L** within some ground model $\mathbf{V} \vDash \mathsf{ZF}$, we only add sets which are definable from already constructed sets M by taking sets of the form

$$x = \left\{ y \in M : (M, \in) \vDash \varphi(y, p_1, \ldots, p_n) \right\},$$

for some formulae φ and parameters $p_1, \ldots, p_n \in M$. The problem that we encounter at this point is that we do not know whether the satisfaction relation $(M, \in) \vDash \varphi(y, p_1, \ldots, p_n)$ can be defined by an $\mathscr{L}_{\mathsf{ST}}$-formula, which is crucial in order to apply the Axiom Schema of Separation to obtain the existence of the set x. To achieve this, we first gödelise $\mathscr{L}_{\mathsf{ST}}$-formulae within ZF in a similar way as we gödelised $\mathscr{L}_{\mathsf{PA}}$-formulae within PA in Chapter 9. However, in Set Theory the gödelisation is much simpler. We first gödelise atomic $\mathscr{L}_{\mathsf{ST}}$-formulae as follows:

$$\ulcorner v_i \in v_j \urcorner := \langle 0, i, j \rangle$$
$$\ulcorner v_i = v_j \urcorner := \langle 1, i, j \rangle$$

Suppose that φ and ψ have already been gödelised. Then we define:

$$\ulcorner \neg\varphi \urcorner := \langle 2, \ulcorner \varphi \urcorner \rangle$$
$$\ulcorner \varphi \wedge \psi \urcorner := \langle 3, \ulcorner \varphi \urcorner, \ulcorner \psi \urcorner \rangle$$
$$\ulcorner \varphi \vee \psi \urcorner := \langle 4, \ulcorner \varphi \urcorner, \ulcorner \psi \urcorner \rangle$$
$$\ulcorner \varphi \rightarrow \psi \urcorner := \langle 5, \ulcorner \varphi \urcorner, \ulcorner \psi \urcorner \rangle$$
$$\ulcorner \exists v_i \varphi \urcorner := \langle 6, i, \ulcorner \varphi \urcorner \rangle$$
$$\ulcorner \forall v_j \varphi \urcorner := \langle 7, i, \ulcorner \psi \urcorner \rangle$$

We can then define the set of all codes of formalised $\mathscr{L}_{\mathsf{ST}}$-formulae by stipulating:

$$\begin{aligned} f \in \mathrm{Fml} : \iff \exists n \in \omega \text{ and } c : n+1 \to \mathrm{V}_\omega \text{ such that } f = c(n) \text{ and} \\ \forall m \leq n \, \exists i, j \in \omega \, \exists k, l < m \, \big(c(m) = \langle 0, i, j \rangle \vee c(m) = \langle 1, i, j \rangle \\ \vee\, c(m) = \langle 2, c(k) \rangle \vee c(m) = \langle 3, c(k), c(l) \rangle \\ \vee\, c(m) = \langle 4, c(k), c(l) \rangle \vee c(m) = \langle 5, c(k), c(l) \rangle \\ \vee\, c(m) = \langle 6, i, c(k) \rangle \vee c(m) = \langle 7, i, c(k) \rangle \big) . \end{aligned}$$

Notice that Fml is a set which belongs to **V**. We leave the proof of the following fact as an exercise to the reader (see EXERCISE 14.5).

FACT 14.6. *For every $\mathscr{L}_{\mathsf{ST}}$-formula φ, we have* $\ulcorner\varphi\urcorner \in \mathrm{Fml}$.

Let M be a class. In abuse of notation, we shall identify the $\mathscr{L}_{\mathsf{ST}}$-structure $(M, \in)$ with M. Furthermore, let $\mathbf{x} = \langle x_0, \dots, x_n \rangle \in M^{n+1}$, i.e., $\mathbf{x}$ is a function with $\operatorname{dom}\mathbf{x} = n+1$ and $\mathbf{x}(i) = x_i$ for all $0 \le i \le n$. Now, we formalise the satisfaction relation for $\mathscr{L}_{\mathsf{ST}}$-formulae with free variables among $\{v_0, \dots, v_n\}$ as follows:

$$M \ulcorner\vDash\urcorner \langle 0, i, j\rangle[\mathbf{x}] :\Longleftrightarrow x_i \in x_j$$

$$M \ulcorner\vDash\urcorner \langle 1, i, j\rangle[\mathbf{x}] :\Longleftrightarrow x_i = x_j$$

$$M \ulcorner\vDash\urcorner \langle 2, f\rangle[\mathbf{x}] :\Longleftrightarrow \neg M \ulcorner\vDash\urcorner f[\mathbf{x}]$$

$$M \ulcorner\vDash\urcorner \langle 3, f, g\rangle[\mathbf{x}] :\Longleftrightarrow M \ulcorner\vDash\urcorner f[\mathbf{x}] \wedge M \ulcorner\vDash\urcorner g[\mathbf{x}]$$

$$M \ulcorner\vDash\urcorner \langle 6, i, f\rangle[\mathbf{x}] :\Longleftrightarrow \exists a \in M \big(M \ulcorner\vDash\urcorner f[\mathbf{x}\tfrac{i}{a}]\big),$$

where for $0 \le k \le n$,

$$(\mathbf{x}\tfrac{i}{a})_k = \begin{cases} x_k & k \neq i, \\ a & k = i. \end{cases}$$

FACT 14.7. *Let M be a class in $\mathbf{V}$ and let $\varphi(\nu_0, \dots, \nu_n)$ be an $\mathscr{L}_{\mathsf{ST}}$-formula. Then for any $x_0, \dots, x_n \in M$ and $\mathbf{x} = \langle x_0, \dots, x_n\rangle$ we have*

$$M \vDash \varphi(x_0, \dots, x_n) \iff M \ulcorner\vDash\urcorner \ulcorner\varphi\urcorner[\mathbf{x}].$$

Let now M be a set in $\mathbf{V}$. We say that a set x is **definable** over M, if there exist an $\mathscr{L}_{\mathsf{ST}}$-formula $\varphi(\nu_0, \dots, \nu_n)$ and $p_1, \dots, p_n \in M$ such that

$$x = \big\{y \in M : M \vDash \varphi(y, p_1, \dots, p_n)\big\} = \big\{y \in M : \varphi^M(y, p_1, \dots, p_n)\big\}.$$

Furthermore, we define

$$\mathrm{Def}(M) := \big\{x \in \mathscr{P}(M) : x \text{ is definable over } M\big\}.$$

Notice that since $M \in \mathbf{V}$, $\mathrm{Def}(M)$ belongs to $\mathbf{V}$ as well. Moreover, by definition we have $\mathrm{Def}(M) \subseteq \mathscr{P}(M)$.

In order to achieve that the set $\mathrm{Def}(M)$ itself is definable within $\mathbf{V}$, we have to use the formalised satisfaction relation. The reason is that we cannot quantify over $\mathscr{L}_{\mathsf{ST}}$-formulae, but need to quantify over elements of Fml instead. By the above definitions we obtain

$$x \in \mathrm{Def}(M) :\Longleftrightarrow$$
$$\exists f \in \mathrm{Fml}\ \exists n \in \omega\ \exists \mathbf{p} \in M^n\ \forall y \in M \big(y \in x \leftrightarrow M \ulcorner\vDash\urcorner f[\langle y, \mathbf{p}\rangle]\big),$$

which shows that we can indeed define the set $\mathrm{Def}(M)$ within $\mathbf{V}$. Thus, by transfinite recursion we can define within $\mathbf{V}$ the **constructible hierarchy** as follows:

$$\mathrm{L}_0 = \emptyset$$

$$\mathrm{L}_\alpha = \bigcup_{\beta\in\alpha} \mathrm{L}_\beta \quad \text{if } \alpha \text{ is a limit ordinal}$$

$$\mathrm{L}_{\alpha+1} = \mathrm{Def}(\mathrm{L}_\alpha)$$

The **constructible universe** is then defined as

$$\mathbf{L} = \bigcup_{\alpha\in\Omega} \mathrm{L}_\alpha.$$

By transfinite recursion one can show that the class **L** is definable within **V**. The goal is now to show that **L** is a model of ZFC. In order to simplify the notation, we will not distinguish between the sets L_α and the corresponding $\mathscr{L}_{\mathsf{ST}}$-structures $(\mathrm{L}_\alpha, \in)$; the same applies to the class **L** and the $\mathscr{L}_{\mathsf{ST}}$-structure $(\mathbf{L}, \in)$.

The following result shows that the structure of **L** is the same as the structure of **V**. In particular, we obtain that **L** contains the same ordinals as **V**.

PROPOSITION 14.8.

(a) *If* $\alpha \in \beta \in \Omega$, *then* $\mathrm{L}_\alpha \subsetneq \mathrm{L}_\beta$.

(b) *For every* $\beta \in \Omega$, L_β *is transitive.*

(c) *For every* $\beta \in \Omega$, $\mathrm{L}_\beta \subseteq \mathrm{V}_\beta$.

(d) *For every* $\beta \in \Omega$, $\mathrm{L}_\beta \cap \Omega = \beta$.

Proof. The proof is by induction on $\beta \in \Omega$. For $\beta = 0$ we have $\mathrm{L}_\beta = \emptyset$ and therefore, the claim is trivial. The limit case follows immediately from the definition of L_β at the limit stage. Hence we may assume that $\beta = \gamma + 1$ is a successor ordinal. For (a) it suffices to check that $\mathrm{L}_\gamma \subseteq \mathrm{L}_\beta$ and $\mathrm{L}_\gamma \in \mathrm{L}_\beta$. For the first claim, let $x \in \mathrm{L}_\gamma$. By transitivity of L_γ we have $x \subseteq \mathrm{L}_\gamma$ and hence

$$x = \{y \in \mathrm{L}_\gamma : (\mathrm{L}_\gamma, \in) \vDash y \in x\} \in \mathrm{Def}(\mathrm{L}_\gamma) = \mathrm{L}_\beta.$$

For the second claim, note that

$$\mathrm{L}_\gamma = \{x \in \mathrm{L}_\gamma : (\mathrm{L}_\gamma, \in) \vDash x = x\} \in \mathrm{L}_\beta.$$

For (b) let $x \in \mathrm{L}_\beta$. Then (a) and the transitivity of L_γ imply that $x \subseteq \mathrm{L}_\gamma \subseteq \mathrm{L}_\beta$. By our induction hypothesis we have $\mathrm{L}_\gamma \subseteq \mathrm{V}_\gamma$ and thus $\mathscr{P}(\mathrm{L}_\gamma) \subseteq \mathscr{P}(\mathrm{V}_\gamma) = \mathrm{V}_\beta$. Since $\mathrm{V}_\beta = \mathrm{Def}(\mathrm{L}_\gamma) \subseteq \mathscr{P}(\mathrm{L}_\gamma)$, (c) holds. For (d), observe first that $\mathrm{L}_\beta \cap \Omega \subseteq \mathrm{V}_\beta \cap \Omega = \beta$. For the reverse inclusion, by induction we have $\mathrm{L}_\gamma \cap \Omega = \gamma$ and therefore, by absoluteness of the formula ordinal(x)', we finally have

$$\gamma = \{\delta \in \mathrm{L}_\gamma : \mathrm{ordinal}(\delta)\} = \{\delta \in \mathrm{L}_\gamma : (\mathrm{L}_\gamma, \in) \vDash \mathrm{ordinal}(\delta)\} \in \mathrm{Def}(\mathrm{L}_\gamma) = \mathrm{L}_\beta\,.$$

⊣

THEOREM 14.9 (LÉVY'S REFLECTION THEOREM). *Let φ be an $\mathscr{L}_{\mathsf{ST}}$-formula and $\alpha \in \Omega$. Then there is $\beta \in \Omega$ with $\beta \geq \alpha$ such that φ is absolute between L_β and $\mathbf{L}$.*

Proof. By THEOREM 1.7, we may assume that φ only contains $\neg$ and $\wedge$ as logical operators and $\exists$ as quantifier. Suppose that $\varphi_0, \dots, \varphi_m$ is a list of all subformulae of φ with the property that all proper subformulae of φ_i occur among $\varphi_0, \dots, \varphi_{i-1}$, i.e., φ is the formula φ_m. Furthermore, assume that all free variables of $\varphi_0, \dots, \varphi_m$ are among $\{v_0, \dots, v_n\}$. For the sake of simplicity, for $\mathbf{x} = \langle x_0, \dots, x_n \rangle \in \mathbf{L}^{n+1}$ we define

$$\varphi(\mathbf{x}) :\equiv \varphi(x_0, \dots, x_n)\,.$$

For every $i \leq m$, we define a class function $F_i : \mathbf{L}^{n+1} \to \Omega$ by stipulating

$$F_i(\mathbf{x}) = \begin{cases} \min_{\in} \left\{\beta \in \Omega : \exists b \in \mathrm{L}_\beta\, \psi^{\mathbf{L}}(\tfrac{\nu}{b}, \mathbf{x})\right\} & \text{if } \varphi_i(\mathbf{x}) \equiv \exists \nu\, \psi(\nu, \mathbf{x}) \\ & \text{and } \exists a \in \mathbf{L}\, \psi^{\mathbf{L}}(\tfrac{\nu}{a}, \mathbf{x}), \\ 0 & \text{otherwise.} \end{cases}$$

Notice that the class function F_i guarantees that if φ_i is of the form $\exists \nu\, \psi(\nu)$ and there is a witness in $\mathbf{L}$ for $\psi(\nu)$, then there is already a witness for $\psi(\nu)$ in L_β.

Now, we recursively define a sequence of ordinals $\langle \beta_k : k \in \omega \rangle$ as follows: Let $\beta_0 := \alpha$. Suppose that β_k is given. Then let

$$\beta_{k+1} := \bigcup \left\{ F_i(\mathbf{x}) : i \leq m \wedge \{x_0, \dots, x_n\} \subseteq \mathrm{L}_{\beta_k} \right\}.$$

Finally, set $\beta := \bigcup_{k \in \omega} \beta_k$. We show by induction that for every $i \leq m$, the formula φ_i is absolute between L_β and $\mathbf{L}$. Suppose that $\{x_0, \dots, x_n\} \subseteq \mathrm{L}_\beta$.

Case 1. φ_i is atomic. Then φ_i is obviously absolute between L_β and $\mathbf{L}$.

Case 2. φ_i is $\neg\varphi_j$ for some $j < i$ and φ_j is absolute between L_β and $\mathbf{L}$. By assumption, we have $\varphi_j^{\mathrm{L}_\beta}(\mathbf{x}) \iff \varphi_j^{\mathbf{L}}(\mathbf{x})$ and hence

$$\varphi_i^{\mathrm{L}_\beta}(\mathbf{x}) \iff \neg\varphi_j^{\mathrm{L}_\beta}(\mathbf{x}) \iff \neg\varphi_j^{\mathbf{L}}(\mathbf{x}) \iff \varphi_i^{\mathbf{L}}(\mathbf{x})\,.$$

Case 3. φ_i is $\varphi_j \wedge \varphi_k$ for $j, k < i$ and φ_j, φ_k are absolute between L_β and $\mathbf{L}$. Then we have

$$\varphi_i^{\mathrm{L}_\beta}(\mathbf{x}) \iff \varphi_j^{\mathrm{L}_\beta}(\mathbf{x}) \wedge \varphi_k^{\mathrm{L}_\beta}(\mathbf{x}) \iff \varphi_j^{\mathbf{L}}(\mathbf{x}) \wedge \varphi_k^{\mathbf{L}}(\mathbf{x}) \iff \varphi_i^{\mathbf{L}}(\mathbf{x})\,.$$

Case 4. φ_i is $\exists \nu \varphi_j$ for some $j < i$ such that φ_j is absolute between L_β and $\mathbf{L}$. Then on the one hand, since $\mathbf{L}$ and $\mathrm{L}_\beta \subseteq \mathbf{L}$, we have

$$\varphi_i^{\mathrm{L}_\beta}(\mathbf{x}) \Longrightarrow \exists \nu \in \mathrm{L}_\beta\, \varphi_j^{\mathrm{L}_\beta}(\mathbf{x}) \Longrightarrow \exists \nu \in \mathbf{L}\, \varphi_j^{\mathbf{L}}(\mathbf{x}) \Longrightarrow \varphi_i^{\mathbf{L}}(\mathbf{x})\,,$$

and on the other hand, by construction of β and since $\{x_0, \dots, x_n\} \subseteq \mathrm{L}_\beta$, we have

$$\varphi_i^{\mathbf{L}}(\mathbf{x}) \Longrightarrow \exists \nu \in \mathbf{L}\, \varphi_j^{\mathbf{L}}(\mathbf{x}) \Longrightarrow \exists \nu \in \mathrm{L}_\beta\, \varphi_j^{\mathrm{L}_\beta}(\mathbf{x}) \Longrightarrow \varphi_i^{\mathrm{L}_\beta}(\mathbf{x}).$$

Hence, φ_i is absolute between L_β and $\mathbf{L}$. ⊣

L ⊨ ZF

Now we are ready to show the following

THEOREM 14.10. *The constructible universe is a model of* ZF, *i.e.,*

$$\mathbf{L} \vDash \mathsf{ZF}.$$

Proof. First notice that since $\emptyset \in \mathbf{L}$, the Axiom of Empty Set holds in $\mathbf{L}$, and since $\mathbf{L}$ is a transitive class, by LEMMA 14.5 also the Axiom of Extensionality and the Axiom of Foundation hold in $\mathbf{L}$.

By applying LEMMA 14.5, we now show that the following five axioms of ZF hold in $\mathbf{L}$:

Axiom of Pairing. Let $a, b \in \mathbf{L}$ and let $\alpha \in \Omega$ be such that $a, b \in \mathrm{L}_\alpha$. Then

$$\{a, b\} = \big\{x \in \mathrm{L}_\alpha : x = a \vee x = b\big\} \in \mathrm{L}_{\alpha+1} \subseteq \mathbf{L}.$$

Axiom of Union. Let $a \in \mathbf{L}$ and let $\alpha \in \Omega$ such that $a \in \mathrm{L}_\alpha$. Since $\mathbf{L}_\alpha$ is transitive, we have $\bigcup a \subseteq \mathrm{L}_\alpha$, and thus,

$$\begin{aligned} \bigcup a &= \big\{x \in \mathrm{L}_\alpha : \exists y (x \in y \wedge y \in a)\big\} \\ &= \big\{x \in \mathrm{L}_\alpha : \mathrm{L}_\alpha \vDash \exists y \in a (x \in y)\big\} \in \mathrm{L}_{\alpha+1}. \end{aligned}$$

Axiom of Infinity. By PROPOSITION 14.8 we have $\omega \in \mathrm{L}_{\omega+1} \subseteq \mathbf{L}$.

Axiom of Power Set. Let $a \in \mathbf{L}$. By the Axiom of Power Set in $\mathbf{V}$ we obtain that $\mathscr{P}(a) \cap \mathbf{L}$ is a set, and thus, there is an $\alpha \in \Omega$ such that $\mathscr{P}(a) \cap \mathbf{L} \subseteq \mathrm{L}_\alpha$. Therefore, we have

$$\mathscr{P}(a) \cap \mathbf{L} = \big\{x \in \mathrm{L}_\alpha : x \subseteq a\big\} = \big\{x \in \mathrm{L}_\alpha : \mathrm{L}_\alpha \vDash x \subseteq a\big\} \in \mathrm{L}_{\alpha+1},$$

since the subset relation is absolute.

It remains to show that the two axiom schema of ZF hold in $\mathbf{L}$ as well:

Axiom Schema of Separation. Let $\varphi(\nu_0, \ldots, \nu_n)$ be an $\mathscr{L}_{\mathsf{ST}}$-formula such that $\mathrm{free}(\varphi) \subseteq \{\nu_0, \ldots, \nu_n\}$, let $\{x, p_1, \ldots, p_n\} \subseteq \mathbf{L}$ and let $\mathbf{p} = \langle p_1, \ldots, p_n\rangle$. It suffices to prove that

$$\big\{y \in x : \mathbf{L} \vDash \varphi(y, \mathbf{p})\big\} \in \mathbf{L}.$$

Let $\alpha \in \Omega$ be such that $\{x, p_1, \ldots, p_n\} \subseteq \mathrm{L}_\alpha$. By LÉVY'S REFLECTION THEOREM, there is a $\beta \in \Omega$ with $\beta \geq \alpha$ such that φ is absolute between L_β and $\mathbf{L}$. Then by transitivity of L_β we have

$$\{y \in x : \mathbf{L} \vDash \varphi(y, \mathbf{p})\} = \{y \in x : \mathrm{L}_\beta \vDash \varphi(y, \mathbf{p})\}$$
$$= \{y \in \mathrm{L}_\beta : \mathrm{L}_\beta \vDash \psi(y, \mathbf{p}, x)\} \in \mathrm{L}_{\beta+1},$$

where $\psi(v_0, \ldots, v_{n+1})$ is the formula $v_0 \in v_{n+1} \wedge \varphi(v_0, \ldots, v_n)$.

Axiom Schema of Replacement. Let φ be an $\mathscr{L}_{\mathsf{ST}}$-formula with $n+2$ free variables and let $\{p_1, \ldots, p_n, A\} \in \mathbf{L}$. Suppose that φ defines a class function in $\mathbf{L}$, i.e.,

$$\forall x \in \mathbf{L}\, \exists! y \in \mathbf{L}\, \varphi(x, y, \mathbf{p}).$$

Consider the function F on A given by

$$F(x) = \min_{\in} \{\beta \in \Omega : \exists y \in \mathrm{L}_\beta\, \varphi^{\mathbf{L}}(x, y, \mathbf{p})\}.$$

Since $A \in \mathbf{L}$, by the Axiom Schema of Replacement applied in $\mathbf{V}$ we have $X := F[A] \in \mathbf{V}$. Now, X is a set of ordinals, and therefore, $\alpha := \bigcup X \in \Omega$. Consider

$$B := \{y \in \mathbf{L} : \exists x \in A\, \varphi^{\mathbf{L}}(x, y, \mathbf{p})\}.$$

Since φ is functional and using the Axiom Schema of Replacement in $\mathbf{V}$, we obtain that B is a set satisfying $B \subseteq \mathrm{L}_\alpha$. Now, by LÉVY'S REFLECTION THEOREM there is an ordinal $\beta \geq \alpha$ such that $A \in \mathrm{L}_\beta$ and φ is absolute between L_β and $\mathbf{L}$. Hence,

$$B = \{y \in \mathbf{L} : \exists x \in A\, \varphi^{\mathbf{L}}(x, y, \mathbf{p})\}$$
$$= \{y \in \mathrm{L}_\beta : \exists x \in A\, \varphi^{\mathrm{L}_\beta}(x, y, \mathbf{p})\} \in \mathrm{L}_{\beta+1}.$$

Therefore, we have shown that $\mathbf{L}$ satisfies all axioms of ZF, i.e., $\mathbf{L} \vDash \mathsf{ZF}$. ⊣

$\mathbf{L} \vDash$ ZFC

In this section, we will show that it is possible to define a class-sized well-ordering of Gödel's model $\mathbf{L}$, from which it follows that $\mathbf{L}$ is in fact a model of the Axiom of Choice. Since our ground model $\mathbf{V}$, in which we carried out the construction of $\mathbf{L}$, was just a model of ZF, it may come as a surprise that $\mathbf{L} \vDash$ ZFC. In particular, we obtain that in *every* model $\mathbf{V}$ of ZF there exists a sub-model of ZFC, no matter whether or not AC holds in $\mathbf{V}$.

The idea of the proof is to show that each level L_α of the constructible hierarchy can be well-ordered, which can be used to construct a well-ordering of $\mathbf{L}$.

Suppose that for some $\alpha \in \Omega$, a well-ordering $\prec_\alpha$ of L_α is given such that for any $\beta \in \Omega$ with $\alpha < \beta$, $\prec_\beta$ is an *end-extension* of $\prec_\alpha$, i.e., $\prec_\beta$ satisfies the following two properties:

- For any $x, y \in \mathrm{L}_\alpha$, if $x \prec_\alpha y$ then $x \prec_\beta y$.
- If $x \in \mathrm{L}_\alpha$ and $y \in \mathrm{L}_\beta \setminus \mathrm{L}_\alpha$, then $x \prec_\beta y$.

Assuming the existence of such a well-ordering $\prec_\alpha$ for every $\alpha \in \Omega$, we obtain a class-sized well-ordering of **L** by stipulating

$$x \prec_{\mathbf{L}} y :\Longleftrightarrow \exists \alpha \in \Omega (x \prec_\alpha y).$$

We will define $\prec_\alpha$ by transfinite recursion. Note that the only non-trivial case will be the successor case. Recall that we have defined $\mathrm{L}_{\alpha+1}$ to be $\mathrm{Def}(\mathrm{L}_\alpha)$, and each element of $\mathrm{Def}(\mathrm{L}_\alpha)$ is of the form $x = D(\alpha, f, \mathbf{p})$, where $f \in \mathrm{Fml}$, $\mathbf{p} \in \mathrm{seq}(\mathrm{L}_\alpha)$ and

$$D(\alpha, f, \mathbf{p}) := \left\{ y \in \mathrm{L}_\alpha : \mathrm{L}_\alpha \ulcorner \vDash \urcorner f[\langle y, \mathbf{p} \rangle] \right\}.$$

Therefore, the task of defining a well-ordering on $\mathrm{L}_{\alpha+1}$ essentially reduces to ordering triples $(\alpha, f, \mathbf{p})$, whereby one has to take into account that different triples can generate the same set. Thus, we will also define recursively a well-ordering $\tilde{\prec}_\alpha$ on triples of the form $(\beta, f, \mathbf{p})$ for $\beta < \alpha$, $f \in \mathrm{Fml}$, and $\mathbf{p} \in \mathrm{seq}(\mathrm{L}_\alpha)$. Now, since the sequence of parameters $\mathbf{p}$ is in L_α, one needs to refer to $\prec_\alpha$ in order to define $\tilde{\prec}_{\alpha+1}$. Hence, we define both well-orderings by a simultaneous recursion.

Moreover, observe that ordering such triples further requires ordering Fml. By construction, we have $\mathrm{Fml} \subseteq \mathrm{V}_\omega$, and thus, Fml is countable. Hence, there is a well-ordering $\prec_{\mathrm{Fml}}$ of Fml. We proceed as follows:

- Let $\prec_0$ be the empty ordering, i.e., $\prec_0 := \emptyset$.
- Suppose that for some $\alpha \in \Omega$, $\prec_\alpha$ has already been defined. We first tackle $\tilde{\prec}_{\alpha+1}$. Let $\beta, \gamma < \alpha$, $f, g \in \mathrm{Fml}$, and let $\mathbf{p}, \mathbf{q} \in \mathrm{seq}(\mathrm{L}_\alpha)$ of length $l_{\mathbf{p}}$ and $l_{\mathbf{q}}$, respectively. Then we define:

$$\begin{aligned} \langle \beta, f, \mathbf{p} \rangle \tilde{\prec}_{\alpha+1} \langle \gamma, g, \mathbf{q} \rangle :\Longleftrightarrow\ & \beta < \gamma \vee (\beta = \gamma \wedge f \prec_{\mathrm{Fml}} g) \\ & \vee (\beta = \gamma \wedge f = g \wedge l_{\mathbf{p}} < l_{\mathbf{q}}) \\ & \vee \Big(\beta = \gamma \wedge f = g \wedge l_{\mathbf{p}} = l_{\mathbf{q}} \wedge \\ & \quad \exists n \in \omega \big(n = \min \{ m < l_{\mathbf{p}}, l_{\mathbf{q}} : \mathbf{p}(m) \neq \mathbf{q}(m) \} \\ & \quad\quad \wedge \mathbf{p}(n) \prec_\alpha \mathbf{q}(n) \big) \Big) \end{aligned}$$

  Now, we are ready to define $\prec_{\alpha+1}$. For $x, y \in \mathrm{L}_{\alpha+1}$, we set

$$x \prec_{\alpha+1} y :\Longleftrightarrow \langle \beta, f, p \rangle \tilde{\prec}_{\alpha+1} \langle \gamma, g, q \rangle,$$

  where $\langle \beta, f, \mathbf{p} \rangle$ and $\langle \gamma, g, \mathbf{q} \rangle$ are $\tilde{\prec}_{\alpha+1}$-minimal triples such that $x = D(\beta, f, \mathbf{p})$ and $y = D(\gamma, g, \mathbf{q})$.

- If α is a limit ordinal, then we set

$$\prec_\alpha := \bigcup_{\beta < \alpha} \prec_\beta .$$

By construction, $\prec_\beta$ is an end-extension of $\prec_\alpha$ for all $\alpha, \beta \in \Omega$ with $\alpha < \beta$. Therefore, the ordering $\prec_{\mathbf{L}}$ as defined above is a well-ordering of the entire constructible universe **L**. Note that the existence of a well-ordering of the whole model, a so-called **global well-ordering**, yields a strengthening of the Axiom of Choice, namely a class-sized choice function which chooses an element from every set.

As a consequence of the existence of a global well-ordering of **L**, we obtain the following

THEOREM 14.11. *The constructible universe is a model of the Axiom of Choice, i.e.,*

$$\mathbf{L} \vDash \mathsf{ZFC}\,.$$

Proof. For every family $\mathscr{F} \in \mathrm{L}$ such that $\emptyset \notin \mathscr{F}$, there is a choice function

$$f : \mathscr{F} \to \bigcup \mathscr{F},$$

where $f(x)$ is the $\prec_{\mathbf{L}}$-minimal element of $x \in \mathscr{F}$. ⊣

In particular, as a consequence of THEOREM 14.11 we obtain that the consistency of ZF implies the consistency of ZFC, i.e.,

$$\mathrm{Con}(\mathsf{ZF}) \Longrightarrow \mathrm{Con}(\mathsf{ZFC}).$$

NOTES

The constructible universe **L** was introduced by Gödel in his 1938 paper [17], in which he proved both that if ZFC is consistent, then **L** is a model of the Axiom of Choice and the Continuum Hypothesis. The construction of **L** presented here is mainly taken from Koepke [29] (see also Kunen [30]). According to Bernays, Gödel originally used the old German script $\mathfrak{L}$ to denote the constructible universe, where $\mathfrak{L}$ is a capital C and not—as one could think—a capital L. In 1963, Cohen developed in [4] and [5] the method of forcing (see, e.g., Halbeisen [21] for an introduction) to prove that both the Axiom of Choice and the Continuum Hypothesis are in fact independent of the axioms of ZF.

EXERCISES

14.0 Prove FACT 14.4, i.e., show that every $\mathscr{L}_{\mathsf{ST}}$-formula which is logically equivalent to a Δ-formula is absolute for M.

14.1 For a cardinal κ, we define the class

$$H_\kappa := \{x : |\,\mathrm{TC}(\{x\})| < \kappa\}.$$

Examine which of the axioms of ZFC hold in the structure $(H_\kappa, \in)$ depending on κ.

14.2 Show that Zermelo's axiom system Z does not imply the Axiom Schema of Replacement.

Hint: Show that $V_{\omega+\omega}$ is a model of Z but the Axiom Schema of Replacement fails in $V_{\omega+\omega}$.

14.3 Show that the axiom system of ZF is not equivalent to a FINITE set of axioms.

Hint: Note that LÉVY'S REFLECTION THEOREM also holds if we replace L_α by V_α. Use this fact and the Axiom of Foundation to show that the assumption that ZF is equivalent to a FINITE axiom system leads to a contradiction.

14.4 Use the fact that $\text{ind}(x)$ can be expressed by a Δ-formula to show that the formula $\varphi(x)$ given by $x = \omega$ is absolute for every model of ZF.

14.5 Prove that for every $\mathscr{L}_{\mathsf{ST}}$-formula φ we have $\ulcorner\varphi\urcorner \in \text{Fml}$.

Hint: Use induction on the construction of φ.

14.6 The Continuum Hypothesis, denoted by CH, is the statement $2^\omega = \omega_1$, and the Generalised Continuum Hypothesis, denoted by GCH, states

$$\forall \alpha \in \Omega \left(2^{\omega_\alpha} = \omega_{\alpha+1}\right).$$

(a) Prove that $\mathbf{L} \vDash \mathsf{GCH}$.

(b) Show that $\text{Con}(\mathsf{ZF}) \implies \text{Con}(\mathsf{ZFC} + \mathsf{GCH})$.

Hint: Use the well-ordering $<_{\mathbf{L}}$ on $\mathbf{L}$.

Chapter 15
Models and Ultraproducts

The goal of this chapter is to show that every consistent $\mathscr{L}$-theory has a model, no matter whether the signature $\mathscr{L}$ is countable or uncountable. In addition, we will show that if a consistent $\mathscr{L}$-theory T has an infinite model, then, on the one hand, T has arbitrarily large models, and on the other hand, T has a model of size at most $\max\{\aleph_0, |\mathscr{L}|\}$.

In order to prove these results, we shall work within a model of ZFC, in particular, we shall make use of the $\mathsf{Axiom\ of\ Choice}$. Therefore, in contrast to the proofs of the corresponding results in Part II, the following proofs are not constructive in general. As a matter of fact, we would like to mention that even though the proofs are carried out in a model of ZFC, in general, they cannot be carried out in ZFC. In fact, we do not work with ZFC as a formal system, but we just take a model of ZFC and use it as a framework in which we carry out the proofs.

Filters and Ultrafilters

Let S be an arbitrary non-empty set and let $\mathscr{P}(S)$ be the *power-set* of S, i.e., the set of all subsets of S. A set $\mathscr{F} \subseteq \mathscr{P}(S)$ is called a **filter** over S, if $\mathscr{F}$ has the following properties:

- $S \in \mathscr{F}$ and $\emptyset \notin \mathscr{F}$
- $(x \in \mathscr{F} \wedge y \in \mathscr{F}) \to (x \cap y) \in \mathscr{F}$
- $(x \in \mathscr{F} \vee y \in \mathscr{F}) \to (x \cup y) \in \mathscr{F}$

In particular, if $x \in \mathscr{F}$ and $x \subseteq y$, then $y \in \mathscr{F}$. Thus, a filter over S is a set of subsets of S which does not contain the empty set and which is closed under intersections and supersets. For example, the set $\{S\}$ is a filter over S.

L. Halbeisen, R. Krapf, *Gödel's Theorems and Zermelo's Axioms*,
https://doi.org/10.1007/978-3-030-52279-7_15

A more interesting example of a filter over S is the set

$$\mathscr{F} := \{x \subseteq S : S \setminus x \text{ is finite}\},$$

which is the so-called *Fréchet-filter*. Now, a set $\mathscr{U} \subseteq \mathscr{P}(S)$ is called an **ultrafilter** over S, if $\mathscr{U}$ is a filter over S and for each $x \in \mathscr{P}(S)$, either $x \in \mathscr{U}$ or $(S \setminus x) \in \mathscr{U}$. In other words, a filter $\mathscr{U}$ is an ultrafilter if $\mathscr{U}$ is not properly contained in any filter. For example, for each $a \in S$, the set

$$\mathscr{U}_a := \{x \subseteq S : a \in x\}$$

is an ultrafilter over S, called *trivial ultrafilter*. In particular, every ultrafilter over a finite set is trivial. It is natural to ask whether there exist also non-trivial ultrafiters, e.g., ultrafilters which contain the Fréchet-filter. Or in general, we can ask whether every filter can be extended to an ultrafiter. This is what the *Ultrafilter Theorem* states:

> Ultrafilter Theorem: If $\mathscr{F}$ is a filter over a set S, then $\mathscr{F}$ can be extended to an ultrafilter.

Surprisingly, we cannot prove the Ultrafilter Theorem without assuming some form of the Axiom of Choice. However, proving the Ultrafilter Theorem within ZFC is not so hard (see EXERCISE 15.0).

Ultraproducts and Ultrapowers

Let $\mathscr{L}$ be an arbitrary but fixed signature, let I be a non-empty set, and for each $\iota \in I$, let $\mathbf{M}_\iota$ be an $\mathscr{L}$-structure with domain A_ι. Furthermore, let $A := \times_{\iota \in I} A_\iota$ be the Cartesian product of the sets A_ι. Below, we shall identify the elements of A with functions $f : I \to \bigcup_{\iota \in I} A_\iota$, where for each $\iota \in I$, $f(\iota) \in A_\iota$. Finally, let $\mathscr{U} \subseteq \mathscr{P}(I)$ be an ultrafilter over I. With respect to $\mathscr{U}$, we define a binary relation $\sim$ on A by stipulating

$$f \sim g \;:\Longleftrightarrow\; \{\iota \in I : f(\iota) = g(\iota)\} \in \mathscr{U}\,.$$

FACT 15.1. *The relation $\sim$ is an equivalence relation.*

Proof. We have to show that $\sim$ is reflexive, symmetric, and transitive.

- For all $f \in A$, we obviously have $f \sim f$.
- For all $f, g \in A$, we obviously have $f \sim g \leftrightarrow g \sim f$.
- Let $f, g, h \in A$ and assume that $f \sim g$ and $g \sim h$. Furthermore, let $x := \{\iota \in I : f(\iota) = g(\iota)\}$ and $y := \{\iota \in I : g(\iota) = h(\iota)\}$. Then $x, y \in \mathscr{U}$, and since $\mathscr{U}$ is a filter, $x \cap y$ as well as every superset of $x \cap y$ belongs to $\mathscr{U}$. Thus,

  $$x \cap y \subseteq \{\iota \in I : f(\iota) = h(\iota)\} \in \mathscr{U},$$

  which shows that $f \sim h$. ⊣

For each $f \in A$, let

$$[f] := \{g \in A : g \sim f\}$$

and let

$$A^* := \{[f] : f \in A\}.$$

We now construct the $\mathscr{L}$-structure $\mathbf{M}^*$ with domain A^* as follows:

- For every constant symbol $c \in \mathscr{L}$, let $f_c \in A$ be defined by stipulating

  $$f_c(\iota) := c^{\mathbf{M}_\iota} \quad \text{for all } \iota \in I,$$

  and let

  $$c^{\mathbf{M}^*} := [f_c].$$

- For every n-ary function symbol $F \in \mathscr{L}$, let $F^{\mathbf{M}^*} : (A^*)^n \to A^*$ be such that

  $$F^{\mathbf{M}^*}\big([f_0], \dots, [f_{n-1}]\big) = [f] \iff \Big\{\iota \in I : F^{\mathbf{M}_\iota}\big(f_0(\iota), \dots, f_{n-1}(\iota)\big) = f(\iota)\Big\} \in \mathscr{U}.$$

- For every n-ary relation symbol $R \in \mathscr{L}$, let $R^{\mathbf{M}^*} \subseteq (A^*)^n$ be such that

  $$\big\langle [f_0], \dots, [f_{n-1}] \big\rangle \in R^{\mathbf{M}^*} \iff \Big\{\iota \in I : \big\langle f_0(\iota), \dots, f_{n-1}(\iota)\big\rangle \in R^{\mathbf{M}_\iota}\Big\} \in \mathscr{U}.$$

FACT 15.2. *The constants $c^{\mathbf{M}^*}$, the functions $F^{\mathbf{M}^*}$, and the relations $R^{\mathbf{M}^*}$ are well-defined.*

Proof. We just show that the functions $F^{\mathbf{M}^*} : (A^*)^n \to A$ are well-defined and leave the proofs for $c^{\mathbf{M}^*}$ and $R^{\mathbf{M}^*}$ as an exercise (see EXERCISE 15.1). Let $F \in \mathscr{L}$ be an n-ary function symbol and let $\langle f_0, \dots, f_{n-1}\rangle$ and $\langle g_0, \dots, g_{n-1}\rangle$ be elements in A^n such that for each $0 \leq i < n$ we have

$$f_i \sim g_i \quad \text{or equivalently} \quad [f_i] = [g_i]\,.$$

Furthermore, we define $f, g \in A$ by stipulating

$$f(\iota) := F^{\mathbf{M}_\iota}\big(f_0(\iota), \dots, f_{n-1}(\iota)\big) \quad \text{and} \quad g(\iota) := F^{\mathbf{M}_\iota}\big(g_0(\iota), \dots, g_{n-1}(\iota)\big)\,.$$

By definition of $\sim$ and since $\mathscr{U}$ is an ultrafilter over I, we have

$$\big\{\iota \in I : f_0(\iota) = g_0(\iota) \wedge \dots \wedge f_{n-1}(\iota) = g_{n-1}(\iota)\big\} \in \mathscr{U}\,,$$

and consequently, we obtain

$$\left\{\iota \in I : F^{\mathbf{M}_\iota}\big(f_0(\iota), \dots, f_{n-1}(\iota)\big) = F^{\mathbf{M}_\iota}\big(g_0(\iota), \dots, g_{n-1}(\iota)\big)\right\} \in \mathscr{U}.$$

Hence, $\{\iota \in I : f(\iota) = g(\iota)\} \in \mathscr{U}$, which shows that $[f] = [g]$ and implies that

$$F^{\mathbf{M}^*}\big([f_0], \dots, [f_{n-1}]\big) = F^{\mathbf{M}^*}\big([g_0], \dots, [g_{n-1}]\big).$$

Therefore, the value of the function $F^{\mathbf{M}^*}$ does not depend on the particular representatives that we choose from the equivalence classes $[f_i]$. ⊣

The $\mathscr{L}$-structure $\mathbf{M}^*$ with domain A^* is called the **ultraproduct** of the $\mathscr{L}$-structures $\mathbf{M}_\iota$ ($\iota \in I$) with respect to the ultrafilter $\mathscr{U}$ over I. If for all $\iota \in I$ we have $\mathbf{M}_\iota = \mathbf{M}$ for some $\mathscr{L}$-structure $\mathbf{M}$, then $\mathbf{M}^*$ is called the **ultrapower** of $\mathbf{M}$ with respect to $\mathscr{U}$.

In the next section, we show that if each $\mathscr{L}$-structure $\mathbf{M}_\iota$ is a model of some $\mathscr{L}$-theory T, then also the ultraproduct $\mathbf{M}^*$ is a model of T.

Łoś's Theorem

As above, let $\mathscr{L}$ be an arbitrary signature, let I be a non-empty set, and for each $\iota \in I$, let $\mathbf{M}_\iota$ be an $\mathscr{L}$-structure with domain A_ι. Finally, let $\mathscr{U}$ be an ultrafilter over I and let $\mathbf{M}^*$ be the ultraproduct of the $\mathscr{L}$-structures $\mathbf{M}_\iota$ ($\iota \in I$) with respect to $\mathscr{U}$. The following result allows us to decide whether a given $\mathscr{L}$-sentence is valid in $\mathbf{M}^*$.

THEOREM 15.3 (ŁOŚ'S THEOREM). *For each $\mathscr{L}$-sentence σ, we have*

$$\mathbf{M}^* \vDash \sigma \iff \{\iota \in I : \mathbf{M}_\iota \vDash \sigma\} \in \mathscr{U}.$$

Proof. By THEOREM 1.7, for every $\mathscr{L}$-sentence σ there is an equivalent $\mathscr{L}$-sentence σ' which contains only $\neg$ and $\wedge$ as logical operators and $\exists$ as quantifier. Therefore, it is enough to prove ŁOŚ'S THEOREM for the $\mathscr{L}$-sentence σ'. The proof is by induction on the number of the symbols $\neg$, $\wedge$, and $\exists$ which appear in the $\mathscr{L}$-sentence σ'.

By construction of $\mathbf{M}^*$, ŁOŚ'S THEOREM holds for atomic $\mathscr{L}$-sentences σ'. Now, assume that $\sigma' \equiv \neg\sigma_0$ and that ŁOŚ'S THEOREM holds for σ_0. Then we have:

$$\begin{aligned}
\mathbf{M}^* \vDash \neg\sigma_0 &\iff \mathbf{M}^* \nvDash \sigma_0 \\
&\iff \{\iota \in I : \mathbf{M}_\iota \vDash \sigma_0\} \notin \mathscr{U} \\
&\iff I \setminus \{\iota \in I : \mathbf{M}_\iota \vDash \sigma_0\} \in \mathscr{U} \\
&\iff \{\iota \in I : \mathbf{M}_\iota \vDash \neg\sigma_0\} \in \mathscr{U}
\end{aligned}$$

Now, assume that $\sigma' \equiv \sigma_1 \wedge \sigma_2$ and that ŁOŚ'S THEOREM holds for σ_1 and σ_2. Then we have:

$$\begin{aligned}
\mathbf{M}^* \vDash \sigma_1 \wedge \sigma_2 &\iff \mathbf{M}^* \vDash \sigma_1 \quad \text{AND} \quad \mathbf{M}^* \vDash \sigma_2 \\
&\iff \underbrace{\{\iota \in I : \mathbf{M}_\iota \vDash \sigma_1\}}_{=:x_1} \in \mathscr{U} \text{ AND } \underbrace{\{\iota \in I : \mathbf{M}_\iota \vDash \sigma_2\}}_{=:x_2} \in \mathscr{U} \\
&\iff x_1 \cap x_2 \in \mathscr{U} \\
&\iff \{\iota \in I : \mathbf{M}_\iota \vDash \sigma_1 \wedge \sigma_2\} \in \mathscr{U}
\end{aligned}$$

Finally, assume that $\sigma' \equiv \exists\nu\sigma_0$ (for some variable ν) and that for any $[g] \in A^*$ we have

$$\mathbf{M}^* \tfrac{[g]}{\nu} \vDash \sigma_0(\nu) \iff \{\iota \in I : \mathbf{M}_\iota \tfrac{g(\iota)}{\nu} \vDash \sigma_0(\nu)\} \in \mathscr{U}.$$

Then we have:

$$\begin{aligned}
\mathbf{M}^* \vDash \exists\nu\sigma_0 &\iff \text{IT EXISTS } [g_0] \text{ IN } A^* : \quad \mathbf{M}^* \tfrac{[g_0]}{\nu} \vDash \sigma_0(\nu) \\
&\iff \text{IT EXISTS } [g_0] \text{ IN } A^* : \underbrace{\{\iota \in I : \mathbf{M}_\iota \tfrac{g_0(\iota)}{\nu} \vDash \sigma_0(\nu)\}}_{=:x} \in \mathscr{U}
\end{aligned}$$

Because $x \subseteq \{\iota \in I : \mathbf{M}_\iota \vDash \exists\nu\sigma_0\}$, we have $\{\iota \in I : \mathbf{M}_\iota \vDash \exists\nu\sigma_0\} \in \mathscr{U}$, which shows that

$$\mathbf{M}^* \vDash \exists\nu\sigma_0 \implies \{\iota \in I : \mathbf{M}_\iota \vDash \exists\nu\sigma_0\} \in \mathscr{U}.$$

In order to show the converse implication, we have to make use of the Axiom of Choice. If, for $\iota \in I$, $\mathbf{M}_\iota \vDash \exists\nu\sigma_0$, then let $a_\iota \in A_\iota$ be such that $\mathbf{M}_\iota \frac{a_\iota}{\nu} \vDash \sigma_0(\nu)$, otherwise, let a_ι be an arbitrary element of A_ι. Now, for the function

$$\begin{aligned}
g_0 : I &\to \bigcup A \\
\iota &\mapsto a_\iota
\end{aligned}$$

we have $\{\iota \in I : \mathbf{M}_\iota \vDash \exists\nu\sigma_0\} = \{\iota \in I : \mathbf{M}_\iota \frac{g_0(\iota)}{\nu} \vDash \sigma_0(\nu)\}$. In particular, if $\{\iota \in I : \mathbf{M}_\iota \vDash \exists\nu\sigma_0\} \in \mathscr{U}$, then also

$$\{\iota \in I : \mathbf{M}_\iota \tfrac{g_0(\iota)}{\nu} \vDash \sigma_0(\nu)\} \in \mathscr{U},$$

which shows that

$$\{\iota \in I : \mathbf{M}_\iota \vDash \exists\nu\sigma_0\} \in \mathscr{U} \implies \mathbf{M}^* \vDash \exists\nu\sigma_0,$$

and consequently, we obtain

$$\mathbf{M}^* \vDash \exists\nu\sigma_0 \iff \{\iota \in I : \mathbf{M}_\iota \vDash \exists\nu\sigma_0\} \in \mathscr{U}.$$

⊣

The Completeness Theorem for Uncountable Signatures

In Chapter 5, we have proven GÖDEL'S COMPLETENESS THEOREM 5.5 (i.e., the COMPLETENESS THEOREM for countable signatures). The proof given there was based on potentially infinite lists, and the metamathematical assumptions we made were very mild. In fact, our proof for GÖDEL'S COMPLETENESS THEOREM 5.5 can be carried out effectively in a kind of *algorithmic* way. In contrast to the proof for countable signatures, the proof of the COMPLETENESS THEOREM for uncountable signatures — which will follow from the semantic form of the COMPACTNESS THEOREM 2.15 — is much more formal. In particular, it makes use of ŁOŚ'S THEOREM 15.3, which is based on the existence of ultrafilters and choice functions, and is carried out in a model of ZFC — but not in ZFC itself.

THEOREM 15.4 (SEMANTIC FORM OF THE COMPACTNESS THEOREM). *Let* T *be an* $\mathscr{L}$*-theory such that for every finite subset* $\Phi \subseteq \mathsf{T}$ *there is an* $\mathscr{L}$*-structure* $\mathbf{M}_\Phi$ *such that* $\mathbf{M}_\Phi \vDash \Phi$. *Then* T *has a model.*

Proof. Let I be the set of all finite subsets of T, i.e.,

$$I := \{\Phi \subseteq \mathsf{T} : \Phi \text{ is finite}\}.$$

For each $\Phi \in I$, let $\mathbf{M}_\Phi$ be an $\mathscr{L}$-structure with domain A_Φ such that $\mathbf{M}_\Phi \vDash \Phi$. Furthermore, for every $\Phi \in I$ let

$$\Delta(\Phi) := \{\Phi' \in I : \Phi \subseteq \Phi'\}.$$

In other words, $\Delta(\Phi)$ is the set of all finite supersets $\Phi' \supseteq \Phi$. In particular, for every $\Phi \in I$ we have $\Phi \in \Delta(\Phi)$ and $\Delta(\Phi) \subseteq I$. Now, for all $\Phi_1, \Phi_2 \in I$ we have $\Delta(\Phi_1) \cap \Delta(\Phi_2) = \Delta(\Phi_1 \cup \Phi_2)$, where $\Phi_1 \cup \Phi_2 \in I$. Therefore, the set

$$\mathscr{F} := \{\Psi \subseteq I : \exists \Phi \in I\big(\Delta(\Phi) \subseteq \Psi\big)\}$$

is a filter over I, which, by the Ultrafilter Theorem, can be extended to an ultrafilter $\mathscr{U}$.

Let $\mathbf{M}^*$ with domain A^* be the ultraproduct of the $\mathscr{L}$-structures $\mathbf{M}_\Phi$ ($\Phi \in I$) with respect to the ultrafilter $\mathscr{U}$ over I, and let $\sigma_0 \in \mathsf{T}$ be an arbitrary $\mathscr{L}$-sentence. Then $\{\sigma_0\} \in I$ and $\mathbf{M}_{\{\sigma_0\}} \vDash \sigma_0$. Moreover, for every $\Phi \in \Delta(\{\sigma_0\})$ we have $\mathbf{M}_\Phi \vDash \sigma_0$. Therefore, we have

$$\Delta(\{\sigma_0\}) = \{\Phi \in I : \sigma_0 \in \Phi\} \subseteq \{\Phi \in I : \mathbf{M}_\Phi \vDash \sigma_0\}.$$

Now, since $\Delta(\{\sigma_0\}) \in \mathscr{F} \subseteq \mathscr{U}$, by ŁOŚ'S THEOREM 15.3 we obtain

$$\mathbf{M}^* \vDash \sigma_0,$$

and since $\sigma_0 \in \mathsf{T}$ was arbitrary, this shows that $\mathbf{M}^* \vDash \mathsf{T}$. Hence, T has a model. ⊣

As a consequence of THEOREM 15.4 and GÖDEL'S COMPLETENESS THEOREM 5.5, we obtain the COMPLETENESS THEOREM for arbitrarily large signatures.

THEOREM 15.5 (COMPLETENESS THEOREM). *If $\mathscr{L}$ is an arbitrary signature and T is a consistent set of $\mathscr{L}$-sentences, then T has a model.*

Proof. Firstly, if T is consistent, then, by the COMPACTNESS THEOREM 2.15, every finite subset $\Phi \subseteq \mathsf{T}$ is consistent. Secondly, as in the proof of GÖDEL'S COMPLETENESS THEOREM 5.5, for every finite subset of $\Phi \subseteq \mathsf{T}$ we can construct an $\mathscr{L}'$-structure $\mathbf{M}'_\Phi$ with domain A_Φ, such that $\mathbf{M}'_\Phi \vDash \Phi$, where $\mathscr{L}'$ is the finite subset of $\mathscr{L}$ consisting of all non-logical symbols which appear in sentences of Φ. Now, we extend each $\mathscr{L}'$-structure $\mathbf{M}'_\Phi$ to an $\mathscr{L}$-structure $\mathbf{M}_\Phi$ with the same domain A_Φ such that $\mathbf{M}_\Phi \vDash \Phi$ (see EXERCISE 3.2). Hence, for every finite subset of $\Phi \subseteq \mathsf{T}$ there is an $\mathscr{L}$-structure $\mathbf{M}_\Phi$ such that $\mathbf{M}_\Phi \vDash \Phi$, and therefore, we can apply THEOREM 15.4 in order to construct a model $\mathbf{M}^* \vDash \mathsf{T}$. ⊣

As an immediate consequence of the COMPLETENESS THEOREM 15.5 and the SOUNDNESS THEOREM 3.8, we obtain the following

COROLLARY 15.6. *For any signature $\mathscr{L}$, a set T of $\mathscr{L}$-sentences has a model if and only if T is consistent.*

The Upward Löwenheim-Skolem Theorem

Next, we will show that every $\mathscr{L}$-theory which has an infinite model has arbitrarily large models.

THEOREM 15.7 (UPWARD LÖWENHEIM-SKOLEM THEOREM). *Let T be an $\mathscr{L}$-theory which has an infinite model, and let κ be an arbitrarily large cardinal. Then there exists a model $\mathbf{M}^* \vDash \mathsf{T}$ with domain A^* such that $|A^*| \geq \kappa$ (i.e., the cardinality of A^* is at least κ).*

Proof. For each $\gamma \in \kappa$, we define a constant symbol c_γ which does not belong to $\mathscr{L}$. Let $\mathscr{L}^* := \mathscr{L} \cup \{c_\gamma : \gamma \in \kappa\}$. Furthermore, let T^* be the $\mathscr{L}^*$-theory consisting of the sentences in T together with the sentences $c_\gamma \neq c_{\gamma'}$ (for any distinct $\gamma, \gamma' \in \kappa$). As in the proof of THEOREM 15.4, let I be the set of all finite subsets of T^*. Now, let $\mathbf{M} \vDash \mathsf{T}$ be a model with infinite domain A. For any $\Phi \in I$, we can extend the $\mathscr{L}$-structure $\mathbf{M}$ to an $\mathscr{L}^*$-structure $\mathbf{M}_\Phi$ such that

$$\mathbf{M}_\Phi \vDash \mathsf{T} + \Phi.$$

In order to see this, notice that the domain A of $\mathbf{M}$ is infinite and that there are just finitely many constant symbols c_γ which appear in Φ. Therefore, we can apply THEOREM 15.4 in order to construct an $\mathscr{L}^*$-structure $\mathbf{M}^*$ with domain A^* such that $\mathbf{M}^* \vDash \mathsf{T}^*$. Finally, by definition of T^*, the elements $c_\gamma^{\mathbf{M}^*}$ in A^* (for $\gamma \in \kappa$) are pairwise distinct, which shows that $|A^*| \geq \kappa$. ⊣

As an immediate consequence of the UPWARD LÖWENHEIM–SKOLEM THEOREM 15.7, we get the following

COROLLARY 15.8. *If an $\mathscr{L}$-theory* T *has a countably infinite model, then* T *also has an uncountable model. In particular,* PA *has an uncountable model.*

As a matter of fact, we would like to mention that the proof of the UPWARD LÖWENHEIM–SKOLEM THEOREM 15.7 can be carried out neither in the formal language of ZFC (since we use an infinite set of constant symbols), nor in the language of metamathematics (since we use THEOREM 15.4, which is based on ŁOŚ'S THEOREM 15.3 and therefore on ultrafilters).

The Downward Löwenheim-Skolem Theorem

The last result of this chapter provides an upper bound for the minimum size of a model of a given theory.

THEOREM 15.9 (DOWNWARD LÖWENHEIM-SKOLEM THEOREM). *If a consistent $\mathscr{L}$-theory* T *has an infinite model, then* T *has a model of size at most* $\max\{\aleph_0, |\mathscr{L}|\}$.

Proof. If the signature $\mathscr{L}$ is countable, then, by GÖDEL'S COMPLETENESS THEOREM 5.5, T has a model, which is — by construction — a countable model. Now, assume that $|\mathscr{L}|$ (i.e., the cardinality of $\mathscr{L}$) is uncountable. First notice that with the signature $\mathscr{L}$ we can build at most $|\mathscr{L}|$ terms. In order to see this, recall that a term is just a special finite string of logical and non-logical symbols, and by FACT 13.9, the cardinality of the set of such strings is $\max\{\aleph_0, |\mathscr{L}|\}$. Now, in order to build a model $\mathbf{M} \vDash \mathsf{T}$ of cardinality at most $\max\{\aleph_0, |\mathscr{L}|\}$, we can essentially follow the proof of GÖDEL'S COMPLETENESS THEOREM 5.5. However, instead of potentially infinite lists we have to work with actual infinite sequences of length at most $|\mathscr{L}|$. At the end of the construction, the domain of $\mathbf{M}$ will be a sequence of length at most $|\mathscr{L}|$ of sequences of length at most $|\mathscr{L}|$. ⊣

As an immediate consequence of the DOWNWARD LÖWENHEIM–SKOLEM THEOREM 15.9, we get the following

COROLLARY 15.10. *If* T *is a consistent $\mathscr{L}$-theory and the signature $\mathscr{L}$ is countable, then* T *has a countable model.*

As a matter of fact, we would like to mention that the proof of the DOWNWARD LÖWENHEIM–SKOLEM THEOREM 15.9 cannot be carried out in the formal language of ZFC either. Otherwise, since the signature of ZFC just contains the single symbol $\in$ and is therefore countable, we would be able to construct a countable model of ZFC within a model of ZFC. In particular, we would be able to prove within ZFC that ZFC is consistent, which obviously contradicts the SECOND INCOMPLETENESS THEOREM.

NOTES

Most of the material of this chapter is taken from Bell and Slomson [3, Ch. 5], where one can find some more historical background. ŁOŚ'S THEOREM, also called the FUNDAMENTAL THEOREM OF ULTRAPRODUCTS, is due to the Polish mathematician Łoś (see, e.g., [32]). A first version of the LÖWENHEIM-SKOLEM THEOREMS was proved by Löwenheim in 1915 (see [33]). Some years later, Skolem generalised Löwenheim's result in [48].

EXERCISES

15.0 Find a proof of the Ultrafilter Theorem within ZFC.

Hint: First take a well-ordering of $\mathscr{P}(S)$, and then extend the filter $\mathscr{F}$ over S to an ultrafilter by transfinite induction.

15.1 Complete the proof of FACT 15.2, i.e., show that the constants $c^{\mathbf{M}^*}$ and the relations $R^{\mathbf{M}^*}$, defined in the construction of the $\mathscr{L}$-structure $\mathbf{M}^*$, are well-defined.

15.2 Prove ŁOŚ'S THEOREM 15.3 for $\mathscr{L}$-sentences σ' which contain only $\neg$ and $\vee$ as logical operators and $\forall$ as quantifier.

Chapter 16
Models of Peano Arithmetic

The Standard Model of Peano Arithmetic in ZF

In this section, we will show that ZF is sufficiently strong to prove that PA is consistent. In fact, within a model $\mathbf{V}$ of ZF we can construct a model $\mathbb{N}_\omega$ of PA with domain ω. The model $\mathbb{N}_\omega$ which we obtain in $\mathbf{V}$ is the *standard model of* PA *with respect to* $\mathbf{V}$. In the case when the model $\mathbf{V}$ is a standard model of ZF, the model $\mathbb{N}_\omega$ is isomorphic to the standard model $\mathbb{N}$ of PA which we constructed in Chapter 7. However, if the model $\mathbf{V}$ is a non-standard model of ZF, then $\mathbb{N}_\omega$ is a non-standard model of PA (i.e., $\mathbb{N}_\omega$ is a model of PA which is not isomorphic to $\mathbb{N}$), and there is no way to obtain the standard model of PA within $\mathbf{V}$. In general, people living in $\mathbf{V}$, no matter whether $\mathbf{V}$ is a standard or a non-standard model of ZF, believe that $\mathbb{N}_\omega$ is the standard model $\mathbb{N}$.

Now, let $\mathbf{V}$ be a model of ZF. Within $\mathbf{V}$, we construct an $\mathscr{L}_{\mathsf{PA}}$-structure $\mathbb{N}_\omega$ with domain ω, and show that $\mathbb{N}_\omega$ is a model of PA. Recall that $\mathscr{L}_{\mathsf{PA}} = \{0, \mathtt{s}, +, \cdot\}$. The $\mathscr{L}_{\mathsf{PA}}$-structure is defined by the following assignments which are based on ordinal arithmetic (see EXERCISE 13.5):

$$\begin{aligned}
0^{\mathbb{N}_\omega} &:= \emptyset \\
\mathtt{s}^{\mathbb{N}_\omega} &: \omega \to \omega \\
&\quad\; n \mapsto n+1 \\
+^{\mathbb{N}_\omega} &: \omega \times \omega \to \omega \\
&\quad\; \langle n, m\rangle \mapsto n+m \\
\cdot^{\mathbb{N}_\omega} &: \omega \times \omega \to \omega \\
&\quad\; \langle n, m\rangle \mapsto n \cdot m
\end{aligned}$$

L. Halbeisen, R. Krapf, *Gödel's Theorems and Zermelo's Axioms*,
https://doi.org/10.1007/978-3-030-52279-7_16

Before we show that the $\mathscr{L}_{\mathsf{PA}}$-structure $\mathbb{N}_\omega$ is a model of Peano Arithmetic, we first recall the axioms of PA:

PA_0: $\neg\exists x(\mathtt{s}x = 0)$

PA_1: $\forall x\forall y(\mathtt{s}x = \mathtt{s}y \to x = y)$

PA_2: $\forall x(x + 0 = x)$

PA_3: $\forall x\forall y(x + \mathtt{s}y = \mathtt{s}(x + y))$

PA_4: $\forall x(x \cdot 0 = 0)$

PA_5: $\forall x\forall y(x \cdot \mathtt{s}y = (x \cdot y) + x)$

If φ is any $\mathscr{L}_{\mathsf{PA}}$-formula with $x \in \text{free}(\varphi)$, then:

PA_6: $\big(\varphi(0) \wedge \forall x(\varphi(x) \to \varphi(\mathtt{s}(x)))\big) \to \forall x\varphi(x)$

Let us now show that $\mathbb{N}_\omega \vDash \mathsf{PA}$:

- PA_0: Since $n + 1 = n \cup \{n\}$ and $n \in \{n\}$ (i.e., $n \cup \{n\} \neq \emptyset$), there is no $n \in \omega$ such that $n + 1 = \emptyset$.
- PA_1: If $n, m \in \omega$ and $n \neq m$, then, by THEOREM 13.1.(c), we have either $n \in m$ or $m \in n$, and in both cases we get $n + 1 \neq m + 1$.
- PA_2 and PA_3: Follow immediately from (a) and (b) of ordinal addition.
- PA_4 and PA_5: Follow immediately from (a) and (b) of ordinal multiplication.
- PA_6: Let φ be an $\mathscr{L}_{\mathsf{PA}}$-formula with $x \in \text{free}(\varphi)$ and assume that

$$\varphi(\emptyset) \wedge \forall n \in \omega\big(\varphi(n) \to \varphi(n + 1)\big).$$

  Furthermore, let $E := \{n \in \omega : \neg\varphi(n)\}$. Obviously, E is a subset of ω. If $E = \emptyset$, then $\forall n \in \omega\big(\varphi(n)\big)$ and we are done. Otherwise, if $E \neq \emptyset$, let m be the $\in$-minimal element of E. Now, m can *neither* be $\emptyset$, since we assumed $\varphi(\emptyset)$, *nor* a successor ordinal (i.e., of the form $n + 1$), since we assumed $\varphi(n) \to \varphi(n + 1)$ which is equivalent to $\neg\varphi(n + 1) \to \neg\varphi(n)$. Thus, there is no $\in$-minimal element of E, which is only possible when $E = \emptyset$.

Thus, $\mathbb{N}_\omega$ is a model of PA with domain ω.

In Chapter 7, we saw that there are non-standard models of PA. However, the existence of these models was obtained by the COMPACTNESS THEOREM 2.15, and the proof cannot be carried out in ZFC. In the next section, we will now give a construction of non-standard models of PA which can be carried out in ZFC. Since the construction uses ultrapowers, it cannot be carried out without the aid of the Axiom of Choice.

A Non-Standard Model of Peano Arithmetic in ZFC

The non-standard model of PA which we now construct is the ultrapower of the standard model $\mathbb{N}_\omega$ with respect to some arbitrary but fixed non-trivial ultrafilter $\mathscr{U}$ over ω. First, let ${}^\omega\omega$ be the set of all functions $f:\omega\to\omega$. With respect to $\mathscr{U}$, we define the binary relation $\sim$ on ${}^\omega\omega$ by stipulating

$$f\sim g \;:\Longleftrightarrow\; \big\{n\in\omega : f(n)=g(n)\big\}\in\mathscr{U}\,.$$

Then the relation $\sim$ is an equivalence relation (see Chapter 15). For each $f\in{}^\omega\omega$, let

$$[f] := \{g\in{}^\omega\omega : g\sim f\},$$

and let

$$\omega^* := \big\{[f] : f\in{}^\omega\omega\big\}.$$

We now construct the $\mathscr{L}_{\mathsf{PA}}$-structure $\mathbb{N}^*_\omega$ with domain ω^* as follows:

- For the constant symbol $\mathsf{0}\in\mathscr{L}_{\mathsf{PA}}$, let $f_{\mathsf{0}}\in{}^\omega\omega$ be defined by stipulating

  $$f_{\mathsf{0}}(n) := 0 \quad \text{for all } n\in\omega,$$

  and let

  $$\mathsf{0}^{\mathbb{N}^*_\omega} := [f_{\mathsf{0}}].$$

- For the unary function symbol s in $\mathscr{L}_{\mathsf{PA}}$, we define $\mathsf{s}(f)$ by stipulating

  $$\mathsf{s}(f)(n) := f(n)+1 \quad \text{for } n\in\omega,$$

  and let

  $$\mathsf{s}^{\mathbb{N}^*_\omega}\big([f]\big) := [\mathsf{s}(f)].$$

- For the binary function symbols $+$ and $\cdot$ in $\mathscr{L}_{\mathsf{PA}}$, we define $f+g$ and $f\cdot g$ (for $f,g\in{}^\omega\omega$) by stipulating for all $n\in\omega$

  $$(f+g)(n) := f(n) +^{\mathbb{N}_\omega} g(n)\,,$$
  $$(f\cdot g)(n) := f(n) \cdot^{\mathbb{N}_\omega} g(n)\,,$$

  and let

  $$[f] +^{\mathbb{N}^*_\omega} [g] := [f+g] \quad\text{and}\quad [f]\cdot^{\mathbb{N}^*_\omega}[g] := [f\cdot g].$$

By ŁOŚ'S THEOREM 15.3, the $\mathscr{L}_{\mathsf{PA}}$-structure $\mathbb{N}^*_\omega$ is a model of PA. In order to see that $\mathbb{N}^*_\omega$ is a non-standard model of PA, first notice that $\mathbb{N}_\omega$ can be embedded into $\mathbb{N}^*_\omega$ by the embedding

$$\begin{aligned}\omega &\to \omega^*\\ k &\mapsto [f_k],\end{aligned}$$

where $f_k(n) := k$ for all $n\in\omega$. This shows that $\mathbb{N}_\omega$ is a substructure of $\mathbb{N}^*_\omega$.

In order to show that the models $\mathbb{N}_\omega$ and $\mathbb{N}^*_\omega$ are not isomorphic, let $g \in {}^\omega\omega$ be such that for all $n \in \omega$,

$$g(n) := n\,.$$

Then for all $k \in \omega$, $[f_k] < [g]$, which shows that the models $\mathbb{N}_\omega$ and $\mathbb{N}^*_\omega$ are not isomorphic, even though they are elementarily equivalent (see EXERCISE 16.0).

EXERCISES

16.0 Show that the $\mathscr{L}_{\mathsf{PA}}$-structures $\mathbb{N}_\omega$ and $\mathbb{N}^*_\omega$ are elementarily equivalent.

Hint: Use ŁOŚ'S THEOREM 15.3.

16.1 Show that the domain ω^* of $\mathbb{N}^*_\omega$ is uncountable. In particular, show that $\mathbb{N}^*_\omega$ is an uncountable model of PA.

16.2 Show that for any $[g], [g'] \in \omega^*$ with $[g] < [g']$, the cardinality of the set

$$\big\{[f] \in \omega^* : [g] < [f] < [g']\big\}$$

is either finite or uncountable.

Chapter 17
Models of the Real Numbers

In this chapter, we will first construct a model of the real numbers using Cauchy sequences of rational numbers. We also present a second model of the real numbers according to A'Campo [1]. This construction has the advantage that it only relies on the integers and not on the rational numbers, and that the definition of the multiplication is much simpler and natural than the classical definition based on equivalence classes of Cauchy sequences. Afterwards, we will show that both constructions yield isomorphic models. The constructions of the real numbers will be quite general, such that — depending on whether we start with the standard or a non-standard model of the natural numbers — we obtain the standard or a non-standard model of the real numbers. We shall conclude this chapter by giving a brief introduction to the so-called Non-Standard Analysis, which is essentially just Analysis in a non-standard model of the reals.

Let us first introduce the axioms R of the real numbers. The language of R is $\mathscr{L}_{\mathsf{R}} = \{\mathsf{0}, \mathsf{1}, +, \cdot, <\}$, where $\mathsf{0}$ and $\mathsf{1}$ are constant symbols, $+$ and $\cdot$ are binary function symbols and $<$ is a binary relation symbol. The first group of axioms are simply the field axioms:

R_0: $\forall x \forall y \forall z\big(x + (y + z) = (x + y) + z\big)$

R_1: $\forall x(x + \mathsf{0} = x)$

R_2: $\forall x \exists y(x + y = 0)$

R_3: $\forall x \forall y(x + x = y + x)$

R_4: $\forall x \forall y \forall z\big(x \cdot (y \cdot z) = (x \cdot y) \cdot z\big)$

R_5: $\forall x(x \cdot \mathsf{1} = x)$

R_6: $\forall x\big(x \neq \mathsf{0} \rightarrow \exists y(x \cdot y = \mathsf{1})\big)$

R_7: $\forall x \forall y(x \cdot y = y \cdot x)$

R_8: $\forall x \forall y \forall z\big(x \cdot (y + z) = (x \cdot y) + (x \cdot z)\big)$

R_9: $\mathsf{0} \neq \mathsf{1}$

L. Halbeisen, R. Krapf, *Gödel's Theorems and Zermelo's Axioms*,
https://doi.org/10.1007/978-3-030-52279-7_17

The second group of axioms are the so-called order axioms:

R_{10}: $\forall x \neg(x < x)$
R_{11}: $\forall x \forall y \forall z(x < y \wedge y < z \rightarrow x < z)$
R_{12}: $\forall x \forall y(x < y \vee x = y \vee y < x)$
R_{13}: $\forall x \forall y \forall z(x < y \rightarrow x + z < y + z)$
R_{14}: $\forall x \forall y \forall z(x < y \wedge z \neq \mathsf{0} \rightarrow x \cdot z < y \cdot z)$

The last axioms together form the Completeness Axiom which is, in contrast to the other axioms, a so-called *second-order axiom* (i.e., a statement, not about the real numbers, but about *sets* of real numbers). The set $\mathcal{N}$ denotes either the standard or a non-standard model of the natural numbers.

R_{15} Every Cauchy sequence of reals converges.
R_{16} If $x > \mathsf{0}$ and $y > 0$ there exists $n \in \mathcal{N}$ such that $nx > y$.

Axiom R_{16} is also called the Archimedian Axiom.

A Model of the Real Numbers

Let $\mathcal{N}$ be either ω or ω^*, where ω^* is the ultrapower of ω with respect to some non-trivial ultrafilter $\mathscr{U} \subseteq \mathscr{P}(\omega)$. In other words, $\mathcal{N}$ is either the domain of the model $\mathbb{N}_\omega$ or of $\mathbb{N}^*_\omega$. Recall that the former model is the standard model of PA within some model of ZF, whereas the latter model is a non-standard model of PA, constructed in a model of ZFC, which is elementarily equivalent to the corresponding model $\mathbb{N}_\omega$. Furthermore, let $\mathbf{N}$ be the structure $(\mathcal{N}, 0, +, \cdot)$, i.e., $\mathbf{N}$ is either $\mathbb{N}_\omega$ or $\mathbb{N}^*_\omega$.

From $\mathbf{N}$, we first construct a model $\mathbb{Z}_\mathcal{N}$ of the integers, then we construct a model $\mathbb{Q}_\mathcal{N}$ of the rationals, and finally we construct a model $\mathbb{R}^{\mathscr{C}}_\mathcal{N}$ of the reals using Cauchy sequences.

A Model of the Integers

On $\mathcal{N}$, we first define the binary function $\dot{-}$ by stipulating

$$x \dot{-} y = z \;:\Longleftrightarrow\; \exists u\big(y + u = x \;\wedge\; z = u\big) \vee \neg\exists u\big(y + u = x \;\wedge\; z = 0\big).$$

Now, we define the set of integers $\mathbb{Z}_\mathcal{N}$ as a subset of $\mathcal{N} \times \mathcal{N}$ by stipulating

$$\mathbb{Z}_\mathcal{N} := \big\{\langle x, 0\rangle : x \in \mathcal{N}\big\} \cup \big\{\langle 0, y\rangle : y \in \mathcal{N}\big\}.$$

We identify the elements $x \in \mathcal{N}$ with integers of the form $\langle x, 0\rangle$.
On $\mathbb{Z}_\mathcal{N}$, we define the two binary functions $+$ and $\cdot$, and the unary function $-$ by stipulating the following:

$$\langle x_0, y_0\rangle + \langle x_1, y_1\rangle = z :\Longleftrightarrow z = \big\langle (x_0 + x_1) \dot{-} (y_0 + y_1),\ (y_0 + y_1) \dot{-} (x_0 + x_1)\big\rangle$$
$$\langle x_0, y_0\rangle \cdot \langle x_1, y_1\rangle = z :\Longleftrightarrow z = \big\langle (x_0 \cdot y_1) + (y_0 \cdot x_1),\ (x_0 \cdot x_1) + (y_0 \cdot y_1)\big\rangle$$
$$-\langle x, y\rangle = z :\Longleftrightarrow z = \langle y, x\rangle$$

In order to simplify the notation, we usually write $\langle x_0, y_0\rangle - \langle x_1, y_1\rangle$ instead of $\langle x_0, y_0\rangle + \big(-\langle x_1, y_1\rangle\big)$. Notice that $\langle x, y\rangle - \langle x, y\rangle = \langle 0, 0\rangle$.

We leave it as an exercise to the reader to check that the structure

$$\mathbb{Z}_{\mathcal{N}} := \big(\mathcal{N} \times \mathcal{N}, \langle 0, 0\rangle, \langle 0, 1\rangle, +, \cdot\big)$$

is a model of the ring of integers, where $\langle 0, 0\rangle$ and $\langle 0, 1\rangle$ are the neutral elements with respect to the binary operations $+$ and $\cdot$, respectively. Notice that if $\mathcal{N}$ is equal to ω^*, then $\mathbb{Z}_{\mathcal{N}}$ is a non-standard model of the integers.

On $\mathbb{Z}_{\mathcal{N}}$, we define the binary relation $<$ and the unary function symbol $|\cdot|$ as follows:

$$\langle x_0, y_0\rangle < \langle x_1, y_1\rangle \quad :\Longleftrightarrow \quad x_0 < x_1 \vee y_0 < y_1$$
$$\big|\langle x, y\rangle\big| = z \quad :\Longleftrightarrow \quad \begin{cases} z = \langle x, y\rangle & \text{if } x = 0 \\ z = \langle y, x\rangle & \text{otherwise} \end{cases}$$

A Model of the Rational Numbers

Let $\mathcal{N}^+ := \mathcal{N} \setminus \{0\}$. On pairs $\langle x_0, y_0\rangle, \langle x_1, y_1\rangle \in \mathbb{Z}_{\mathcal{N}} \times \mathcal{N}^+$ we define an equivalence relation $\sim$ by stipulating

$$\langle x_0, y_0\rangle \sim \langle x_1, y_1\rangle :\Longleftrightarrow x_0 \cdot y_1 = x_1 \cdot y_0.$$

Now, we denote the equivalence classes by

$$\tfrac{x}{y} := \big[\langle x, y\rangle\big] = \big\{\langle x', y'\rangle \in \mathbb{Z}_{\mathcal{N}} \times \mathcal{N}^+ : \langle x', y'\rangle \sim \langle x, y\rangle\big\}$$

and call $\frac{x}{y}$ a **rational number**. Let $\mathbb{Q}_{\mathcal{N}}$ denote the set of all rational numbers, i.e.,

$$\mathbb{Q}_{\mathcal{N}} := \big\{\tfrac{x}{y} : x \in \mathbb{Z}_{\mathcal{N}}, y \in \mathcal{N}^+\big\}.$$

We can now introduce the two binary functions $+$ and $\cdot$ by

$$\tfrac{x_0}{y_0} + \tfrac{x_1}{y_1} := \tfrac{x_0 y_1 + y_0 x_1}{y_0 y_1},$$
$$\tfrac{x_0}{y_0} \cdot \tfrac{x_1}{y_1} := \tfrac{x_0 \cdot x_1}{y_0 \cdot y_1}.$$

We leave it as an exercise for the reader to check that these functions are well-defined and that the structure $\mathbb{Q}_{\mathcal{N}} = (\mathbb{Q}_{\mathcal{N}}, \frac{0}{1}, \frac{1}{1}, +, \cdot)$ satisfies the field axioms. As in the case of the integers, if $\mathcal{N}$ is ω^*, then $\mathbb{Q}_{\mathcal{N}}$ is a non-standard model of the rational numbers.

On $\mathbb{Q}_{\mathcal{N}}$, we define the binary relation $<$ and the unary function symbol $|\cdot|$ as follows:

$$\frac{x_0}{y_0} < \frac{x_1}{y_1} \quad :\Longleftrightarrow \quad x_0 \cdot y_1 < x_1 \cdot y_0$$

$$\left|\frac{x}{y}\right| = z \quad :\Longleftrightarrow \quad \begin{cases} z = \frac{x}{y} & \text{if } x \geq 0 \\ z = \frac{-x}{y} & \text{otherwise} \end{cases}$$

Again, it is easy to check that the order $<$ and the absolute value function are well-defined and satisfy the usual properties.

A Model of the Real Numbers using Cauchy Sequences

Let $\mathbb{Q}_{\mathcal{N}}^+$ denote the positive rational numbers, i.e., those $p \in \mathbb{Q}_{\mathcal{N}}$ that satisfy $p > 0$. We define a sequence (a_n) of rational numbers to be a **Cauchy sequence**, if for every $\varepsilon \in \mathbb{Q}_{\mathcal{N}}^+$ there is an $N \in \mathcal{N}$ such that for all $m, n \in \mathcal{N}$ with $m, n \geq N$, $|a_n - a_m| < \varepsilon$. We denote the set of all Cauchy sequences of rationals by $\mathscr{C}$. Two Cauchy sequences $(a_n), (b_n) \in \mathscr{C}$ are said to be **equivalent**, denoted by $(a_n) \approx (b_n)$, if for each positive rational number $\varepsilon \in \mathbb{Q}_{\mathcal{N}}^+$ there is an $N \in \mathcal{N}$ such that for all $n \in \mathcal{N}$ with $n \geq N$, $|a_n - b_n| < \varepsilon$. In order to simplify the notation, we shall write $\lim_{n\to\infty}(a_n - b_n) = 0$. Notice that the meaning of $\lim_{n\to\infty}$ depends on whether $\mathcal{N} = \omega$ or $\mathcal{N} = \omega^*$.

It is obvious that the relation $\approx$ is reflexive and symmetric. Moreover, it is also transitive, since for Cauchy sequences $(a_n), (b_n)$ and (c_n) such that $(a_n) \approx (b_n)$ and $(b_n) \approx (c_n)$, it follows that

$$\lim_{n\to\infty}(a_n - c_n) = \lim_{n\to\infty}(a_n - b_n) + \lim_{n\to\infty}(b_n - c_n) = 0.$$

Therefore, $\approx$ is an equivalence relation on $\mathscr{C}$, and the equivalence classes with respect to $\approx$ are given by

$$[(a_n)] := \{(b_n) \in \mathscr{C} : (b_n) \approx (a_n)\}.$$

Let $\mathbb{R}_{\mathcal{N}}^{\mathscr{C}}$ denote the set of all equivalence classes of rational Cauchy sequences, i.e.,

$$\mathbb{R}_{\mathcal{N}}^{\mathscr{C}} := \{[(a_n)] : (a_n) \in \mathscr{C}\}.$$

The elements of $\mathbb{R}_{\mathcal{N}}^{\mathscr{C}}$ are called **real numbers**.

In order to obtain a model of the real numbers, we need to define the functions addition $+$ and multiplication $\cdot$ on $\mathbb{R}_{\mathcal{N}}^{\mathscr{C}}$, including the neutral elements $0_{\mathscr{C}}$ and $1_{\mathscr{C}}$, respectively; then we have to define a linear ordering $<$, on $\mathbb{R}_{\mathcal{N}}^{\mathscr{C}}$ and finally, we check that the structure $\mathbb{R}_{\mathcal{N}}^{\mathscr{C}} = (\mathbb{R}_{\mathcal{N}}^{\mathscr{C}}, 0_{\mathscr{C}}, 1_{\mathscr{C}}, +, \cdot, <)$ thus obtained satisfies all axioms of the real numbers.

Addition and multiplication: For $r, s \in \mathbb{R}^{\mathscr{C}}_{\mathcal{N}}$, represented by $(a_n), (b_n) \in \mathscr{C}$ we define:

$$\begin{aligned} r + s &:= [(a_n + b_n)] \\ r \cdot s &:= [(a_n \cdot b_n)] \end{aligned}$$

LEMMA 17.1. *Addition and multiplication of reals are well-defined, i.e., if $r, s \in \mathbb{R}^{\mathscr{C}}_{\mathcal{N}}$ such that r is represented by $(a_n), (a'_n)$ and s is represented by $(b_n), (b'_n)$, then $(a_n + b_n)$ and $(a_n b_n)$ are again Cauchy sequences such that $(a_n + b_n) \approx (a'_n + b'_n)$ and $(a_n b_n) \approx (a'_n b'_n)$.*

Proof. In order to verify that $(a_n + b_n)$ is a Cauchy sequence, let $\varepsilon \in \mathbb{Q}^+_{\mathcal{N}}$. By assumption, there are $N_1, N_2 \in \mathcal{N}$ such that for all $n, m \geq N := \max\{N_1, N_2\}$ we have $|a_m - a_n| < \frac{\varepsilon}{2}$ and $|b_m - b_n| < \frac{\varepsilon}{2}$. Then it follows

$$|(a_n + b_n) - (a_m + b_m)| \leq |a_n - a_m| + |b_n - b_m| < \tfrac{\varepsilon}{2} + \tfrac{\varepsilon}{2} = \varepsilon$$

for all $n, m \geq N$.

Next, we prove that $(a_n b_n)$ is a Cauchy sequence. Since Cauchy sequences are bounded, there is a $C \in \mathcal{N}$ such that $|a_n|, |b_n| \leq C$ for all $n \in \mathcal{N}$. Now we can choose $M_1, M_2 \in \mathcal{N}$ such that for all $n, m \geq M := \max\{M_1, M_2\}$ we have $|a_n - a_m| < \frac{\varepsilon}{2C}$ and $|b_n - b_m| < \frac{\varepsilon}{2C}$ for all $n, m \geq M$. Consequently, we obtain:

$$\begin{aligned} |a_n b_n - a_m b_m| &= |a_n(b_n - b_m) + b_m(a_n - a_m)| \\ &\leq |a_n| \cdot |b_n - b_m| + |b_m| \cdot |a_n - a_m| \\ &\leq C \cdot \tfrac{\varepsilon}{2C} + C \cdot \tfrac{\varepsilon}{2C} \\ &= \varepsilon \end{aligned}$$

Hence, $(a_n b_n)$ is a Cauchy sequence. The second part uses similar arguments. ⊣

Furthermore, we define the neutral elements $0_{\mathscr{C}}$ and $1_{\mathscr{C}}$ in the following way:

$$\begin{aligned} 0_{\mathscr{C}} &:= [(a_n)] \quad \text{where } a_n = 0 \text{ for all } n \in \mathcal{N} \\ 1_{\mathscr{C}} &:= [(b_n)] \quad \text{where } b_n = 1 \text{ for all } n \in \mathcal{N} \end{aligned}$$

Linear ordering: Let $r, s \in \mathbb{R}^{\mathscr{C}}_{\mathcal{N}}$ such that $r = [(a_n)]$ and $s = [(b_n)]$. Then we define:

$$r < s :\Longleftrightarrow \exists \varepsilon \in \mathbb{Q}^+_{\mathcal{N}}\, \exists N \in \mathcal{N}\, \forall n \geq N (b_n - a_n > \varepsilon)$$

Again, we have to verify that this definition is well-defined.

THEOREM 17.2. *The structure* $\mathbb{R}^{\mathscr{C}}_{\mathcal{N}} = (\mathbb{R}^{\mathscr{C}}_{\mathcal{N}}, 0_{\mathscr{C}}, 1_{\mathscr{C}}, +, \cdot, <)$ *is a model of the axioms of the real numbers.*

Proof. The only non-trivial axioms are the existence of a multiplicative inverse and the completeness axiom. Suppose that $r \neq 0_{\mathscr{C}}$ is a real number represented by (a_n). Then we define $r^{-1} = [(\tilde{a}_n)]$, where

$$\tilde{a}_n := \begin{cases} \frac{1}{a_n} & \text{if } a_n \neq 0, \\ 1 & \text{otherwise.} \end{cases}$$

Since (a_n) is a Cauchy sequence such that $[(a_n)] \neq 0_{\mathscr{C}}$, only for finitely many $n \in \mathcal{N}$ we have $a_n = 0$. Thus, $\lim_{n\to\infty}(a_n\tilde{a}_n - 1) = 0$ and hence $r \cdot r^{-1} = 1_{\mathscr{C}}$.

In order to prove that $\mathbb{R}^{\mathscr{C}}_{\mathcal{N}}$ is complete, we first verify R_{15}. Suppose that (r_n) is a Cauchy sequence of real numbers and let r_n be represented by (a^n_k), where (a^n_k) is a Cauchy sequence of rational numbers. Since (a^n_k) is a Cauchy sequence, for every $n \in \mathcal{N}$ there is $N_n \in \mathcal{N}$ such that

$$\forall k, l \geq N_n \big(|a^n_k - a^n_l| < \tfrac{1}{n}\big).$$

Now we consider the diagonal sequence (d_n) with $d_n := a^n_{N_n}$ for every $n \in \mathcal{N}$.

CLAIM. (d_n) *is a Cauchy sequence of rationals which represents a real number* $r = [(d_n)]$ *such that* $\lim_{n\to\infty} r_n = r$.

Proof of Claim. First we show that (d_n) is a Cauchy sequence, and then we prove that it represents a limit of the sequence (r_n) of reals.

(d_n) *is a Cauchy sequence*: Suppose that $\varepsilon \in \mathbb{Q}^+_{\mathcal{N}}$. Note that since (r_n) is a Cauchy sequence of reals, there exists $N \in \mathcal{N}$ with $N \geq \frac{3}{\varepsilon}$ such that

$$\forall m, n \geq N \big(|r_m - r_n| < \tfrac{\varepsilon}{3}\big).$$

In particular, this implies that for all $m, n \in \mathcal{N}$ with $m, n \geq N$, there is $N_{m,n}$ such that

$$\forall k \geq N_{m,n} \big(|a^m_k - a^n_k| < \tfrac{\varepsilon}{3}\big).$$

Now let $m, n \in \mathcal{N}$ such that $n, m \geq N$. We have to verify that $|d_m - d_n| < \varepsilon$. Choose $k \in \mathcal{N}$ with $k \geq N_{m,n}$. We have

$$\begin{aligned} |d_m - d_n| &= |a^m_{N_m} - a^n_{N_n}| \\ &\leq \underbrace{|a^m_{N_m} - a^m_k|}_{<\frac{1}{m}\leq\frac{1}{N}<\frac{\varepsilon}{3}} + \underbrace{|a^m_k + a^n_k|}_{<\frac{\varepsilon}{3}} + \underbrace{|a^n_k - a^n_{N_n}|}_{<\frac{1}{n}\leq\frac{1}{N}<\frac{\varepsilon}{3}}. \end{aligned}$$

Hence we have $|d_n - d_n| < \varepsilon$, which proves that (d_n) is a Cauchy sequence.

(r_n) *converges to* $r = [(d_n)]$: Suppose that $e \in \mathbb{R}^{\mathscr{C}}_{\mathcal{N}}$ is a positive real number, i.e., $e > 0_{\mathscr{C}}$. We need to find $N \in \mathcal{N}$ and $\varepsilon \in \mathbb{Q}^+_{\mathcal{N}}$ such that $|r_n - r| < e$ for all $n \geq N$. Choose a rational Cauchy sequence (b_n) representing e. Since $e > 0_{\mathscr{C}}$, the definition of our linear ordering yields $\delta \in \mathbb{Q}^+_{\mathcal{N}}$ and $N_0 \in \mathcal{N}$ such that for all $n \geq N_0$ we have $b_n > \delta$. Moreover, since (d_n) is a Cauchy sequence, there is $N_1 \in \mathcal{N}$ such that for all $m, n \geq N_1$ we have $|d_m - d_n| < \frac{\delta}{3}$. Now let $N \in \mathcal{N}$ be defined by $N := \max\{N_0, N_1, \frac{3}{\delta}\}$ and let $n \geq N$. We prove that $|r_n - r| < e$, i.e., we show that there is $\varepsilon \in \mathbb{Q}^+_{\mathcal{N}}$ and $N' \in \mathcal{N}$ such that

$$\forall k \geq N' \left(b_k - |a^n_k - d_k| > \varepsilon\right).$$

Let $\varepsilon := \frac{\delta}{3}$ and $N' = \max\{N, N_n\}$. Then for each $k \geq N'$, we have

$$\begin{aligned} |a^n_k - d_k| &\leq |a^n_k - d_n| + |d_n - d_k| \\ &= \underbrace{|a^n_k - a^n_{N_n}|}_{<\frac{1}{n}\leq\frac{1}{N}\leq\frac{\delta}{3}} + \underbrace{|d_n - d_k|}_{<\frac{\delta}{3}}, \end{aligned}$$

and hence $|a^n_k - d_k| < \frac{2\delta}{3}$. Since $b_k < \delta$, we further obtain

$$b_k - |a^n_k - d_k| > \delta - \tfrac{2\delta}{3} = \varepsilon.$$

Therefore, we have $\lim_{n\to\infty} r_n = r$. $\dashv_{\text{Claim}}$

Moreover, the Archimedian Axiom R_{16} holds as a consequence of the fact that Cauchy sequences are always bounded: If $r, s \in \mathbb{R}^{\mathscr{C}}_{\mathcal{N}}$ such that $0_{\mathscr{C}} < r < s$, then $\frac{s}{r}$ is again a real number represented by some Cauchy sequence (a_n). Since (a_n) is bounded (as a sequence of rational numbers), there is a natural number $N \in \mathcal{N}$ such that $|a_n| < N$ for all $n \in \mathcal{N}$. Hence, $\frac{s}{r} < N + 1$ and $r(N+1) > s$. $\dashv$

A Natural Construction of the Real Numbers

In this section, we construct a model of the real numbers in which the real numbers are equivalence classes of certain functions, so-called *slopes*, from $\mathbb{Z}_{\mathcal{N}}$ to $\mathbb{Z}_{\mathcal{N}}$. In fact, from $\mathbf{N}$ we first construct a model $\mathbb{Z}_{\mathcal{N}}$ of the integers, and from $\mathbb{Z}_{\mathcal{N}}$ we directly construct a model $\mathbb{R}^{\mathscr{S}}_{\mathcal{N}}$ of the real numbers. It will turn out that the models $\mathbb{R}^{\mathscr{S}}_{\mathcal{N}}$ and $\mathbb{R}^{\mathscr{C}}_{\mathcal{N}}$ are isomorphic, no matter whether $\mathcal{N} = \omega$ or $\mathcal{N} = \omega^*$.

A **slope** is a function $\lambda : \mathbb{Z}_{\mathcal{N}} \to \mathbb{Z}_{\mathcal{N}}$ for which there exists an $M_\lambda \in \mathcal{N}$ such that for all $n, m \in \mathbb{Z}_{\mathcal{N}}$ we have

$$\left|\lambda(n+m) - \left(\lambda(n) + \lambda(m)\right)\right| \leq M_\lambda .$$

Roughly speaking, a slope is an almost linear function from $\mathbb{Z}_{\mathcal{N}}$ to $\mathbb{Z}_{\mathcal{N}}$. Let $\mathscr{S}$ denote the set of all slopes. We say that two slopes $\lambda, \lambda' \in \mathscr{S}$ are **equivalent**, denoted by $\lambda \sim \lambda'$, if there exists an $M \in \mathcal{N}$ such that for all $n \in \mathbb{Z}_{\mathcal{N}}$ we have

$$\left|\lambda(n) - \lambda'(n)\right| \leq M .$$

Obviously, the relation $\sim$ is reflexive and symmetric, and if for all $n \in \mathbb{Z}_{\mathcal{N}}$,

$$\big|\lambda(n) - \lambda'(n)\big| \;\le\; M_{\lambda,\lambda'}$$

and

$$\big|\lambda'(n) - \lambda''(n)\big| \;\le\; M_{\lambda',\lambda''},$$

then

$$\big|\lambda(n) - \lambda''(n)\big| \;\le\; (M_{\lambda,\lambda'} + M_{\lambda',\lambda''}),$$

which shows that $\sim$ is also transitive. Therefore, $\sim$ is an equivalence relation on $\mathscr{S}$. For a slope $\lambda \in \mathscr{S}$, let

$$[\lambda] := \big\{\lambda' \in \mathscr{S} : \lambda' \sim \lambda\big\}.$$

Let $\mathbb{R}^{\mathscr{S}}_{\mathcal{N}}$ denote the set of equivalence classes of slopes, i.e.,

$$\mathbb{R}^{\mathscr{S}}_{\mathcal{N}} = \big\{[\lambda] : \lambda \in \mathscr{S}\big\}.$$

The elements of $\mathbb{R}^{\mathscr{S}}_{\mathcal{N}}$ are denoted by letters like $r, s, t, \dots$ and are called **real numbers**.

In what follows, we shall first define two binary functions addition $+$ and multiplication $\cdot$ on $\mathbb{R}^{\mathscr{S}}_{\mathcal{N}}$, including the neutral elements $0_{\mathscr{S}}$ and $1_{\mathscr{S}}$, respectively; then we introduce the binary relation $<$; and finally, we shall define an isomorphism between the structures $\mathbb{R}^{\mathscr{C}}_{\mathcal{N}} = \big(\mathbb{R}^{\mathscr{C}}_{\mathcal{N}}, 0_{\mathscr{C}}, 1_{\mathscr{C}}, +, \cdot, <\big)$ and $\mathbb{R}^{\mathscr{S}}_{\mathcal{N}} = \big(\mathbb{R}^{\mathscr{S}}_{\mathcal{N}}, 0_{\mathscr{S}}, 1_{\mathscr{S}}, +, \cdot, <\big)$, which implies that $\mathbb{R}^{\mathscr{S}}_{\mathcal{N}}$ is a model of the reals.

Addition: Let $r, s \in \mathbb{R}^{\mathscr{S}}_{\mathcal{N}}$ be two reals. Then there are slopes $\lambda, \lambda' \in \mathscr{S}$, such that $r = [\lambda]$ and $s = [\lambda']$. We define $r + s$ by stipulating

$$r + s := [\lambda + \lambda'],$$

where

$$\begin{aligned} \lambda + \lambda' : \mathbb{Z}_{\mathcal{N}} &\to \mathbb{Z}_{\mathcal{N}} \\ n &\mapsto \lambda(n) + \lambda'(n). \end{aligned}$$

It is easy to see that $\lambda + \lambda'$ is a slope and that $r + s$ is independent of the choice of representatives λ, λ'. Furthermore, we define

$$0_{\mathscr{S}} := [\lambda_0] \quad \text{where } \lambda_0(n) := 0 \text{ for all } n \in \mathbb{Z}_{\mathcal{N}}.$$

We obviously have that $0_{\mathscr{S}}$ is a neutral element with respect to addition. For a real $r = [\lambda]$, let

$$-r := [-\lambda] \quad \text{where } (-\lambda)(n) := -\lambda(n) \text{ for all } n \in \mathbb{Z}_{\mathcal{N}}.$$

For all reals r, we obviously have $r + (-r) = 0_{\mathscr{S}}$, where $r + (-r)$ is usually written as $r - r$.

Multiplication: Let $r, s \in \mathbb{R}^{\mathscr{S}}_{\mathcal{N}}$ be two reals, and let $\lambda, \lambda' \in \mathscr{S}$ be the corresponding slopes. We define $r \cdot s$ by stipulating

$$r \cdot s := [\lambda \circ \lambda'],$$

where

$$\begin{array}{rccc} \lambda\circ\lambda' : & \mathbb{Z}_{\mathcal{N}} & \to & \mathbb{Z}_{\mathcal{N}} \\ & n & \mapsto & \lambda\big(\lambda'(n)\big) . \end{array}$$

Furthermore, we define

$$1_{\mathscr{S}} := [\lambda_1] \quad \text{where } \lambda_1(n) = n \text{ for all } n \in \mathbb{Z}_{\mathcal{N}} .$$

We obviously have that $1_{\mathscr{S}}$ is a neutral element with respect to multiplication. However, we have to show that the composition $\lambda\circ\lambda'$ of two slopes is again a slope, and that $r \cdot s$ is independent of the choice of representatives λ, λ'.

In order to simplify the notation, for a slope $\lambda \in \mathscr{S}$ and any $n, m \in \mathbb{Z}_{\mathcal{N}}$, we say that $\lambda(n+m)$ and $\lambda(n) + \lambda(m)$ are **similar (with respect to $\boldsymbol{\lambda}$)**, denoted by

$$\lambda(n+m) \underset{\lambda}{\approx} \lambda(n) + \lambda(m) .$$

In fact, $\lambda(n+m) \underset{\lambda}{\approx} \lambda(n)+\lambda(m)$ just means that the absolute value of the difference of $\lambda(n+m)$ and $\lambda(n)+\lambda(m)$ is uniformly bounded (i.e., independently of $n, m \in \mathbb{Z}_{\mathcal{N}}$). Notice that by definition, for each $u \in \mathbb{Z}_{\mathcal{N}}$ we have

$$\lambda(n+m+u) \underset{\lambda}{\approx} \lambda(n) + \lambda(m) . \tag{$*$}$$

LEMMA 17.3. *Let the slopes λ, λ' represent $r \in \mathbb{R}^{\mathscr{S}}_{\mathcal{N}}$, and let the slopes μ, μ' represent $s \in \mathbb{R}^{\mathscr{S}}_{\mathcal{N}}$. Then the compositions $\lambda\circ\mu$ and $\lambda'\circ\mu'$ are equivalent slopes.*

Proof. We first show that $\lambda\circ\mu$ is a slope, i.e., there exists an $M_{\lambda\circ\mu}$ such that for all $n, m \in \mathbb{Z}_{\mathcal{N}}$,

$$\big|\lambda\circ\mu(n+m) - \big(\lambda\circ\mu(n) + \lambda\circ\mu(m)\big)\big| \;\le\; M_{\lambda\circ\mu} .$$

Since μ and λ are both slopes, we have the following two relations:

$$\mu(n+m) \underset{\mu}{\approx} \mu(n) + \mu(m)$$

$$\lambda\big(\underbrace{\mu(n)}_{n'} + \underbrace{\mu(m)}_{m'}\big) \;\underset{\lambda}{\approx}\; \lambda\circ\underbrace{\mu(n)}_{n'} + \lambda\circ\underbrace{\mu(m)}_{m'}$$

By the former relation, for all $n, m \in \mathbb{Z}_{\mathcal{N}}$ there exists a $u_{n,m}$ with

$$|u_{n,m}| \le M_{\mu} ,$$

such that

$$\lambda\big(\mu(n+m)\big) = \lambda\big(\mu(n) + \mu(m) + u_{n,m}\big) ,$$

and therefore, by $(*)$ we obtain

$$\lambda\big(\mu(n+m)\big) \underset{\lambda}{\approx} \lambda\big(\mu(n) + \mu(m)\big) .$$

Thus, by the latter relation we obtain

$$\lambda\circ\mu(n+m) \underset{\lambda}{\approx} \lambda\circ\mu(n) + \lambda\circ\mu(m)\,,$$

which shows that $\lambda\circ\mu$ (as well as $\lambda'\circ\mu'$) is a slope.

In order to see that $\lambda\circ\mu$ and $\lambda'\circ\mu'$ are equivalent slopes, first notice that

$$\lambda\circ\mu(n+m) \underset{\lambda}{\approx} \lambda\big(\mu(n)+\mu(m)\big) \underset{\lambda}{\approx} \lambda\big(\mu'(n)+\mu'(m)\big)\,.$$

Similarly, we have $\lambda'\circ\mu(n+m) \underset{\lambda'}{\approx} \lambda'\big(\mu'(n)+\mu'(m)\big)$, and since $\lambda \sim \lambda'$, there is an $M_{\lambda,\lambda'} \in \mathcal{N}$ such that for all $n \in \mathbb{Z}_{\mathcal{N}}$,

$$\big|\lambda\circ\mu(n) - \lambda'\circ\mu'(n)\big| \le M_{\lambda,\lambda'}\,,$$

which shows that the slopes $\lambda\circ\mu$ and $\lambda'\circ\mu'$ are equivalent. ⊣

Linear ordering: In order to define the binary relation $<$, we first define the unary relation $\operatorname{pos}(\cdot)$ on $\mathscr{S}$ by stipulating

$$\operatorname{pos}(\lambda) \;:\Longleftrightarrow\; \forall N \in \mathcal{N}\, \exists m \in \mathcal{N}\big(\lambda(m) > N\big)\,.$$

Now, for any slopes $\lambda, \lambda' \in \mathscr{S}$, we define

$$\lambda < \lambda' \;:\Longleftrightarrow\; \operatorname{pos}(\lambda' - \lambda)\,.$$

Notice that the relation $<$ is transitive and that $\operatorname{pos}(\lambda)$ if and only if $0_{\mathscr{S}} < \lambda$.

In order to show that the structures $\mathbb{R}^{\mathscr{C}}_{\mathcal{N}}$ and $\mathbb{R}^{\mathscr{S}}_{\mathcal{N}}$ are isomorphic – which implies that $\mathbb{R}^{\mathscr{S}}_{\mathcal{N}}$ is a model of the reals –, we first prove the following

FACT 17.4. *Let $\lambda \in \mathscr{S}$ and $M_\lambda \in \mathcal{N}$ be such that for all $n, m \in \mathbb{Z}_{\mathcal{N}}$ we have*

$$\big|\lambda(n+m) - \lambda(n) - \lambda(m)\big| \;\le\; M_\lambda\,.$$

Then for all $n, m \in \mathbb{Z}_{\mathcal{N}}$ we have

$$\big|\lambda(n)\cdot m - \lambda(n\cdot m)\big| \;\le\; (m+1)\cdot M_\lambda\,.$$

Proof. Notice that for each $n \in \mathbb{Z}_{\mathcal{N}}$ we have $|\lambda(0)| \le M_\lambda$, since

$$M_\lambda \ge \big|\lambda(n+0) - \lambda(n) - \lambda(0)\big| = |-\lambda(0)| = |\lambda(0)|\,.$$

The proof is now by induction on m. If $m = 0$, then for all $n \in \mathbb{Z}_{\mathcal{N}}$ we have

$$\big|\lambda(n)\cdot 0 - \lambda(n\cdot 0)\big| = |-\lambda(0)| = |\lambda(0)| \le M_\lambda.$$

Assume that for some $m \ge 0$ and for all $n \in \mathbb{Z}_{\mathcal{N}}$, we have

$$\big|\lambda(n)\cdot m - \lambda(n\cdot m)\big| \;\le\; (m+1)\cdot M_\lambda\,.$$

Then, for all $n \in \mathbb{Z}_{\mathcal{N}}$ we have:

$$\begin{aligned}\big|\lambda(n)\cdot(m+1)-\lambda(n\cdot(m+1))\big| &= \big|\lambda(n)\cdot m+\lambda(n)-\lambda(n\cdot m+n)\big|\\ &\le \big|\lambda(n)\cdot m+\lambda(n)-\lambda(n\cdot m)-\lambda(n)\big|+M_\lambda\\ &= \big|\lambda(n)\cdot m-\lambda(n\cdot m)\big|+M_\lambda\\ &\le (m+1)\cdot M_\lambda+M_\lambda\\ &= (m+2)\cdot M_\lambda\end{aligned}$$

This obviously completes the proof. ⊣

Now, we are ready to prove that $\mathbb{R}^{\mathscr{S}}_{\mathcal{N}}$ is a model of the reals.

PROPOSITION 17.5. *The two structures*

$$\mathbb{R}^{\mathscr{C}}_{\mathcal{N}} = (\mathbb{R}^{\mathscr{C}}_{\mathcal{N}}, 0_{\mathscr{C}}, 1_{\mathscr{C}}, +, \cdot, <) \quad \textit{and} \quad \mathbb{R}^{\mathscr{S}}_{\mathcal{N}} = (\mathbb{R}^{\mathscr{S}}_{\mathcal{N}}, 0_{\mathscr{S}}, 1_{\mathscr{S}}, +, \cdot, <)$$

are isomorphic.

Proof. First, we define a mapping $\gamma : \mathscr{S} \to \mathscr{C}$ which maps each slope $\lambda \in \mathscr{S}$ to a Cauchy sequence $\gamma(\lambda)$. Then we show that $\lambda \sim \lambda'$ if and only if $\gamma(\lambda) \approx \gamma(\lambda')$. With γ, we then define a bijection $\Gamma : \mathbb{R}^{\mathscr{S}}_{\mathcal{N}} \to \mathsf{RC}$ which induces an isomorphism between $\mathbb{R}^{\mathscr{S}}_{\mathcal{N}}$ and $\mathbb{R}^{\mathscr{C}}_{\mathcal{N}}$.

Let $\gamma : \mathscr{S} \to \mathscr{C}$ be defined by stipulating $(a^\lambda_n) := \gamma(\lambda)$, where for each $n \in \mathcal{N}$ we have

$$a^\lambda_n = \begin{cases} 0 & \text{if } n = 0,\\ \frac{\lambda(n)}{n} & \text{otherwise.}\end{cases}$$

We have to show that γ is well-defined (i.e., (a^λ_n) is a Cauchy sequence). For this, let $\lambda \in \mathscr{S}$ and consider $\frac{\lambda(n)}{n} - \frac{\lambda(m)}{m}$ for some $n, m \in \mathcal{N}^+$. Notice that

$$\frac{\lambda(n)}{n} - \frac{\lambda(m)}{m} = \frac{\lambda(n)\cdot m - \lambda(m)\cdot n}{n\cdot m},$$

and that by FACT 17.4 we have

$$\big|\lambda(n)\cdot m-\lambda(n\cdot m)\big| \le (m+1)\cdot M_\lambda \quad \text{and} \quad \big|\lambda(m)\cdot n-\lambda(m\cdot n)\big| \le (n+1)\cdot M_\lambda\,.$$

Hence,

$$\left|\frac{\lambda(n)\cdot m - \lambda(m)\cdot n}{n\cdot m}\right| \le \frac{n+m+2}{n\cdot m}\cdot M_\lambda\,,$$

and since M_λ is fixed, for every $\varepsilon \in \mathbb{Q}^+_{\mathcal{N}}$ we find an $N \in \mathcal{N}$ such that for all $n, m \in \mathcal{N}$ with $m, n \ge N$,

$$\left|\frac{\lambda(n)}{n} - \frac{\lambda(m)}{m}\right| \le \frac{n+m+2}{n\cdot m}\cdot M_\lambda \le \varepsilon\,,$$

which shows that (a^λ_n) is a Cauchy sequence.

Now we show that for any slopes $\lambda, \lambda' \in \mathscr{S}$, if $\lambda \sim \lambda'$ then $(a_n^{\lambda}) \approx (a_n^{\lambda'})$. For this purpose, recall that $\lambda \sim \lambda'$ if and only if there exists an $M \in \mathcal{N}$ such that for all $n \in \mathbb{Z}_{\mathcal{N}}$ we have $|\lambda(n) - \lambda'(n)| \le M$. With respect to the corresponding Cauchy sequences (a_n^{λ}) and $(a_n^{\lambda'})$, this gives us

$$\left|(a_n^{\lambda}) - (a_n^{\lambda'})\right| = \left|\frac{\lambda(n)}{n} - \frac{\lambda'(n)}{n}\right| = \left|\frac{\lambda(n) - \lambda'(n)}{n}\right| \le \frac{M}{n},$$

which shows that $(a_n^{\lambda}) \approx (a_n^{\lambda'})$.

Let us define the function $\Gamma : \mathbb{R}_{\mathcal{N}}^{\mathscr{S}} \to \mathbb{R}_{\mathcal{N}}^{\mathscr{C}}$ by stipulating

$$\Gamma\big([\lambda]\big) := \big[\gamma(\lambda)\big].$$

By the above result, the function Γ is well-defined. In order to show that the structures $\mathbb{R}_{\mathcal{N}}^{\mathscr{S}}$ and $\mathbb{R}_{\mathcal{N}}^{\mathscr{C}}$ are isomorphic, we have to show that Γ is a bijection. For this, we show that Γ is surjective and injective.

Γ is surjective: Let $(a_n) \in \mathscr{C}$ be a Cauchy sequence. With respect to (a_n), let $k_1 < k_2 < \dots$ be a strictly increasing sequence in $\mathcal{N}$ such that for every $n \in \mathcal{N}^+$ we have

$$\forall m_1, m_2 \ge k_n \left(\big|\lfloor n \cdot a_{m_1} \rfloor - \lfloor n \cdot a_{m_2} \rfloor\big| \le 1\right),$$

where for a rational $\frac{p}{q}$, $\lfloor \frac{p}{q} \rfloor := \max\{z \in \mathbb{Z}_{\mathcal{N}} : z \le \frac{p}{q}\}$. In order to see that such a sequence $k_1 < k_2 < \dots$ exists, notice that since $(a_n) \in \mathscr{C}$, for every $n \in \mathcal{N}^+$ we find a $k_n \in \mathcal{N}$ such that

$$\forall m_1, m_2 \ge k_n \left(\big|a_{m_1} - a_{m_2}\big| < \frac{1}{n^2}\right).$$

Hence, for $n \in \mathcal{N}^+$ and all $m_1, m_2 \ge k_n$ we obtain

$$\big|n \cdot a_{m_1} - n \cdot a_{m_2}\big| < \frac{1}{n}$$

and therefore

$$\big|\lfloor n \cdot a_{m_1} \rfloor - \lfloor n \cdot a_{m_2} \rfloor\big| \le 1.$$

Now, we define $\lambda : \mathbb{Z}_{\mathcal{N}} \to \mathbb{Z}_{\mathcal{N}}$ with respect to (a_n) by stipulating

$$\lambda(n) = \begin{cases} \lfloor n \cdot a_{k_n} \rfloor & \text{for } n \in \mathcal{N}^+, \\ 0 & \text{for } n = 0, \\ \lfloor n \cdot a_{k_{-n}} \rfloor & \text{otherwise.} \end{cases}$$

Notice that for all $n \in \mathbb{Z}_{\mathcal{N}}$, we have $\lambda(-n) = -\lambda(n)$. therefore, in order to show that $\lambda \in \mathscr{S}$ is a slope, it is enough to show that $u_{n,m} := |\lambda(n+m) - \lambda(n) - \lambda(m)|$ is bounded for $n, m \in \mathcal{N}$. Now, for all $n, m \in \mathcal{N}$ we have:

$$\begin{aligned}
u_{n,m} &= |\lambda(n+m) - \lambda(n) - \lambda(m)| \\
&= \big|\lfloor (n+m)\cdot a_{k_{n+m}}\rfloor - \lfloor n\cdot a_{k_n}\rfloor - \lfloor m\cdot a_{k_m}\rfloor\big| \\
&\le \big|\lfloor n\cdot a_{k_{n+m}}\rfloor + \lfloor m\cdot a_{k_{n+m}}\rfloor - \lfloor n\cdot a_{k_n}\rfloor - \lfloor m\cdot a_{k_m}\rfloor\big| + 1 \\
&= \big|\lfloor n\cdot a_{k_{n+m}}\rfloor - \lfloor n\cdot a_{k_n}\rfloor + \lfloor m\cdot a_{k_{n+m}}\rfloor - \lfloor m\cdot a_{k_m}\rfloor\big| + 1 \\
&\le \big|\lfloor n\cdot a_{k_{n+m}}\rfloor - \lfloor n\cdot a_{k_n}\rfloor\big| + \big|\lfloor m\cdot a_{k_{n+m}}\rfloor - \lfloor m\cdot a_{k_m}\rfloor\big| + 1 \\
&\le 1+1+1 \\
&= 3
\end{aligned}$$

This shows that λ is a slope. Moreover, for $k_0 := 0$ and $a_0 := 0$, we obtain $\gamma(\lambda) \approx (a_{k_n}) \approx (a_n)$, and since $(a_n) \in \mathscr{C}$ was arbitrary, this implies that Γ is surjective.

Γ is injective: We have to show that for any slopes $\lambda, \lambda' \in \mathscr{S}$, if $\lambda \not\sim \lambda'$ then $\gamma(\lambda) \not\approx \gamma(\lambda')$. Let $(a_n) = \gamma(\lambda)$ and $(b_n) = \gamma(\lambda')$. Then $a_0 = b_0 = 0$ and for all $n \in \mathcal{N}^+$, $a_n = \frac{\lambda(n)}{n}$ and $b_n = \frac{\lambda'(n)}{n}$. Since $\lambda, \lambda' \in \mathscr{S}$, there are $M_\lambda, M_{\lambda'} \in \mathcal{N}$ such that for all $n, m \in \mathcal{N}$,

$$|\lambda(2n) - 2\lambda(n)| \le M_\lambda \quad \text{and} \quad |\lambda'(2n) - 2\lambda'(n)| \le M_{\lambda'}\,.$$

Assume that $\lambda \not\sim \lambda'$. Then for each $M \in \mathcal{N}$, there is an $n \in \mathcal{N}$ such that $|\lambda(n) - \lambda'(n)| > M$. Let

$$M_0 := M_\lambda + M_{\lambda'} + 1$$

and let $n_0 \in \mathcal{N}^+$ be such that

$$|\lambda(n_0) - \lambda'(n_0)| > M_0\,.$$

Now, since

$$\big|\big(\lambda(2n_0) - 2\lambda(n_0)\big) - \big(\lambda'(2n_0) - 2\lambda'(n_0)\big)\big| \le M_\lambda + M_{\lambda'},$$

we obtain

$$\big|\lambda(2n_0) - \lambda'(2n_0)\big| > 2M_0 - (M_\lambda + M_{\lambda'}) = M_\lambda + M_{\lambda'} + 2\,.$$

Similarly, we obtain

$$\big|\lambda(4n_0) - \lambda'(4n_0)\big| > 2(M_\lambda + M_{\lambda'} + 2) - (M_\lambda + M_{\lambda'}) = M_\lambda + M_{\lambda'} + 4\,,$$

and in general, we have

$$\big|\lambda(2^k n_0) - \lambda'(2^k n_0)\big| > M_\lambda + M_{\lambda'} + 2^k\,.$$

For the corresponding Cauchy sequences (a_n) and (b_n), we therefore have

$$|a_{2^k n_0} - b_{2^k n_0}| = \left|\frac{\lambda(2^k n_0)}{2^k n_0} - \frac{\lambda'(2^k n_0)}{2^k n_0}\right| > \frac{M_\lambda + M_{\lambda'} + 2^k}{2^k n_0} \ge \frac{1}{n_0}\,,$$

which shows that $(a_n) \not\approx (b_n)$ and completes the proof that $\Gamma : \mathbb{R}_{\mathscr{N}}^{\mathscr{S}} \to \mathbb{R}_{\mathscr{N}}^{\mathscr{C}}$ is a bijection.

The proof that the structures $\mathbb{R}_{\mathscr{N}}^{\mathscr{S}}$ and $\mathbb{R}_{\mathscr{N}}^{\mathscr{C}}$ are isomorphic is left as an exercise to the reader (see EXERCISE 17.0). ⊣

Non-Standard Models of the Reals

In the previous section, starting with either $\mathscr{N} = \omega$ or $\mathscr{N} = \omega^*$, we have constructed four models of the real numbers, namely $\mathbb{R}_{\omega}^{\mathscr{C}}$, $\mathbb{R}_{\omega^*}^{\mathscr{C}}$, $\mathbb{R}_{\omega}^{\mathscr{S}}$, $\mathbb{R}_{\omega^*}^{\mathscr{S}}$, and we have shown in PROPOSITION 17.5 that $\mathbb{R}_{\omega}^{\mathscr{C}} \cong \mathbb{R}_{\omega}^{\mathscr{S}}$ and $\mathbb{R}_{\omega^*}^{\mathscr{C}} \cong \mathbb{R}_{\omega^*}^{\mathscr{S}}$. The models $\mathbb{R}_{\omega}^{\mathscr{C}}$ and $\mathbb{R}_{\omega}^{\mathscr{S}}$ correspond to the standard model $\mathbb{R}$ (with respect to some model of ZF), whereas $\mathbb{R}_{\omega^*}^{\mathscr{C}}$ and $\mathbb{R}_{\omega^*}^{\mathscr{S}}$ are isomorphic non-standard models of the reals (constructed in some model of ZFC), denoted by $\mathbb{R}_{\omega^*}$.

Other non-standard models of the reals are obtained by an ultrapower of the standard model $\mathbb{R}$ with respect to some non-trivial ultrafilters $\mathscr{U} \subseteq \mathscr{P}(\omega)$. The models which we obtain with this construction are denoted by $\mathbb{R}^*$. By ŁOŚ'S THEOREM 15.3 we know that all these models $\mathbb{R}^*$ are elementarily equivalent to $\mathbb{R}$, independent of the choice of the ultrafilter $\mathscr{U}$. Therefore, beside the non-standard models $\mathbb{R}_{\omega^*}$ as constructed above, we also have the non-standard models $\mathbb{R}^*$. It is natural to ask whether the models $\mathbb{R}_{\omega^*}$ are also elementarily equivalent to the standard model $\mathbb{R}$. This is indeed the case. Moreover, if we use the same ultrafilter to construct ω^* (from ω) and $\mathbb{R}^*$ (from $\mathbb{R}$), then the models $\mathbb{R}_{\omega^*}$ and $\mathbb{R}^*$ are isomorphic (see EXERCISE 17.1).

A Brief Introduction to Non-Standard Analysis

The idea of Non-Standard Analysis is that we work simultaneously with two models of the real numbers. One model, let us call it the *ground model*, takes the role of the standard model $\mathbb{R}$, and the other model, denoted by $\mathbb{R}^*$, is in the view of $\mathbb{R}$ a non-standard model which is elementarily equivalent to $\mathbb{R}$. Now, we take the standpoint that proper Analysis takes place in the model $\mathbb{R}^*$, but — as people living in $\mathbb{R}$ — we can only "see" the standard part of the reals in $\mathbb{R}^*$. Even though we have quite a restricted view to proper Analysis from $\mathbb{R}$, by the fact that the models $\mathbb{R}$ and $\mathbb{R}^*$ are elementarily equivalent, we cannot detect any difference between the two models on the formal level. In fact, all that we can prove in one model is also valid for the other model. For example, in order to solve a problem in $\mathbb{R}$, we can carry out our calculations in $\mathbb{R}^*$, where we can use reals in $\mathbb{R}^*$ which do not exist in $\mathbb{R}$, and at the end we simply "project" the result to $\mathbb{R}$ again.

Let us now have a closer look at the models $\mathbb{R}$ and $\mathbb{R}^*$, and let us fix some notation: The domain of $\mathbb{R}$ is denoted by $\mathbb{R}$ with the natural numbers $\mathbb{N}$, and the domain of $\mathbb{R}^*$ is denoted by $\mathbb{R}^*$ with the natural numbers $\mathbb{N}^*$. Furthermore, let $\bar{\mathbb{R}}$ be the set of all reals $r^* \in \mathbb{R}^*$ such that $s_1 \leq r^* \leq s_2$ for some $s_1, s_2 \in \mathbb{R}$. We obviously have $\mathbb{R} \subseteq \bar{\mathbb{R}} \subseteq \mathbb{R}^*$. For any number $c \in \mathbb{N}^* \setminus \mathbb{N}$, we have that $\delta_0 := \frac{1}{c}$ belongs to $\mathbb{R}^*$, and from the viewpoint of $\mathbb{R}^*$, δ_0 is just a positive real, in fact a positive rational. However, from the viewpoint of $\mathbb{R}$, δ_0 does not exist, since it would be an infinitely small real number, a so-called **infinitesimal** (i.e., a non-zero real number whose absolute value is smaller than $\frac{1}{n}$ for any $n \in \mathbb{N}$). We say that $r^*, s^* \in \mathbb{R}^*$ are **infinitely close**, denoted by $r^* \approx s^*$, if $r^* - s^*$ is infinitesimal. Note that $\approx$ defines an equivalence relation on $\mathbb{R}^*$. The following fact states that the reals in $\bar{\mathbb{R}}$ are exactly those reals which can be "projected" to $\mathbb{R}$:

FACT 17.6. *For each $r^* \in \bar{\mathbb{R}}$, there is a unique real $r \in \mathbb{R}$ such that $r^* \approx r$.*

Proof. Uniqueness is obvious, since if there are $r_1, r_2 \in \mathbb{R}$ such that $r^* \approx r_1$ and $r^* \approx r_2$, then by transitivity we have $r_1 \approx r_2$. Since $r_1, r_2 \in \mathbb{R}$, it follows that $r_1 - r_2 = 0$ and thus $r_1 = r_2$.

For the existence, suppose that $s, t \in \mathbb{R}$ are such that $s \leq r^* \leq t$. We construct sequences (s_n) and (t_n) in $\mathbb{R}$ as follows: Let $s_0 := s$ and $t_0 := t$. If s_n and t_n are given, set

$$s_{n+1} = \begin{cases} s_n & \text{if } r^* \leq \frac{s_n+t_n}{2}, \\ \frac{s_n+t_n}{2} & \text{otherwise,} \end{cases}$$

and

$$t_{n+1} = s_{n+1} + \frac{t_n - s_n}{2}.$$

By construction, (s_n) and (t_n) are Cauchy sequences, $\lim_{n\to\infty}(t_n - s_n) = 0$, and $s_n \leq r^* \leq t_n$ for all $n \in \mathbb{N}$. By the Completeness Axiom, there exists a limit $r \in \mathbb{R}$ of (s_n). Suppose towards a contradiction that $r^* \not\approx r$. Then there is an $n \in \mathbb{N}$ such that $r^* - r \geq \frac{1}{n}$. Since $\lim_{n\to\infty}(t_n - s_n) = 0$, there is $N \in \mathbb{N}$ such that $t_N - s_N < \frac{1}{n}$ and hence,

$$r^* \geq r + \frac{1}{n} > r_N + \frac{1}{n} > t_N,$$

which is a contradiction. ⊣

We call the unique $r \in \mathbb{R}$ such that $r^* \approx r$ the **standard part** of $r^* \in \bar{\mathbb{R}}$, denoted by $\mathrm{st}(r^*)$. Obviously, the standard part of δ_0 is 0, i.e., for people living in $\mathbb{R}$, $\delta_0 \approx 0$. Similarly, $2 + \delta_0 \approx 2$, or more formally, $\mathrm{st}(2 + \delta_0) = 2$.

For example, $2 + \delta_0$ belongs to $\bar{\mathbb{R}}$, but $c = \frac{1}{\delta_0}$ does not belong to $\bar{\mathbb{R}}$, since there is no $s \in \mathbb{R}$ such that $\frac{1}{\delta_0} \leq s$. The set $\bar{\mathbb{R}}$, as a subset of $\mathbb{R}^*$, is linearly ordered by $<^*_{\mathbb{R}}$. Notice that $<^*_{\mathbb{R}}$ restricted to $\mathbb{R}$ is just the usual linear ordering on $\mathbb{R}$ (which follows from the fact that $\mathbb{R}$ is a submodel of $\mathbb{R}^*$ and that $\mathbb{R}$ and $\mathbb{R}^*$ are elementarily equivalent). The following figure visualises some calculations in $\bar{\mathbb{R}}$.

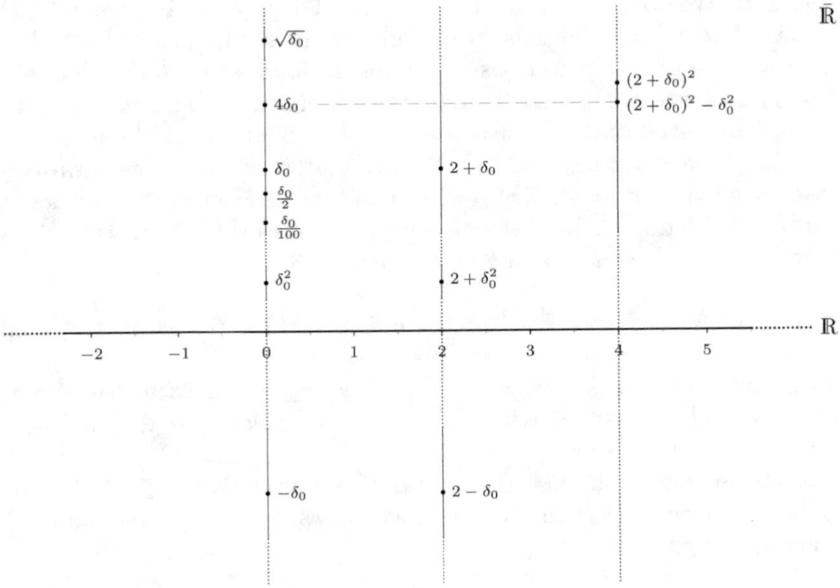

In Non-Standard Analysis, it is quite easy to define, e.g., the derivative of a function at some point (see EXERCISE 17.2) or definite integrals (see EXERCISE 17.3) without using any limits. Other applications of Non-Standard Analysis are given by the following three examples.

Example 17.7. As a first example, we prove **L'Hospital's Rule**: Let f and g be two real-valued functions which are derivable at $x_0 \in \mathbb{R}$, where $f(x_0) = g(x_0) = 0$. Furthermore, let ε be an infinitesimal. Then

$$\frac{f(x_0)}{g(x_0)} = \operatorname{st}\Big(\frac{f(x_0)}{g(x_0)}\Big),$$

and since

$$\operatorname{st}\big(f(x_0)\big) = \operatorname{st}\big(f(x_0+\varepsilon) - f(x_0)\big) = 0\,,$$

(and similarly for g), we have:

$$\begin{aligned}
\operatorname{st}\Big(\frac{f(x_0)}{g(x_0)}\Big) &= \operatorname{st}\Big(\frac{f(x_0+\varepsilon)-f(x_0)}{g(x_0+\varepsilon)-g(x_0)}\Big)\\
&= \operatorname{st}\Big(\frac{f(x_0+\varepsilon)-f(x_0)}{g(x_0+\varepsilon)-g(x_0)}\cdot\frac{\varepsilon}{\varepsilon}\Big)\\
&= \operatorname{st}\Big(\frac{f(x_0+\varepsilon)-f(x_0)}{\varepsilon}\cdot\frac{\varepsilon}{g(x_0+\varepsilon)-g(x_0)}\Big)\\
&= \operatorname{st}\Big(\frac{f(x_0+\varepsilon)-f(x_0)}{\varepsilon}\Big)\cdot\operatorname{st}\Big(\frac{\varepsilon}{g(x_0+\varepsilon)-g(x_0)}\Big)
\end{aligned}$$

Therefore, by EXERCISE 17.2, we finally have

$$\frac{f(x_0)}{g(x_0)} = \frac{f'(x_0)}{g'(x_0)}.$$

Example 17.8. Let us now consider the **Dirac Delta Function**: For this, let g be a real-valued function which is continuous at 0, and let δ be a positive infinitesimal. With respect to δ, we define the function

$$f_\delta(x) = \begin{cases} \frac{x}{\delta^2} + \frac{1}{\delta} & \text{for } -\delta < x \leq 0, \\ -\frac{x}{\delta^2} + \frac{1}{\delta} & \text{for } 0 < x < \delta, \\ 0 & \text{otherwise.} \end{cases}$$

Then

$$\int_{\mathbb{R}} g(x) \cdot f_\delta(x)\, dx \;=\; \int_{-\delta}^{\delta} g(x) \cdot f_\delta(x)\, dx\,,$$

and since g is continuous at 0, we obtain

$$\mathrm{st}\Bigg(\int_{-\delta}^{\delta} g(x) \cdot f_\delta(x)\, dx\Bigg) \;=\; g(0)\,.$$

Computing the Fourier coefficients of f_δ, we obtain $a_0 := \frac{1}{\pi}$, and for all positive $n \in \mathbb{N}^*$ we have $b_n = 0$ and

$$a_n := \frac{2}{\pi n^2 \delta^2}\big(1 - \cos(n\delta)\big).$$

Notice that for all $n \in \mathbb{N}$ we have $\mathrm{st}(a_n) = \frac{1}{\pi}$.

Example 17.9. In this last example, we define a non-vanishing function which vanishes on $\mathbb{R}$: Let μ be a positive infinitesimal and let c be a positive real in $\mathbb{R}$. Consider the real-valued function

$$f_\mu(x) := \begin{cases} \dfrac{\mu}{\sqrt{1 - (\frac{x}{c})^2}} & \text{for } |x| < c, \\ 0 & \text{otherwise.} \end{cases}$$

If we consider the function f_μ as a function $\mathbb{R} \to \mathbb{R}^*$, then $\mathrm{st}\big(f_\mu(x)\big) = 0$ for all $x \in \mathbb{R}$. However, if we consider the function f_μ as a function $\bar{\mathbb{R}} \to \mathbb{R}^*$, then there are some $x \in \bar{\mathbb{R}}$ such that $\mathrm{st}\big(f_\mu(x)\big) \neq 0$. For example, for $c = 17^2$ and $x_0 = c - \mu^2$, we have $\mathrm{st}\big(f_\mu(x_0)\big) = \frac{17}{\sqrt{2}}$. Therefore, f_μ is a real-valued function whose standard part vanishes on $\mathbb{R}$, but takes non-zero values for some $x \in \bar{\mathbb{R}}$.

Notes

The natural construction of a model of the real numbers is due to A'Campo [1], who proved all results directly from the properties of slopes — without using Cauchy sequences. The structure of non-standard models of A'Campo's construction of the reals was first studied by Mizrahi [35]. Non-Standard Analysis was developed in the early 1960s by Robinson. Even though the idea of working with infinitesimals can be traced back to Leibniz and L'Hospital, it was Robinson who laid the logical foundations for infinitesimals and infinite numbers (see, e.g., [44]).

Exercises

17.0 (a) Show that for any slopes $\lambda, \mu \in \mathscr{S}$ we have $\Gamma\big([\lambda]+[\mu]\big) = \Gamma\big([\lambda]\big) + \Gamma\big([\mu]\big)$.

(b) Show that $\Gamma(0_{\mathscr{S}}) = 0_{\mathscr{C}}$.

(c) Show that for any slopes $\lambda, \mu \in \mathscr{S}$ we have $\Gamma\big([\lambda]\cdot[\mu]\big) = \Gamma\big([\lambda]\big) \cdot \Gamma\big([\mu]\big)$.

Hint: Notice that $\frac{\lambda(\mu(n))}{n} = \frac{\lambda(\mu(n))}{\mu(n)} \cdot \frac{\mu(n)}{n}$.

(d) Show that $\Gamma(1_{\mathscr{S}}) = 1_{\mathscr{C}}$.

(e) Show that for any slopes $\lambda, \mu \in \mathscr{S}$ we have $[\lambda] < [\mu] \iff \Gamma\big([\lambda]\big) < \Gamma\big([\mu]\big)$.

17.1 Asume that all the ultrapowers are constructed with respect to the same non-trivial ultrafilter $\mathscr{U} \subseteq \mathscr{P}(\omega)$.

(a) Show that $\mathbb{Z}_{\omega^*} \cong \mathbb{Z}_\omega^*$.

(b) Show that $\mathbb{Q}_{\omega^*} \cong \mathbb{Q}_\omega^*$.

(c) Let $\overline{\mathbb{Q}_\omega^*}$ be the completion of the ultrapower of $\mathbb{Q}_\omega^*$ with Cauchy sequences, and let $(\overline{\mathbb{Q}_\omega})^*$ be the ultrapower of the completion of $\overline{\mathbb{Q}_\omega}$ with Cauchy sequences. Show that $\overline{\mathbb{Q}_\omega^*} \cong (\overline{\mathbb{Q}_\omega})^*$.

(d) Show that $\mathbb{R}^{\mathscr{C}}_{\omega^*} \cong \overline{\mathbb{Q}_\omega^*}$ and that $(\mathbb{R}^{\mathscr{C}}_\omega)^* = (\overline{\mathbb{Q}_\omega})^*$.

(e) Show that the six models $\mathbb{R}^{\mathscr{C}}_\omega, (\mathbb{R}^{\mathscr{C}}_\omega)^*, \mathbb{R}^{\mathscr{C}}_{\omega^*}, \mathbb{R}^{\mathscr{S}}_\omega, (\mathbb{R}^{\mathscr{S}}_\omega)^*, \mathbb{R}^{\mathscr{S}}_{\omega^*}$ are pairwise elementarily equivalent.

17.2 Let f be a real-valued function which is defined at $x_0 \in \mathbb{R}$, and let δ be a positive infinitesimal. For every non-zero ε in the interval $[-\delta, \delta]$, let

$$\Delta_f(x_0, \varepsilon) := \frac{f(x_0+\varepsilon) - f(x_0)}{\varepsilon}.$$

Show that if there is an $a \in \mathbb{R}$ such that for all non-zero $\varepsilon \in [-\delta, \delta]$ we have

$$\operatorname{st}\big(\Delta_f(x_0, \varepsilon)\big) = a,$$

then, in $\mathbb{R}$, $f'(x_0) = a$, i.e.,

$$f'(x_0) = \operatorname{st}\Big(\frac{f(x_0+\varepsilon) - f(x_0)}{\varepsilon}\Big) \quad \text{for some infinitesimal } \varepsilon.$$

17.3 Let $b \in \mathbb{R}$, let f be a real-valued function which is continuous on the interval $[0, b]$, and let $N \in \mathbb{N}^* \setminus \mathbb{N}$. Show that in $\mathbb{R}$,

$$\int_0^b f(x)\,dx = \operatorname{st}\left(\frac{b}{N}\sum_{j=1}^{N} f\Big(\frac{jb}{N}\Big)\right).$$

17.4 Let $N \in \mathbb{N}^* \setminus \mathbb{N}$. Show that for all $x \in \mathbb{R}$,

$$\left(1 + \frac{x}{N}\right)^N \approx e^x .$$

17.5 The following exercise is taken from Robert [43], where one can find many more applications of Non-Standard Analysis.

Show that for $a^2 \neq 1$,

$$\int_{x=0}^{\pi} \ln\left(1 - 2a\cos(x) + a^2\right) dx = \begin{cases} 0 & \text{if } |a| < 1, \\ \pi \ln(a^2) & \text{if } |a| > 1. \end{cases}$$

Hint: Let $N \in \mathbb{N}^* \setminus \mathbb{N}$ and first show that

$$\begin{aligned} \int_0^{\pi} \ln\left(1 - 2a\cos(x) + a^2\right) dx &= \operatorname{st}\left(\frac{\pi}{N} \sum_{k=0}^{N-1} \ln\left(1 - 2a\cos\left(\frac{k\pi}{N}\right) + a^2\right)\right) \\ &= \operatorname{st}\left(\frac{\pi}{N} \ln\left(\prod_{k=0}^{N-1} \underbrace{\left(1 - a\left(e^{ik\pi/N} + e^{-ik\pi/N}\right) + a^2\right)}_{(a - e^{ik\pi/N})(a - e^{-ik\pi/N})}\right)\right). \end{aligned}$$

Tautologies

In this section we give a list of some of the most important tautologies. Many of them have been used explicitly and implicitly in several formal proofs.

(A.0) $\vdash \varphi \rightarrow \varphi$
(A.1) $\vdash \varphi \leftrightarrow \varphi$

(B) $\{\psi, \varphi\} \vdash \varphi \wedge \psi$

(C) $\vdash (\psi \rightarrow \varphi) \rightarrow (\psi \rightarrow \forall\nu\varphi)$ [for $\nu \notin \text{free}(\psi)$]

(D.0) $\{\varphi_1 \rightarrow \varphi_2, \varphi_2 \rightarrow \varphi_3\} \vdash \varphi_1 \rightarrow \varphi_3$
(D.1) $\{\varphi_1 \rightarrow \psi, \varphi_2 \rightarrow \psi\} \vdash (\varphi_1 \vee \varphi_2) \rightarrow \psi$
(D.2) $\{\psi \rightarrow \varphi_1, \psi \rightarrow \varphi_2\} \vdash \psi \rightarrow (\varphi_1 \wedge \varphi_2)$

(E) $\vdash \varphi \rightarrow \big(\psi \rightarrow (\varphi \wedge \psi)\big)$

(F) $\vdash \varphi \leftrightarrow \neg\neg\varphi$

(G) $\vdash (\varphi \rightarrow \psi) \leftrightarrow (\neg\psi \rightarrow \neg\varphi)$

(H.0) $\{\varphi \leftrightarrow \psi\} \vdash \neg\varphi \leftrightarrow \neg\psi$
(H.1) $\{\varphi \leftrightarrow \varphi', \psi \leftrightarrow \psi'\} \vdash (\varphi \rightarrow \psi) \leftrightarrow (\varphi' \rightarrow \psi')$
(H.2) $\{\varphi \leftrightarrow \varphi', \psi \leftrightarrow \psi'\} \vdash (\varphi \vee \psi) \leftrightarrow (\varphi' \vee \psi')$
(H.3) $\{\varphi \leftrightarrow \varphi', \psi \leftrightarrow \psi'\} \vdash (\varphi \wedge \psi) \leftrightarrow (\varphi' \wedge \psi')$

(I.1) $\vdash (\varphi_1 \wedge \varphi_2) \leftrightarrow (\varphi_2 \wedge \varphi_1)$
(I.2) $\vdash \big((\varphi_1 \wedge \varphi_2) \wedge \varphi_3\big) \leftrightarrow \big(\varphi_1 \wedge (\varphi_2 \wedge \varphi_3)\big)$

(J.0) $\vdash (\varphi_1 \vee \varphi_2) \leftrightarrow (\varphi_2 \vee \varphi_1)$

L. Halbeisen, R. Krapf, *Gödel's Theorems and Zermelo's Axioms*,
https://doi.org/10.1007/978-3-030-52279-7

(J.1) $\vdash \big((\varphi_1 \vee \varphi_2) \vee \varphi_3\big) \leftrightarrow \big(\varphi_1 \vee (\varphi_2 \vee \varphi_3)\big)$

(K) $\vdash (\varphi \to \psi) \leftrightarrow (\neg\varphi \vee \psi)$

(L.0) $\vdash \neg(\varphi \wedge \psi) \leftrightarrow (\neg\varphi \vee \neg\psi)$
(L.1) $\vdash \neg(\varphi \vee \psi) \leftrightarrow (\neg\varphi \wedge \neg\psi)$

(M.0) $\vdash (\varphi_1 \wedge \varphi_2) \vee \varphi_3 \leftrightarrow (\varphi_1 \vee \varphi_3) \wedge (\varphi_2 \vee \varphi_3)$
(M.1) $\vdash (\varphi_1 \vee \varphi_2) \wedge \varphi_3 \leftrightarrow (\varphi_1 \wedge \varphi_3) \vee (\varphi_2 \wedge \varphi_3)$

(N.0) $\vdash \nu = \nu' \leftrightarrow \nu' = \nu$
(N.1) $\vdash (\nu = \nu' \wedge \nu' = \nu'') \to \nu = \nu''$

(O.0) $\vdash \varphi(\nu) \leftrightarrow \varphi(\nu')$ [if ν' does not appear in $\varphi(\nu)$]
(O.1) $\vdash \exists\nu\varphi(\nu) \leftrightarrow \exists\nu'\varphi(\nu')$ [if ν' does not appear in $\varphi(\nu)$]
(O.2) $\vdash \forall\nu\varphi(\nu) \leftrightarrow \forall\nu'\varphi(\nu')$ [if ν' does not appear in $\varphi(\nu)$]

(P.0) $\{\varphi \leftrightarrow \psi\} \vdash \forall\nu\varphi \leftrightarrow \forall\nu\psi$
(P.1) $\{\varphi \leftrightarrow \psi\} \vdash \exists\nu\varphi \leftrightarrow \exists\nu\psi$

(Q.0) $\vdash \neg\exists\nu\varphi \leftrightarrow \forall\nu\neg\varphi$
(Q.1) $\vdash \neg\forall\nu\varphi \leftrightarrow \exists\nu\neg\varphi$

(R) $\vdash \forall\nu\varphi \leftrightarrow \neg\exists\nu\neg\varphi$

(S.0) $\vdash \exists\nu\exists\nu'\varphi \leftrightarrow \exists\nu'\exists\nu\varphi$
(S.1) $\vdash \forall\nu\forall\nu'\varphi \leftrightarrow \forall\nu'\forall\nu\varphi$
(S.2) $\vdash \rlap{\forall}\exists\nu\exists\nu\varphi \leftrightarrow \exists\nu\varphi$
(S.3) $\vdash \rlap{\forall}\exists\nu\forall\nu\varphi \leftrightarrow \forall\nu\varphi$

(T.0) $\vdash \exists\nu\varphi \wedge \exists\nu'\psi \leftrightarrow \exists\nu\exists\nu'(\varphi \wedge \psi)$ [for $\nu \notin \text{free}(\psi)$, $\nu' \notin \text{free}(\varphi)$]
(T.1) $\vdash \forall\nu\varphi \wedge \forall\nu'\psi \leftrightarrow \forall\nu\forall\nu'(\varphi \wedge \psi)$ [for $\nu \notin \text{free}(\psi)$, $\nu' \notin \text{free}(\varphi)$]
(T.2) $\vdash \exists\nu\varphi \wedge \psi \leftrightarrow \exists\nu(\varphi \wedge \psi)$ [for $\nu \notin \text{free}(\psi)$]
(T.3) $\vdash \forall\nu\varphi \wedge \psi \leftrightarrow \forall\nu(\varphi \wedge \psi)$ [for $\nu \notin \text{free}(\psi)$]

(U.0) $\vdash \exists\nu\varphi \vee \exists\nu'\psi \leftrightarrow \exists\nu\exists\nu'(\varphi \vee \psi)$ [for $\nu \notin \text{free}(\psi)$, $\nu' \notin \text{free}(\varphi)$]
(U.1) $\vdash \forall\nu\varphi \vee \forall\nu'\psi \leftrightarrow \forall\nu\forall\nu'(\varphi \vee \psi)$ [for $\nu \notin \text{free}(\psi)$, $\nu' \notin \text{free}(\varphi)$]
(U.2) $\vdash \exists\nu\varphi \vee \psi \leftrightarrow \exists\nu(\varphi \vee \psi)$ [for $\nu \notin \text{free}(\psi)$]
(U.3) $\vdash \forall\nu\varphi \vee \psi \leftrightarrow \forall\nu(\varphi \vee \psi)$ [for $\nu \notin \text{free}(\psi)$]

References

1. NORBERT A'CAMPO, *A natural construction for the real numbers*, ***arXiv.org,*** math/0301015 (2003), pp. 10.
2. ARISTOTLE, ***Topics***, Athens, *published by Andronikos of Rhodos around* 40 B.C.
3. JOHN L. BELL AND ALAN B. SLOMSON, ***Models and Ultraproducts: An Introduction***, North-Holland, Amsterdam, 1969.
4. PAUL J. COHEN, *The independence of the continuum hypothesis I.*, ***Proceedings of the National Academy of Sciences (U.S.A.)***, vol. 50 (1963), 1143–1148.
5. ———, *The independence of the continuum hypothesis II.*, ***Proceedings of the National Academy of Sciences (U.S.A.)***, vol. 51 (1964), 105–110.
6. RICHARD DEDEKIND, ***Was sind und was sollen die Zahlen***, Friedrich Vieweg & Sohn, Braunschweig, 1888 (see also [7, pp. 335–390]).
7. ———, ***Gesammelte mathematische Werke III***, edited by R. Fricke, E. Noether, and Ö. Ore, Friedrich Vieweg & Sohn, Braunschweig, 1932.
8. HERBERT ENDERTON, ***A mathematical introduction to logic***, Academic Press, New York-London, 1972.
9. EUCLID, ***The Thirteen Books of the Elements***, Volume I: Books I & II [translated with introduction and commentary by Sir Thomas L. Heath], Dover, 1956.
10. ADOLF FRAENKEL, *Zu den Grundlagen der Cantor-Zermeloschen Mengenlehre*, ***Mathematische Annalen***, vol. 86 (1922), 230–237.
11. GERHARD GENTZEN, *Untersuchungen über das logische Schließen I*, ***Mathematische Zeitschrift***, vol. 39 (1935), 176–210.
12. ———, *Untersuchungen über das logische Schließen II*, ***Mathematische Zeitschrift***, vol. 39 (1935), 405–431.
13. ———, *Die Widerspruchsfreiheit der reinen Zahlentheorie*, ***Mathematische Annalen***, vol. 112 (1936), 493–565.
14. KURT GÖDEL, *Über die Vollständigkeit des Logikkalküls*, ***Dissertation*** (1929), University of Vienna (Austria), reprinted and translated into English in [18].

L. Halbeisen, R. Krapf, *Gödel's Theorems and Zermelo's Axioms*,
https://doi.org/10.1007/978-3-030-52279-7

15. ______, *Die Vollständigkeit der Axiome des logischen Funktionenkalküls*, ***Monatshefte für Mathematik und Physik***, vol. 37 (1930), 349–360 (see [55, 18] for a translation into English).
16. ______, *Über formal unentscheidbare Sätze der Principia Mathematica und verwandter Systeme*, ***Monatshefte für Mathematik und Physik***, vol. 38 (1931), 173–198 (see [55, 18] for a translation into English).
17. ______, *The consistency of the axiom of choice and of the generalized continuum-hypothesis*, ***Proceedings of the National Academy of Sciences (U.S.A.)***, vol. 24 (1938), 556–557 (reprinted in [19]).
18. ______, ***Collected Works, Volume I: Publications 1929–1936***, edited by S. Feferman (Editor-in-chief), J. W. Dawson, Jr., S. C. Kleene, G. H. Moore, R. M. Solovay, J. van Heijenoort, Oxford University Press, New York, 1986.
19. ______, ***Collected Works, Volume II: Publications 1938–1974***, edited by S. Feferman (Editor-in-chief), J. W. Dawson, Jr., S. C. Kleene, G. H. Moore, R. M. Solovay, J. van Heijenoort, Oxford University Press, New York, 1990.
20. Hermann Grassmann, ***Lehrbuch der Arithmetik für höhere Lehranstalten***, Th. Chr. Fr. Enslin, Berlin, 1861.
21. Lorenz J. Halbeisen, ***Combinatorial Set Theory, with a gentle introduction to forcing***, 2nd ed., Springer Monographs in Mathematics, Springer, London, 2017.
22. Leon Henkin, *The completeness of the first-order functional calculus*, ***The Journal of Symbolic Logic***, vol. 14 (1949), 159–166.
23. ______, *A problem concerning provability*, ***Journal of Symbolic Logic***, vol. 17 (1952), 160.
24. ______, *The discovery of my completeness proofs*, ***The Bulletin of Symbolic Logic***, vol. 2 (1996), 127–158.
25. David Hilbert, *Mathematische Probleme*, ***Vortrag, gehalten auf dem internationalen Mathematiker-Kongreß zu Paris 1900*** (1900), 253–297.
26. ______, *Die Grundlagen der Mathematik. Vortrag, gehalten auf Einladung des Mathematischen Seminars im Juli 1927 in Hamburg.*, ***Abhandlungen aus dem mathematischen Seminar der Hamburgischen Universität***, vol. 6 (1928), 65–85 (see [55] for a translation into English).
27. David Hilbert and Paul Bernays, ***Grundlagen der Mathematik***, Vol. II, Springer, Berlin, 1939.
28. Richard Kaye, ***Models of Peano Arithmetic***, [Oxford Logic Guides 15], The Clarendon Press Oxford University Press, New York, 1991.
29. Peter Koepke, *Models of Set Theory I*, Lecture Notes, University of Bonn (Germany), 2009.
30. Kenneth Kunen, ***Set Theory, an Introduction to Independence Proofs***, [Studies in Logic and the Foundations of Mathematics 102], North-Holland, Amsterdam, 1983.

31. MARTIN HUGO LÖB, *Solution of a problem of Leon Henkin*, ***The Journal of Symbolic Logic***, vol. 20 (1955), 115–118.
32. JERZY ŁOŚ, *Quelques remarques, théorèmes et problèmes sur les classes définissables d'algèbres*, ***Mathematical interpretation of formal systems***, North-Holland Publishing Co., Amsterdam, 1955, pp. 98–113.
33. LEOPOLD LÖWENHEIM, *Über Möglichkeiten im Relativkalkül*, ***Mathematische Annalen***, vol. 76 (1915), 447–470.
34. ELLIOTT MENDELSON, ***Introduction to Mathematical Logic***, 6th ed., [Discrete Mathematics and its Applications], CRC Press, Boca Raton, FL, 2015.
35. LEILA MIZRAHI, *Thoroughly formalizing an uncommon construction of the real numbers*, ***Master Thesis*** (2015), University of Zürich (Switzerland).
36. ROMAN MURAWSKI, *Undefinability of truth. The problem of priority: Tarski vs. Gödel*, ***History and Philosophy of Logic***, vol. 9 (1998), 153–160.
37. JOHN VON NEUMANN, *Eine Axiomatisierung der Mengenlehre*, ***Journal für die Reine und Angewandte Mathematik***, vol. 154 (1925), 219–240 (see [55] for a translation into English).
38. ———, *Die Axiomatisierung der Mengenlehre*, ***Mathematische Zeitschrift***, vol. 27 (1928), 669–752.
39. ———, *Über eine Widerspruchfreiheitsfrage in der axiomatischen Mengenlehre*, ***Journal für die Reine und Angewandte Mathematik***, vol. 160 (1929), 227–241.
40. LAWRENCE PAULSON, *A machine-assisted proof of Gödel's incompleteness theorems for the theory of hereditarily finite sets*, ***Review of Symbolic Logic***, vol. 7 (2014), 484–498.
41. GIUSEPPE PEANO, ***Arithmetices principia, nova methoda exposita***, Fratres Bocca, Torino, 1889 (see [55] for a translation into English).
42. MOJŻESZ PRESBURGER, *Über die vollständigkeit eines gewissen systems der arithmetik ganzer zahlen, in welchem die addition als einzige operation hervortritt*, ***Comptes Rendus I Congès des Mathémathiciens des Pays Slaves*** 92–101, Zusatz ebenda, 395 (1930).
43. ALAIN M. ROBERT, ***Nonstandard Analysis***, Dover Publications, Mineola, New York, 2003.
44. ABRAHAM ROBINSON, ***Non-standard analysis***, North-Holland Publishing Co., Amsterdam, 1966.
45. RAPHAEL M. ROBINSON, *An Essentially Undecidable Axiom System*, ***Proceedings of the International Congress of Mathematics*** (1950), 729–730.
46. BARKLEY ROSSER, *Extensions of some theorems of Gödel and Church*, ***The Journal of Symbolic Logic***, vol. 1 (1936), no. 03, 87–91.
47. JOSEPH R. SHOENFIELD, ***Mathematical Logic***, Addison-Wesley Publishing Co., Reading, Mass.-London-Don Mills, Ont., 1967.
48. THORALF SKOLEM, *Logisch-kombinatorische Untersuchungen über die Erfüllbarkeit oder Beweisbarkeit mathematischer Sätze nebst einem*

Theorem über dichte Mengen, ***Krist. Vid. Selsk. Skr. I, 1920***, Nr. 4, 36 S., (1922).

49. ______, *Einige Bemerkungen zur axiomatischen Begründung der Mengenlehre*, ***Matematikerkongressen i Helsingfors den 4–7 Juli 1922***, *Den femte skandinaviska matematikerkongressen*, Akademiska Bokhandeln Helsingfors, 1923, pp. 217–232 (see [55] for a translation into English).
50. ______, *Über einige Satzfunktionen in der Arithmetik*, ***Skrifter Vitenskapsakademiet i Oslo***, vol. 7 (1931), 1–28.
51. ______, *Über die Unmöglichkeit einer vollständigen Charakterisierung der Zahlenreihe mittels eines endlichen Axiomensystems*, ***Fundamenta Mathematicae***, vol. 23 (1934), 150–161.
52. Raymond M. Smullyan, ***A beginner's guide to mathematical logic***, Dover Publications, Inc., Mineola, NY, 2014.
53. Stanisław S. Świerczkowski, *Finite sets and Gödel's incompleteness theorems*, ***Dissertationes Mathematicae***, vol. 422 (2003), 1–58.
54. Alfred Tarski, *Der Wahrheitsbegriff in den formalisierten Sprachen*, ***Studia Philosophica***, vol. 1 (1936), 261–405.
55. Jean van Heijenoort, ***From Frege to Gödel. A Source Book in Mathematical Logic, 1879–1931***, [Source Books in the History of Science], Harvard University Press, Cambridge, Massachusetts, 1967.
56. Alfred North Whitehead and Bertrand Russell, ***Principia Mathematica, Vol. I–III***, Cambridge University Press, Cambridge, 1910–1913.
57. Ernst Zermelo, *Beweis, dass jede Menge wohlgeordnet werden kann*, ***Mathematische Annalen***, vol. 59 (1904), 514–516 (see [55, 61] for a translation into English).
58. ______, *Neuer Beweis für die Möglichkeit einer Wohlordnung*, ***Mathematische Annalen***, vol. 65 (1908), 107–128 (see [55, 61] for a translation into English).
59. ______, *Untersuchungen über die Grundlagen der Mengenlehre. I.*, ***Mathematische Annalen***, vol. 65 (1908), 261–281 (see [55, 61] for a translation into English).
60. ______, *Über Grenzzahlen und Mengenbereiche. Neue Untersuchungen über die Grundlagen der Mengelehre*, ***Fundamenta Mathematicae***, vol. 16 (1930), 29–47 (see [61] for a translation into English).
61. ______, ***Collected Works / Gesammelte Werke***, *Volume I: Set Theory, Miscellanea / Band I: Mengenlehre, Varia*, [Schriften der Mathematisch-naturwissenschaftlichen Klasse der Heidelberger Akademie der Wissenschaften, Nr. 21 (2010)], edited by Heinz-Dieter Ebbinghaus, Craig G. Fraser, and Akihiro Kanamori, Springer-Verlag, Berlin · Heidelberg, 2010.

Symbols

L. Halbeisen, R. Krapf, *Gödel's Theorems and Zermelo's Axioms*,
https://doi.org/10.1007/978-3-030-52279-7

Persons

L. Halbeisen, R. Krapf, *Gödel's Theorems and Zermelo's Axioms*,
https://doi.org/10.1007/978-3-030-52279-7

Subjects

L. Halbeisen, R. Krapf, *Gödel's Theorems and Zermelo's Axioms*,
https://doi.org/10.1007/978-3-030-52279-7